D1285901

Genomics and Evolution of Microbial Eukaryotes

Genomics and Evolution of Microbial Eukaryotes

EDITED BY

Laura A. Katz
Smith College, MA, USA

Debashish Bhattacharya
University of Iowa, IA, USA

OXFORD

UNIVERSITY PRESS

QR74·5
·G46
2006

OXFORD
UNIVERSITY PRESS

Great Clarendon Street, Oxford OX2 6DP

Oxford University Press is a department of the University of Oxford.
It furthers the University's objective of excellence in research, scholarship,
and education by publishing worldwide in

Oxford New York

Auckland Cape Town Dar es Salaam Hong Kong Karachi
Kuala Lumpur Madrid Melbourne Mexico City Nairobi
New Delhi Shanghai Taipei Toronto

With offices in

Argentina Austria Brazil Chile Czech Republic France Greece
Guatemala Hungary Italy Japan Poland Portugal Singapore
South Korea Switzerland Thailand Turkey Ukraine Vietnam

Oxford is a registered trade mark of Oxford University Press
in the UK and in certain other countries

Published in the United States
by Oxford University Press Inc., New York

© Oxford University Press 2006

The moral rights of the authors have been asserted
Database right Oxford University Press (maker)

First published 2006

All rights reserved. No part of this publication may be reproduced,
stored in a retrieval system, or transmitted, in any form or by any means,
without the prior permission in writing of Oxford University Press,
or as expressly permitted by law, or under terms agreed with the appropriate
reprographics rights organization. Enquiries concerning reproduction
outside the scope of the above should be sent to the Rights Department,
Oxford University Press, at the address above

You must not circulate this book in any other binding or cover
and you must impose the same condition on any acquirer

British Library Cataloguing in Publication Data
Data available

Library of Congress Cataloging in Publication Data
Genomics and evolution of microbial eukaryotes / edited by Laura A.
   Katz, Debashish Bhattacharya.
      p. cm.
   Includes index.
   ISBN-13: 978–0–19–856974–9 (alk. paper)
   ISBN-10: 0–19–856974–2 (alk. paper)
   1. Protista. 2. Microbial genomics. I. Olson, Laura Katz, 1945– II.
Bhattacharya, Debashish, 1959–
   QR74.5.G46 2006
   579—dc22   2006017209

Typeset by Newgen Imaging Systems (P) Ltd., Chennai, India
Printed in Great Britain
on acid-free paper by
Antony Rowe, Chippenham

ISBN 0–19–856974–2   978–0–19–856974–9

10 9 8 7 6 5 4 3 2 1

To my mentor and friend John Tyler Bonner

L.A.K

To Susanne

D.B

University Libraries
Carnegie Mellon University
Pittsburgh, PA 15213-3890

Contents

List of contributors   vii

Introduction   1
    Laura A. Katz and Debashish Bhattacharya

Part I   5
1   Current perspectives on high-level groupings of protists   7
    Alastair G.B. Simpson and David J. Patterson

Part II Evolutionary genomics of eukaryotic microbes   31
2   Comparative genomics of *Plasmodium* species   33
    Jane M. Carlton

3   The genomes of dinoflagellates   48
    Jeremiah D. Hackett and Debashish Bhattacharya

4   Ciliate genome evolution   64
    Casey McGrath, Rebecca Zufall, and Laura A. Katz

5   Molecular evolution of Foraminifera   78
    Samuel S. Bowser, Andrea Habura, and Jan Pawlowski

6   Photosynthetic organelles and endosymbiosis   94
    M. S. Sommer, S. B. Gould, O. Kawach, C. Klemme, C. Voß, U.-G. Maier, and S. Zauner

7   Genome evolution of anaerobic protists: metabolic adaptation
   via gene acquisition   109
    Jan O. Andersson

8   Horizontal and intracellular gene transfer in the Apicomplexa: The scope and
   functional consequences   123
    Jinling Huang and Jessica C. Kissinger

Part III Analyses of complete genomes   137
9   The nuts and bolts of sequencing protist genomes   139
    Daniella C. Bartholomeu, Neil Hall, and Jane M. Carlton

10   **Comparative genomics of the trypanosomatids**   **155**
     Kenneth D. Stuart and Peter J. Myler

11   **The genome of *Entamoeba histolytica***   **169**
     C. Graham Clark

12   **Genome reduction in Microsporidia**   **181**
     Patrick J. Keeling

13   **Nucleomorphs: remnant nuclear genomes**   **192**
     Oliver Kawach, Maik S. Sommer, Sven B. Gould, Christine Voß, Stefan Zauner, and Uwe-G. Maier

14   **Genomic insights into diatom evolution and metabolism**   **201**
     E. Virginia Armbrust, Tatiana A. Rynearson, and Bethany D. Jenkins

15   **The *Dictyostelium* genome – a blueprint for a multicellular protist**   **214**
     Pauline Schaap

Index   **227**

List of contributors

Jan O. Andersson
Institute of Cell and Molecular Biology
Uppsala University
Biomedical Center
Box 596
S-751 24, Uppsala
Sweden

E. Virginia Armbrust
Marine Molecular Biotechnology Laboratory
School of Oceanography
University of Washington
Seattle, Washington 98195
USA

Daniella Bartholomeu
The Institute for Genomic Research
9712 Medical Center Drive
Rockville, MD 20850
USA

Debashish Bhattacharya
Department of Biological Sciences and
Roy J. Carver Center for Comparative Genomics
446 Biology Building
University of Iowa
Iowa City, IA 52242
USA

Samuel S. Bowser
Laboratory of Cell Regulation
Wadsworth Center
New York State Department of Health
P.O. Box 509
Albany, NY 12201
USA

Jane Carlton
The Institute for Genomic Research
9712 Medical Center Drive
Rockville, MD 20850
USA

C. Graham Clark
Department of Infectious and Tropical Diseases
London School of Hygiene and Tropical Medicine
Keppel Street, London, WC1E 7HT
UK

Sven Gould
University of Marburg
Cell Biology, Karl-von-Frisch Strasse
D-35032 Marburg
Germany

Andrea Habura
Laboratory of Cell Regulation
Wadsworth Center
New York State Department of Health
P.O. Box 509
Albany, NY 12201
USA

Jeremiah D. Hackett
Woods Hole Oceanographic Institution
Biology Department
Mail Stop 32
Woods Hole, MA 02543
USA

Neil Hall
The Institute for Genomic Research
9712 Medical Center Drive
Rockville, MD 20850
USA

Jinling Huang
Department of Biology
East Carolina University
Greenville, NC 27858
USA

Bethany D. Jenkins
Department of Cell and Molecular Biology
Graduate School of Oceanography
University of Rhode Island
316 Morrill Hall
45 Lower College Road
Kingston, Rhode of Island 02881
USA

Laura A. Katz
Department of Biological Sciences
Smith College
Northampton, MA 01063
USA

Oliver Kawach
University of Marburg
Cell Biology, Karl-von-Frisch Strasse
D-35032 Marburg
Germany

Patrick J. Keeling
Canadian Institute for Advanced Research
Botany Department
University of British Columbia
3529–6270 University Boulevard
Vancouver, BC, V6T 1Z4
Canada

Jessica Kissinger
Department of Genetics and Center for
Tropical and Emerging Global Diseases
Paul D. Coverdell Center
500 D.W. Brooks Drive
Athens, GA 30602–7394
USA

C. Klemme
University of Marburg
Cell Biology, Karl-von-Frisch Strasse
D-35032 Marburg
Germany

Uwe-G. Maier
University of Marburg
Cell Biology, Karl-von-Frisch Strasse
D-35032 Marburg
Germany

Casey Mc Grath
Department of Biological Sciences
Smith College
Northampton, MA 01063
USA

Peter J. Myler
Seattle Biomedical Research Institute
307 Westlake Ave N, Suite 500
Seattle, WA, 98109–5219
and
Department of Pathobiology and
Division of Biomedical and
Health Informatics
University of Washington
Seattle, WA 98195
USA

David J. Patterson
Marine Biological Laboratory
Woods Hole, MA 02543
USA

Jan Pawlowski
Department of Zoology and Animal Biology
University of Geneva
Sciences III 30, Quai Ernest-Ansermet
CH-1211 Geneva 4
Switzerland

Tatiana A. Rynearson
University of Rhode Island
South Ferry Road
Narrangan Sett, RI, 02882
USA

Pauline Schaap
School of Life Sciences
University of Dundee
MSI/WTB complex
Dow Street, Dundee,
DD1 5EH
UK

Alastair G. B. Simpson
Canadian Institute for Advanced Research
Program in Evolutionary Biology and
Department of Biology
Dalhousie University
Halifax, Nova Scotia, B3H 4J1
Canada

Maik Sommer
University of Marburg
Cell Biology, Karl-von-Frisch Strasse
D-35032 Marburg
Germany

Kenneth D. Stuart
Seattle Biomedical Research Institute
307 Westlake Ave N, Suite 500
Seattle,
WA, 98109–5219
and
Department of Pathobiology
University of Washington
Seattle, WA 98195
USA

Christine Voß,
University of Marburg
Cell Biology, Karl-von-Frisch Strasse

D-35032 Marburg
Germany

Stefan Zauner
University of Marburg
Cell Biology, Karl-von-Frisch Strasse
D-35032 Marburg
Germany

Rebecca Zufall
Department of Biology and Biochemistry
University of Houston
Houston, TX 77204

Introduction

Laura A. Katz[1] and Debashish Bhattacharya[2]

[1] Smith College, MA, USA
[2] University of Iowa, IA, USA

Microbial eukaryotes (i.e. protists), defined as eukaryotes that are neither plants, animals, nor fungi, are essential to many aspects of human health and the environment. These organisms represent at least 1500 million years of evolution, comprise the bulk of eukaryotic lineages, and include members with many key evolutionary innovations. Furthermore, because plants, animals, and fungi arose independently from within microbial groups, elucidating the origins and genetic developments within eukaryotic micro-organisms are necessary to understand fully the evolution of these macroscopic lineages. Hence, genomic studies of eukaryotic microbes are vital for deciphering the biodiversity of life on Earth.

Discoveries from eukaryotic microbes challenge the principles of genome evolution that have emerged mainly from studies of a few plant and animal lineages. Examples of such discoveries, many of which are documented in this volume, include absolute strand polarity in kinetoplastids (including the parasites *Trypanosoma* and *Leishmania*), chromosomal processing in ciliates, and extrachromosomal rDNAs in *Entamoeba* (the causative agent of amoebic dysentery). As genomes have been studied from only a tiny fraction of the diverse tree of eukaryotic microbes, we are confident that many more novel molecular features are yet to be discovered. A wealth of knowledge will result from uncovering both the mechanisms underlying, and the implications of, genome evolution in these taxa.

Elucidating the principles of genome evolution in microbial eukaryotes is, however, a daunting task because so little is known about their basic biology. This challenge is being met, in part, by advances in DNA sequencing technologies that have made the often large genomes of eukaryotic microbes more accessible. High-throughput genomics alleviates the reliance on model systems to understand basic biology, and technological advances in genomics enables the full sequencing of numerous genomes from microbial eukaryotes. Whereas early studies largely focused on parasites with relatively reduced genome sizes, larger genomes from free-living microbial eukaryotes are now (or will soon be) complete (e.g. *Tetrahymena thermophila* (ciliate), *Thalassiosira pseudonana* (diatom), *Dictyostelium discoideum* (cellular slime mold), *Cyanidioschyzon merolae* (red alga)). The availability of these data promises to revolutionize our understanding of eukaryotic genomes, as well as the physiology, systematics, and basic biology of eukaryotes in general.

Goals of this volume

Our intention with this edited volume is to convey a basic understanding of genomics and evolution in microbial eukaryotes to a wide scientific audience, including genome scientists, evolutionary biologists, microbiologists, and ecologists. The chapters in this book focus on evolutionary genomics, but also provide a basic understanding of microbial diversity. The contributing authors were chosen because of their leadership in their fields. Many of the authors contributed to a National Science Foundation- funded workshop entitled "Frontiers in Genomics: Insights into Protist Evolutionary Biology" that was convened at the University of Iowa from June 19 to 21, 2004, and also presented at an ensuing symposium entitled 'Genome Evolution in Microbial Eukaryotes', on June 3, 2004, at the annual meeting of the International Society of Protistologists. This volume is therefore the culmination of a concerted effort by the participants to popularize and focus attention on the impressive achievements in genomic studies of microbial eukaryotes.

Overview of the chapters

This volume is divided into three main sections: (1) an overview of the diversity of eukaryotes (Simpson and Patterson, Chapter 1); (2) a sampling of the diversity of evolutionary genomics in eukaryotic microbes through examination of a limited number of groups (Table 1); (3) analyses of completed genomes (Table 2), including an overview of the approaches used for sequencing protist genomes (Bartholomeu *et al.*, Chapter 9).

The first section, which is represented by a single comprehensive chapter, presents current hypotheses regarding eukaryotic diversity. This section highlights both what is known about eukaryotic diversity, and identifies some of the many remaining questions that remain unanswered.

The second section, on the diversity of evolutionary genomics, highlights a number of groups of eukaryotes (Table 1). Whereas the level of knowledge varies tremendously across eukaryotic clades, unusual and interesting genomic features are emerging from many groups. On the one hand, such features can be exemplified through detailed analyses of the numerous studied genomes from the malaria parasite, *Plasmodium falciparum* and its relatives (Chapter 2, Carlton). At the same time, the tremendous diversity of life cycles and nuclear features within relatively understudied clades such as dinoflagellates (Chapter 3, Hackett and Bhattacharya), ciliates (Chapter 4, McGrath, Zufall, and Katz), and Foraminifera (Chapter 5, Bowser, Habura, and Pawlowski) indicate additional novelties yet to be discovered. Further, evidence from plastid genomes is providing hypotheses on the pattern of symbioses within eukaryotes, while also highlighting areas of uncertainty (Chapter 6, Sommer *et al.*). Finally, this section ends with two chapters that delve into the impact of lateral gene transfers on eukaryotic

Table 1 Exemplar features of genome architecture within clades of eukaryotic microbes

Clades	Feature	Chapter
Comparative genomics of *Plasmodium* species	Conserved synteny coupled with species-specific pathogenicity genes and paralog expansions	2
The genomes of Dinoflagellates	Unusual chromosome structure in both nucleus and plastid, a diverse clade that is still relatively unexplored	3
Ciliates	Genome architecture, including chromosomal processing and amitosis, drives patterns of molecular evolution	4
Foraminifera	Despite limited data, indications are that genomes will be very diverse within these fascinating organisms	5
Photosynthetic organelles and endosymbiosis	Although most plastid genomes share similar architecture, uncovering the history of plastid transfer through endosymbioses remains a major challenge	6
Anaerobic Eukaryotes (multiple clades)	At least some of the metabolism in these lineages has been acquired on multiple occasions through lateral gene transfer	7
Apicomplexan parasites	There have been multiple acquisitions of pathogenicity genes within the Apicomplexa through lateral transfer from diverse hosts	8

Table 2 Some highlights from completed genomes

Clade	Highlight	Chapter
Kinetoplastids	Absolute strand polarity, with long arrays of cotranscribed genes; conserved synteny between species	10
Entamoeba histolytica	High levels of repetitive sequences (rRNAs, tRNAs, transposons), no evidence of telomere repeats	11
Microsporidia	Dramatic reductions, of non-coding and coding material typify this parasitic microbe	12
Nucleomorphs	Impressive parallel reductions in independently derived nucleomorphs (products of secondary endosymbioses)	13
Diatoms	Rich in metabolic genes, intraspecific genomic variation	14
Dictyostelium	Gene-dense, high numbers of repetitive elements (transposons, simple repeats)	15

genomes, first in anaerobic eukaryotes (Chapter 7, Andersson) and second in apicomplexan parasites (Chapter 8, Huang and Kissinger).

The third section of the book provides detailed synthesis of discoveries in some of the completed genomes from eukaryotic microbes (Table 2). This section starts with a presentation of approaches for characterizing genomes from eukaryotic microbes (Chapter 9, Bartholomeu, Hall, and Carlton). From here, discoveries from completed genomes are presented, including: absolute strand polarity in kinetoplastids (Chapter 10, Stuart and Myler); the expansion of repetitive elements in *Entamoeba* (Chapter 11, Clark); the dramatic reduction of genome features (i.e. intervening sequences, introns, genes) in microsporidians (Chapter 12, Keeling) and in nucleomorphs (Chapter 13, Kawach *et al.*); the diversity of metabolic genes in the diatom *Thalassiosira* (Chapter 14, Armbrust, Rynearson, and Jenkins); and the high gene richness in the slime mold *Dictyostelium* (Chapter 15, Schaap).

Future directions

As genomes have been characterized from only a small fraction of eukaryotic diversity, we are confident that additional novel genomic properties will emerge from future studies. Genomes are virtually unexplored in some groups of eukaryotes, including the proposed supergroup "Cercozoa" and many of the diverse amoeboid groups (i.e. Radiolaria, Foraminifera, testate amoebae). We anticipate that existing books on genome evolution will have to be thoroughly revised to incorporate the diverse features that will emerge from studies of eukaryotic microbes.

Acknowledgements

We are very grateful to three institutions for providing financial support for this endeavor. We received a grant from the National Science Foundation for both a workshop on Protist Genomics and a follow-up symposium at the 2004 meetings of the International Society of Protistology. The latter symposium also received generous support from both the Burroughs–Wellcome Fund and the Sloan Foundation – in addition to supporting the symposium speakers, these organizations provided travel funds for numerous students and postdoctoral fellows.

PART I

CHAPTER 1

Current perspectives on high-level groupings of protists

Alastair G. B. Simpson[1] and David J. Patterson[2]

[1] Department of Biology, Dalhousie University, Halifax, Nova Scotia, Canada
[2] Marine Biological Laboratory, Woods Hole, MA, USA

Introduction

Most of the major branches of eukaryotic diversification are protistan. The modern understanding of protistan diversity and evolutionary relationships is based on morphological studies, especially by transmission electron microscopy, and on molecular phylogenetic studies, initially dominated by small subunit ribosomal RNA but increasingly involving multigene datasets. Morphological studies define approximately 60 robust lineages of eukaryotes. Current syntheses, primarily based on molecular phylogenetic studies, place these lineages in six to eight "supergroups". However, this catalogue is incomplete as the definition of most supergroups is not clear and their extent and compositions are disputed, and a further 25 lineages, for which there are electron microscopical data and/or molecular data, are not currently assigned to any supergroup. We review the current perspective on eukaryote diversity and phylogeny, emphasizing protistan groups of economic, ecological, or medical significance, and, especially, those of evolutionary importance, and/or of uncertain placement within eukaryotes.

Understanding the diversity and interrelationships amongst protists (eukaryotes other than animals, land plants, and true fungi) is critical to understanding the evolution and full capabilities of the eukaryotic cells and their genomes. By any measure, protists are highly diverse. There are about 200 000 described species of protists – about the same number as for plants or fungi.

About 60 major types of protists can be identified, and the familiar multicellular groups – animals, true fungi, and land plants – all evolved independently from protistan ancestors. Protists also exhibit tremendous variety in their genome architecture. The smallest reported eukaryotic nuclear genomes are the nucleomorphs of the photosynthetic cryptophyte and chlorarachniophyte protists at ~0.5 MB (Gilson and McFadden 2002), while the largest, an estimated 10^6 MB, have been reported in some amoebae (Graur and Li 1999; McGrath and Katz 2004). Chromosome numbers may be double the number found in multicellular organisms, non-standard genetic codes abound, and genome organizations include the histone-depauperate chromosomes of dinoflagellates and nuclear dualism in ciliates and foraminifera (Graur and Li 1999; Hausmann *et al.* 2003; McGrath and Katz 2004).

A general evolutionary schema is a statement of the range and scope of what can be done by eukaryotic cells, and helps reveal the role of less familiar evolutionary drivers such as symbiosis and lateral gene transfer. This chapter summarizes the current thinking about high-level groupings among protists.

A brief history of understanding protistan diversity and phylogeny

Our understanding of the diversity and phylogeny of protists has progressed in three major phases

7

Table 1.1 A hierarchical arrangement of eukaryote groups based on the list of "ultrastructural identities" from Patterson (1999), now located in the context of emerging supergroups and other higher-level groupings

Opisthokonts			
	Animals (Metazoa)		
	Choanoflagellates		
	Ichthyosporea/		
	Mesomycetozoea/DRIPs		
	Ministeria		
	Capsaspora		
	Fungi (including chytrids and microsporidia)		
	Nucleariidae		
Ramicristates / "Amoebozoa"			
	Lobose amoebae		
	Mycetozoa		
	Pelobionts and entamoebae		
"Plantae"/ Archaeplastida			
	Glaucophytes		
	Viridaeplantae ("green algae", including "prasinophytes", and plants)		
	Rhodophytes		
"Stramenopiles and alveolates"			
	Stramenopiles		
		Bicosoecids	
		Actinophryids	
		Oomycetes	
		Stramenochromes	
		Slopalinids	
	Alveolates		
		Dinoflagellates	
		Apicomplexa	
		Ciliates	
		Colpodellidae	
		Perkinsids	
"Rhizaria"			
	Cercozoa		
		core Cercozoa	
			Cercomonads
			Thaumatomonads
			Euglyphids
			Gymnophryids
			Chlorarachniophytes
			Cryothecomonas
			Spongomonads?
			Phaeodarea
			Desmothoracids
		Plasmodiophorids/ Phytomyxea	
	Granuloreticulosea (~Foraminifera)		
	"Radiolaria"		
		Acantharea	
		Polycystinea	
		Sticholonche	

Table 1.1 (*Contd.*)

	Haplosporids		
Excavates			
	Preaxostyla		
		Oxymonads	
		Trimastix	
	Malawimonas		
	Excavate Clade 1 / "Trichozoa"		
		Carpediemonas	
		Diplomonads	
		Retortamonads	
		Parabasalids	
	Excavate Clade 2		
		Jakobids	
		Euglenozoa	
			Euglenids
			Kinetoplastids
			Diplonemids
			Postgaardi
		Heterolobosea	
Residual UIs			
	Ancyromonas		
	Apusomonads		
	Biomyxa		
	Centroheliozoa		
	Coelosporidium		
	Copromyxids		
	Cryptomonads and Kathablepharids		
	Dimorphids		
	Discocelis		
	Ebriids		
	Fonticula		
	Gymnosphaerida		
	Haptophytes		
	Komokiacea		
	Luffisphaera		
	Multicilia		
	Nephridiophagids		
	Paramyxea		
	Phagodinium		
	Phalansterium		
	Pseudospora		
	Schizoclades		
	Spironemidae		
	Stephanopogon		
	Telonema		
	Vampyrellids		
Plus 200 or so residual genera			

Inverted commas indicate names of an ambiguous nature or a provisional nature, or groups of questionable robustness. For other details see Patterson 1999 and micro*scope (http://microscope.mbl.edu).

with successive and cumulative insights. Each is defined by the application of a new technology.

The first period was driven by light microscopy. The first observations of microbial protists by Leeuwenhoek and Hook in the latter seventeenth century were followed about a hundred years later by O. F. Müller's first catalog of microbial diversity. Much of the descriptive work that followed improvements in lens design over the first half of the nineteenth century was assembled by Bütschli, who more or less defined our understanding from the 1880s until the 1960s. Light microscopy has been the principal contributor to the description of about 200 000 species of protists.

The second phase began in the 1960s and reflects the influence of transmission electron microscopy. As of 1999, about 70 types of protists were distinguished on the basis of their sub-cellular morphology, or "ultrastructure" (Patterson 1999). Members of each type share a distinctive cellular architecture that takes into account the structure of the flagella (if present), their anchorage and elaborations, other cytoskeletal elements, plastids (chloroplasts and homologous organelles), mitochondria, nuclei and nuclear division processes, extrusomes (organelles that rapidly discharge material from the cell, often for prey capture or defense against predation), and so on. Types defined on the basis of ultrastructural distinctiveness have proven to be robust – although some have since been brought together into higher-level groupings by molecular phylogenetic studies, none or one only have been found to be based on false homology (i.e. convergent evolution). Table 1.1 contains an updated listing of those types, referred to here as ultrastructural identity (UI) lineages. Previously undocumented UI lineages, which undoubtedly exist, are described rarely.

The third phase reflects the introduction of molecular phylogenetics. In the case of protists, this phase really began with comparisons of small subunit ribosomal RNA (SSU rRNA) sequences in the early 1980s (McCarroll *et al.* 1983). By the 1990s, molecular techniques had confirmed many of the UI lineages and were identifying new deep relationships, for example uniting dinoflagellates, apicomplexan parasites, and ciliates as the alveolates (Gajadhar *et al.* 1991). The breadth of the

SSU rRNA dataset has increased dramatically since the mid-1990s, greatly improving our understanding of eukaryote phylogeny.

Molecular approaches were initially expected to reveal relationships amongst protists with little ambiguity, but this has not always been the case. In the late 1980s, SSU rRNA trees were characterized by a ladder-like series of very deep branches at the base of the eukaryotes, with the deepest branches being protists that lacked mitochondria (Sogin 1989; Sogin *et al.* 1989). This topology defined our understanding of eukaryotic evolution for nearly a decade and led to evolutionary models in which mitochondria were a relatively late acquisition within eukaryotes. However, this ladder-like tree is now attributed to artifacts caused by an inability of early phylogenetic methods to accommodate drastically unequal rates of sequence evolution across taxa (Philippe *et al.* 2000) and to poor taxon sampling.

The loss of this framework revealed a need for further change, such as more sophisticated analytical techniques or the use of multiple molecular markers. The last decade has seen the emergence of substantial datasets for additional genes, such as large subunit ribosomal RNA, nuclear protein-coding genes, and mitochondrial genomes. Increasingly, multiple genes are combined into large datasets and analyzed together. In parallel, discrete molecular features, such as large insertions/deletions in protein sequences, lateral gene transfers or gene replacements, and gene fusions/fragmentations, are bring employed as molecular apomorphies (characters derived from a single evolutionary event) to define events in eukaryotic history (Philippe *et al.* 2000; Stechmann and Cavalier-Smith 2002; Richards and Cavalier-Smith 2005). The greatest impact is still to come as genomic approaches are extended to sample across eukaryotic diversity.

Emerging supergroups

Current views of eukaryote phylogeny generally place protistan diversity in a few very large groupings informally labeled "supergroups" (Baldauf 2003; Simpson and Roger 2004c; Adl *et al.* 2005). Six supergroups seem to be fairly widely adopted (though see below), are exclusive of each other, and are based on various data including molecular phylogenies, morphological data, and discrete molecular characters (Fig. 1.1). The current supergroup-based accounts are not comprehensive catalogues of eukaryote diversity, as over a third of the UI lineages of protists identified by Patterson (1999) are not yet included in these supergroups, sometimes in spite of the availability of useful molecular sequence data. Many supergroup concepts are imprecise and some are actively disputed, especially excavates and the territory around stramenopiles and alveolates (Fig. 1.1).

Defining high level groups

Taxonomic treatments, especially high-level ones such as those dealing with protists, are characterized by considerable differences of opinion. There are at least four distinct sources of these disputes: (i) disagreements about the actual phylogeny of the organisms concerned; (ii) different views about the relationship between phylogeny and taxonomy; (iii) different preferences about which taxon names to present in a summary taxonomy; and (iv) the use of the same name to refer to different groups or concepts.

Our discussion here is concerned only with the last area – what the supergroup names are used to refer to. Our communications would be much improved if each name would have a single meaning, and retain that meaning through time. At present, this does not happen. Taxon names may be defined in several ways: (i) by reference to the content (composition) of the taxon referred to; (ii) by the general appearance of contained organisms (circumscription); (iii) by the evolutionary innovation (apomorphy) that distinguishes the lineage; or (iv) with reference to a branching point within a tree structure (Patterson 1999). The first two promote instability, because in the first case names would have very short lives, and in the second case the definition would be ambiguous. By contrast, the latter two are less ambiguous, although the last approach makes naming contingent on particular trees.

Names defined by reference to evolutionary innovation (apomorphies) seem the simplest to apply and understand if an apomorphy can be identified. In the absence of clear apomorphies (e.g. in many new groups proposed on the basis of molecular phylogenies), names are often defined by composition based on proposed evolutionary relationships and/or vague circumscriptions. In these cases, changes in the understanding of evolutionary relationships (and other factors) require a choice as to which of several legitimate group concepts (if any) should inherit the old name. Thus names such as Rhizaria have referred to quite different things in different publications (Cavalier-Smith 2002 vs Cavalier-Smith 2004) such that, despite efforts to promote stability through a "node-based" definition (Nikolaev *et al.* 2004), we are left with uncertainty as to what is intended when the name is used. Where possible, we provide definitions for other major taxon names to clarify current usage.

The diversity of protists

The remainder of this chapter is an account of the higher-level diversity of protistan eukaryotes. One object is to place the groups whose genome biology is described in subsequent chapters into phylogenetic context. A second object is to highlight groups that appear distinctive by one or more criteria, but whose phylogenetic position within eukaryotes remains uncertain. This account is updated and expanded from Patterson (1999), to which the reader is directed for more thorough reference to the pre-1999 literature. Readers interested the lower-level diversity of protists are directed to the dynamic treatment available at http://microscope.mbl.edu. While attempting to keep this account up-to-date, we have also been conservative when there is strong ongoing dispute about evolutionary history (e.g. the chromalveolate hypothesis), or where inferences are based on limited data and await repetition (e.g. Paramyxea as possible Haplosporidia). In the account of diversity of higher-level protists we have used the following protocol for describing the hierarchy: groups are headed by large, bold font; sub-groups are headed by small, bold font; sub-sub-groups are headed by italics; and sub-sub-sub-groups are bulleted.

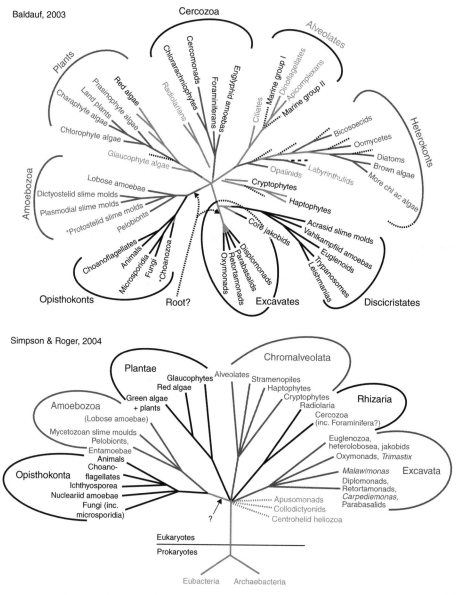

Figure 1.1 Two recently published diagrams that summarize the contemporary understanding of the deep-level diversity of eukaryotes. The schemes are similar with minor differences in emphasis. Baldauf (2003) is agnostic concerning the monophyly of chromalveolates and excavates (considering the possibility of eukaryotes being rooted within excavates), while Simpson and Roger (2004) show both of these groups as monophyletic. These differences illustrate the current uncertainty and incomplete consensus concerning the supergroups. The core jakobids of Baldauf are equivalent to jakobids of Simpson and Roger, and heterokonts approximate to stramenopiles. "Euglenoids", leishmanias, and trypanosomes are subgroups of Euglenozoa, while acrasid slime molds and vahlkampfid amoebae are subtypes of Heterolobosea. The reference to "Choanozoa" in the upper figure perhaps refers collectively to Ichthyosporea and nucleariids (Choanozoa is a rarely-used paraphyletic taxon that includes the various non-animal, non-fungal opisthokonts). Asterisks in the upper diagram represent likely paraphyletic groups. Reproduced from: Baldauf, S. L. (2003). The deep roots of eukaryotes. *Science*, **300**, 1703–1706. Copyright (2003), with permission from the American Association for the Advancement of Science; and from: Simpson, A. G. B. and Roger, A. J. (2004). The real "kingdoms" of eukaryotes. *Current Biology*, **14**, R693–696. Copyright (2004), with permission from Elsevier.

Opisthokonts

A group that includes animals and true fungi and their protist relatives such as choanoflagellates, to which have more recently been added Ichthyosporea (DRIPs) and nucleariids (Fig. 1.3). Although the group embraces taxa that are diverse in morphology and life history, cells of undisputed opisthokonts have a single emergent flagellum which projects behind the cells while they swim. This is treated as defining the opisthokont ancestry. Numerous molecular phylogenies support the monophyly of animals and fungi, and taxon-rich rRNA phylogenies unite all opisthokonts (Cavalier-Smith and Chao 2003a; Ruiz-Trillo et al. 2004; see example SSU rRNA tree in Fig. 1.2). *Ministeria* and *Capsaspora* are small pseudopodia-forming protists of uncertain placement within opisthokonts.

Animals (Metazoa): Animals, including sponges, arose from a common multicellular ancestor. While overlapping in size with large protists (e.g. ciliates, Foraminifera), animals monopolize the large "consumer" niches on earth. Myxozoa, animal parasites that produce spores with several differentiated cells, and previously treated as a group of protozoa, are derived animals (Cnidaria).

Choanoflagellates: Solitary or colonial flagellates that capture small particles using a collar of fine pseudopodia that surrounds the flagellum. Acanthoecid choanoflagellates are numerically significant in marine plankton and produce elaborate siliceous loricae (a lorica is an extracellular organic and/or mineralized structure that encloses a cell – "tests" are similar). Choanoflagellates are of evolutionary interest as the closest relatives of animals. The isolated parasite *Corallochytrium* is argued to be related to choanoflagellates (Cavalier-Smith and Chao 2003a).

Ichthyosporea (also referred to as **Mesomycetozoea** or **DRIPs**): This group contains an array of protistan parasites that form a clade in SSU rRNA trees. Most infect marine animals. The original name "DRIPs" was an acronym based on the names of included genera, and has been replaced by Ichthyosporea. Mesomycetozoea is a later synonym (Mendoza et al. 2002).

Fungi: The colloquial term "fungus" refers to a polyphyletic group that includes oomycetes along with the "true" fungi. The true fungi are a clade within the opisthokonts that have the chytrids as a sister or basal taxon. True fungi usually absorb dissolved nutrients through a system of elongate hyphae, and often have elaborate differentiated structures for sexual reproduction. True fungi are of major importance in the environment, especially on land, both as free-living organisms and as symbionts and parasites of plants (and to a lesser extent, animals). Chytrids produce flagellated dispersal stages and are included within the fungi or treated as the sister group. Microsporidia are a significant group of spore-forming intracellular parasites that used to be regarded as primitive protists but are now seen as a derived type of fungus (Van de Peer et al. 2000) with tiny genomes (Keeling, Chapter 12, this book).

Nucleariidae: A small group of amoebae with fine, extrusome-lacking, "filose" pseudopodia. There is no known flagellate phase. Some rRNA trees indicate a relationship to fungi (Ruiz-Trillo et al. 2004).

Ramicristates ="Amoebozoa" *sensu* Cavalier-Smith

The name "ramicristates" refers to the distinctive, branched mitochondrial cristae morphology typical for this group. Most ramicristates are amoebae and most classical "lobose" amoebae fall within this assemblage. Some groups have a flagellate phase in their life cycle (some are only flagellates) (Fig. 1.4). One large group – mycetozoan slime molds – exhibit limited multicellularity. Early SSU rRNA analyses suggested that both lobose amoebae and Mycetozoa were polyphyletic assemblages, however protein phylogenies (actin, elongation factors) and recent SSU rRNA studies suggest that they are all closely related (Baldauf and Doolittle 1997; Fahrni et al. 2003). Relationships between the amitochondriate "pelobionts plus entamoebae" assemblage and Mycetozoa and/or lobose amoebae are more recent proposals, supported through concatenated multigene phylogenies and taxon-rich SSU rRNA analyses (Bapteste et al. 2002). Some SSU rRNA trees or single morphological characters suggest that the flagellates *Phalansterium*

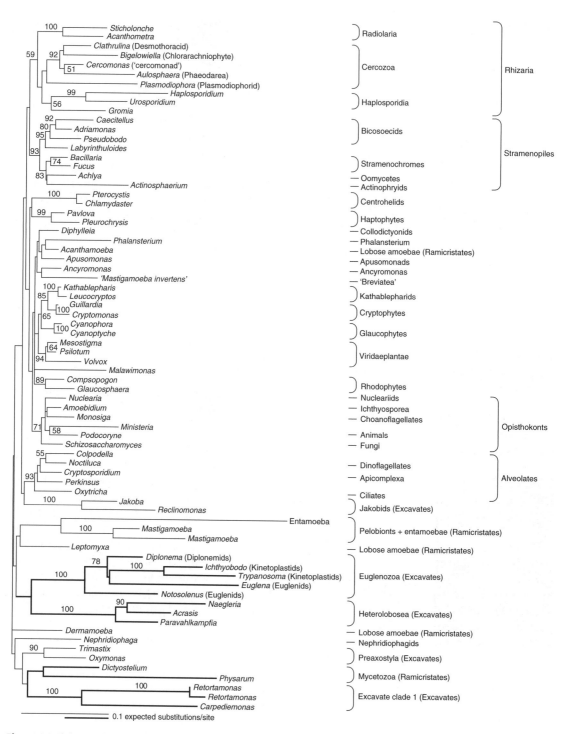

Figure 1.2 Phylogeny of eukaryotic small subunit ribosomal RNA genes. Several "long-branching" taxa have not been included. The phylogeny is estimated from maximum likelihood distances from 1165 well-aligned sites under a TrN + Γ + I model by minimum evolution (50 random addition sequences + TBR rearrangements). Bootstrap support values are calculated based on 500 replicates (three addition sequences + TBR rearrangements). Bootstrap values <50% are not shown. Lengths of thick branches are depicted at 50% of the scale of the rest of the tree. Note that not all relationships reported in the text as supported by previous SSU rRNA phylogenies are recovered in this one tree.

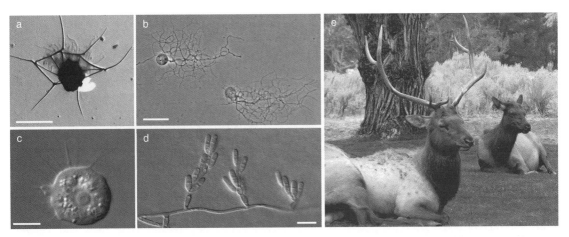

Figure 1.3 Some opisthokonts. (a) The choanoflagellate *Campyloacantha* with cell body surrounded by siliceous lorica; (b) *Chytriomyces*, a chytrid; (c) *Nuclearia*, a filose amoeba; (d) the fungus *Dactylaria* with hyphae and spores; (e) *Cervus*, one of the charismatic megabiota within animals. All scale bars = 10 μm. All images, D. J. Patterson and provided by micro*scope (http://microscope.mbl.edu).

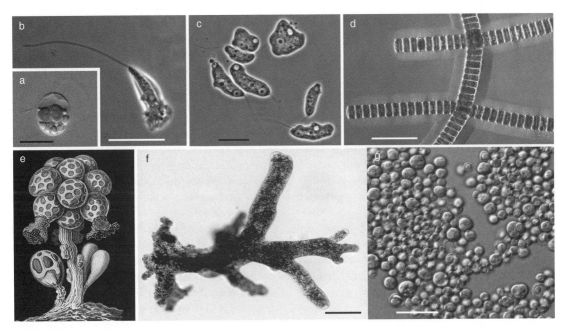

Figure 1.4 Some members of Amoebozoa and "Plantae". (a) *Cyanophora*, a glaucophyte; (b) *Mastigamoeba*, a pelobiont flagellate; (c) mostly flagellated forms of the mycetozoan slime mold, *Planoprotostelium*; (d) *Desmidium*, a filamentous desmid (streptophyte green alga) with cells surrounded by a layer of mucus; (e) fruiting body of the mycetozoan myxogastriid slime mold *Trichia*; (f) *Chaos*, a lobose amoeba; (g) *Cyanidium*, a coccoid red alga from hyperthermal sites. All scale bars = 10 μm, excepting (d) and (f) which represent 100 μm. All images, D. J. Patterson, excepting (e) Ernst Haeckel, and provided by micro*scope (http://microscope.mbl.edu).

and *Multicilia* are ramicristates (see below), but more data are required to confirm this.

Lobose amoebae: Lobose amoebae *sensu lato* are unicellular amoebae that include species that vary in size over three orders of magnitude. They lack a flagellate phase. Some, such as the arcellids, produce a test. SSU rRNA and actin phylogenies identify several major lineages within lobose amoebae, but tend to indicate that the group is paraphyletic (i.e. gave rise to other Amoebozoa – Fahrni *et al.* 2003; Smirnov *et al.* 2005).

Mycetozoa: There are several types of slime molds, two major groups of which (myxogastriids and dictyostelids) are often placed together within the contemporary concept of Mycetozoa. All slime molds occur as amoebae while feeding, but form stalked structures that bear spores or cysts for dispersal. This general life history has evolved multiple times (see "Heterolobosea"). Contemporary molecular phylogenies generally unite myxogastriids and dictyostelids (Baldauf and Doolittle 1997; Fahrni *et al.* 2003). In dictyostelids individual amoebae aggregate and some individuals sacrifice themselves to form a stalk, while the rest encyst. Myxogastriids, also referred to as myxomycetes, grow into large diploid plasmodia that differentiate into stalked fruiting bodies and produce numerous haploid spores. A poorly understood assemblage, "protostelids", may have given rise to both dictyostelids and myxogastriids. *Dictyostelium* is widely used as an experimental model system (see Schaap, Chapter 15, this book).

Pelobionts and entamoebae: Pelobionts are amoeboid organisms, mostly free-living, and usually with one long flagellum. Entamoebae are mostly endobiotic amoebae of animals (usually intestinal). The serious human pathogen *Entamoeba histolytica* is quite intensively studied (see Clark, Chapter 11, this book). Pelobionts and entamoebae lack classical mitochondria. Mitosomes of entamoebae are regarded as remnant mitochondria but do not function in energy metabolism (Tovar *et al.* 1999). A relationship between pelobionts and entamoebae was confirmed by SSU rRNA analyses and concatenated protein trees (Silberman *et al.* 1999; Bapteste *et al.* 2002), but the internal relationships are poorly

understood. Pelobionts plus entamoebae are sometimes referred to as "Archamoebae", but this term has been applied to many different groupings and group concepts over time, and so is ambiguous.

"Plantae" (Archaeplastida)

Primary endosymbiosis describes the conversion of a prokaryotic cell into a dependent organelle of a eukaryote. Three main groups of eukaryotes have primary plastids: glaucophytes, red algae, and "green algae plus land plants" (here called Viridaeplantae). Plastid gene trees and genome organization show plastid genomes as a monophyletic group within cyanobacteria, consistent with the proposal that these three groups descended from a common plastid-bearing ancestor (see Sommer *et al.*, Chapter 6, this book). While there is little other morphological similarity between these three groups, and SSU rRNA analyses tend to place them as separate lineages, large-scale concatenated gene phylogenies unite red algae, Viridaeplantae, and glaucophytes as a clade within a more limited sampling of other eukaryotes (Moreira *et al.* 2000; Rodriguez-Ezpeleta *et al.* 2005). The use of the name "Plantae" for this group by many protistologists causes some confusion because this name is usually used in botanical sciences for narrower groups (the group here called "Viridaeplantae" or for land plants/embryophytes). The alternative name Archaeplastida has recently been introduced (Adl *et al.* 2005).

Glaucophytes (Glaucocystophytes): A small group of unicellular or colonial freshwater algae. Of seemingly minor ecological importance, they are of evolutionary interest because their plastids retain many prokaryote-like features such as the peptidoglycan wall and phycobilisomes (large complexes of accessory pigments attached along the internal thylakoid membranes within the plastid).

Rhodophytes (red algae): Most members of this large group are sessile multicellular algae with complex sexual life cycles. There are some unicellular forms, including the possibly basal Cyanidiales. Many are parasites of other rhodophytes. Their plastids retain phycobilisomes. Red algae completely lack flagella. Several red algae are of commercial importance (producing agar, carrageenan, nori, etc.).

Viridaeplantae (Chloroplastida): This group includes taxa with cellulosic cell walls and plastids with chlorophylls *a* and *b* – the only primary plastids to do so. The group embraces the land plants (embryophytes), and the "green algae" inclusive of the "prasinophytes". The earliest-diverging lineages are the "prasinophytes" – mostly scale-bearing unicellular flagellates from marine environments. Unique among the "Plantae", some prasinophytes might be capable of phagocytosis (see Graham and Wilcox 2000). (Almost) all other Viridaeplantae form two apparently monophyletic groups: chlorophytes (sensu stricto) and streptophytes. The chlorophyte green algae include unicells (e.g. *Chlamydomonas*), several multicellular groups, and the algal components of many lichens. The streptophytes are mostly multicellular (desmids and relatives being the major exceptions), and include land plants, which have a close evolutionary relationship with highly differentiated freshwater algae – "charophytes" (see Lewis and McCourt 2004). The name Viridaeplantae is not commonly used by botanists (see "Plantae" above) and the name "Chloroplastida" was suggested recently as a replacement (Adl *et al.* 2005).

Stramenopiles and alveolates

The stramenopiles and alveolates are both diverse and species rich groups within eukaryotes (Fig. 1.5; Fig. 1.6. Several combined protein phylogenies, and some SSU rRNA analyses have been used to argue that they are related to the exclusion of other well-studied groups (Van de Peer and De Wachter 1997; Baldauf *et al.* 2000; Bapteste *et al.* 2002; Simpson and Roger 2004b; Harper *et al.* 2005). Supporting arguments are based on the presence in both lineages of similar "chromalveolate" plastids (with chlorophylls *a* and *c*, and 3–4 bounding membranes). Similar plastids are also present in haptophytes and cryptomonads (although the cryptomonad plastid also retains phycobilins and a nucleomorph). Many consider that chromalveolate plastids are the result of secondary symbioses

(the engulfment of a eukaryote with a primary plastid), but the number of symbioses is disputed. Concatenated plastid protein trees indicate a common ancestry for the plastids of stramenopiles, haptophytes, and cryptomonads at least (Yoon *et al.* 2002). Furthermore, all four lineages share a distinctive plastid-targeted glyceraldehyde-3-phosphate dehydrogenase (GAPDH) gene that is related to cytosolic forms, rather than the plastid-targeted GAPDH of Plantae (Fast *et al.* 2001, Harper and Keeling 2003). A second similar gene replacement involving fructose-1,6-bisphosphate isomerase (FBA) was reported recently (Patron *et al.* 2004). These data are have been used to underpin an argument for the single acquisition of chromalveolate-type plastids by a common ancestor of stramenopiles, alveolates, haptophytes, and cryptomonads (Fast *et al.* 2001). This is termed the "chromalveolate hypothesis" (Cavalier-Smith 1999). It implies a plastid-bearing origin for many heterotrophic eukaryotes, including ciliates, and oomycetes (water molds). Unfortunately for this hypothesis, no nuclear gene phylogeny unites haptophytes and cryptomonads with stramenopiles and alveolates (Harper *et al.* 2005). Molecular phylogenies generally place plastid-lacking groups in basal positions within cryptomonads, stramenopiles, and alveolates, and this is consistent with independent acquisitions of plastids within each lineage. Plastid remnants have not been found in some well-studied plastid-lacking alveolates/stramenopiles, such as oomycetes or ciliates. Other explanations have been presented to explain the GAPDH data and FBA data that do not require the chromalveolate hypothesis but invoke "tertiary symbioses" (i.e. the acquisition of plastids by engulfment of a eukaryote with a secondary plastid), within stramenopiles, cryptomonads, haptophytes, and/or alveolates (Bodyl 2005).

Some workers group stramenopiles, cryptomonads, and haptophytes as "chromists" or "Chromista", reflecting the idea that they arose from a common secondary endosymbiosis (see Cavalier-Smith 1989). "Chromista" is distinct from the newer "chromalveolate" concept because it excludes alveolates. If a relationship is proven between stramenopiles and alveolates, it would imply that "Chromista" is not monophyletic.

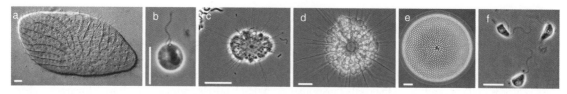

Figure 1.5 Some members of the stramenopiles. (a) *Opalina*, an endobiotic opaline flagellate found in the rectal regions of anuran amphibia; (b) *Ochromonas*, with a short and a long flagellum, and a plastid; (c) *Ciliophrys*, with stiff radiating arms and a single flagellum with a slow figure-of-eight beating motion; (d) *Actinophrys*, a heliozoon; (e) *Coscinodiscus*, shell of a centric diatom; (f) *Cafeteria*, a heterotrophic flagellate. All scale bars = 10 μm. All images, D. J. Patterson and provided by micro*scope (http://microscope.mbl.edu).

Stramenopiles: A large and diverse group defined as organisms with stiff tripartite hairs (i.e. with a distinct base, a long hollow shaft, and terminal filaments) attached to one flagellum, or descendents of such organisms (Fig. 1.5). The group embraces some of the largest multicellular organisms (kelp), the most species rich of the protists (the diatoms) as well as many photosynthetic flagellates, mixotrophs (used here to refer to organisms capable both of photosynthesis and heterotrophic nutrition by phagotrophy), free-living and parasitic heterotrophs, and fungal-like oomycete protists. Recent culture-based studies and environmental PCR studies reveal several additional poorly understood marine stramenopile lineages thought, or known, to be heterotrophic flagellates (Massana *et al.* 2004). There are a number of other names that refer to overlapping groups of organisms – such as "heterokonts" and "chromists" (see above). The name stramenopiles is explicitly defined by the synapomorphy of tripartite flagellar hairs, which are retained by most groups, but have been secondarily lost several times.

Bicosoecids are small unicellular plastid-lacking bacterivorous stramenopiles probably of ecological importance, but understudied. Ultrastructural studies and SSU rRNA phylogenies indicate that some bacterivorous flagellates, such as *Caecitellus* and pseudodendromonads, without flagellar hairs and previously considered separate lineages (Patterson 1999), are derived bicosoecids (O'Kelly and Nerad 1998; Karpov *et al.* 2001; and see Fig. 1.2) or sister groups to them.

Actinophryids: Traditionally, one of the types of protists with a rounded body and stiff radiating pseudopods (axopods) that were grouped as "heliozoa". Morphological and molecular studies indicate that the heliozoan morphology arose multiple times (Smith and Patterson 1986; Nikolaev *et al.* 2004). In actinophryids, the axonemes that support the pseudopodia often start against the nuclear envelope. A position within stramenopiles is supported by morphological arguments and SSU rRNA analyses (see Fig. 1.2 and Nikolaev *et al.* 2004).

Oomycetes: These "water molds" (and related taxa) produce hyphae, and so resemble true fungi. Unlike true fungi, they produce life-history stages that are biflagellated with tripartite flagellar hairs. The group includes several serious plant pathogens (e.g. *Phytophthora infestans* – the cause of potato blight). Oomycetes are the best-studied non-photosynthetic stramenopiles in terms of genomic data and cell biology. Other vaguely fungal-like stramenopiles include hyphochytrids, labyrinthulids (slime-nets), and thraustochytrids.

Stramenochromes ("~heterokont algae"): This includes those stramenopiles with plastids or derived from such organisms. Photosynthetic stramenopiles are extremely diverse. Phycologists currently recognize around a dozen major groups (most are usually ranked as "classes") several of which have been identified since 1999 (see Andersen 2004, for review). Briefly (and not exhaustively), chrysophycea and the related synurophycea are mostly free-swimming and are most abundant (and sometimes dominating) in freshwater. Raphidophyceae are often large free-swimming algae while Dictyochophyceae are flagellated but tend to produce fine pseudopodia. Chrysophycea and dictyochophycea are often mixotrophic (or purely phagotrophic). The closely related xanthophycea, phaeothamniophycea, and phaeophycaea (brown algae) tend to be non-motile and often filamentous, or, in the case of many brown algae, multicellular organisms of large or huge size and several differentiated tissue types (kelp). The most important photosynthetic stramenopiles are the diatoms (Bacillariophyta/Bacillariophyceae) which dominate the >20 μm fraction of the phytoplankton in many aquatic systems. Diatom cells (other than some gametes) lack flagella and are enclosed by the frustule – essentially a lidded box made of highly perforated silica. Their cell biology and genome biology are increasingly investigated (see Armbrust *et al.*, Chapter 14, this book). Other groups, mostly of small or otherwise inconspicuous cells, include eustigmatophycea, pelagophycea, and bolidophycea (the latter group is actually closely related to diatoms). The chromalveolate hypothesis holds that the ancestral stramenopile had a plastid, and therefore would not predict that photosynthetic stramenopiles would form a clade within stramenopiles. However, well-sampled SSU rRNA phylogenies almost always recover a monophyletic "stramenochromes" grouping deeply nested within stramenopiles, though rarely with strong statistical support (Leipe *et al.* 1996; Kostka *et al.* 2004). Studies of other comparable data are required.

Slopalines: Slopalines are multiflagellated intestinal endobionts. Proteromonads have two to four flagella, while the true opalines have flagella arranged in rows and so look superficially like ciliates. The opalines lack tripartite flagellar hairs, although proteromonads have such hairs on the cell surface. Early SSU rRNA phylogenies included sequences referred to as opalines but that had been misidentified. Recent studies unite opalines and proteromonads with *Blastocystis* (a non-flagellated anaerobe) as a clade within stramenopiles (Kostka *et al.* 2004).

Alveolates: This is a species rich group composed primarily of ciliates, dinoflagellates, and Apicomplexa (Fig. 1.6). These groups were united by SSU rRNA phylogenies (Gajadhar *et al.* 1991), and by the shared presence of cortical alveoli – a system of flattened membrane vesicles underneath the cell membrane (Cavalier-Smith 1991). This synapomorphy defines the group. The alveoli-bearing predatory flagellate *Colponema* is considered by some as an alveolate, but this is not consistent with structural data and there are as yet no molecular data.

Dinoflagellates: a diverse assemblage of photosynthetic, heterotrophic, and mixotrophic protists. They include many parasites, and the photosymbionts of corals. Dinoflagellate nuclear DNA is permanently condensed but is not conventionally packaged using histones. This "dinokaryon" structure is usually considered to be apomorphic, and defines the group. Most dinoflagellates are biflagellated and are unicellular or form small colonies. Many dinoflagellates specialize in consuming other eukaryotic cells and even photosynthetic forms are often predatory. Most photosynthetic dinoflagellates have a plastid with the distinctive accessory pigment peridinin in addition to chlorophylls *a* and *c*, and some groups have acquired other types of plastids from other photosynthetic eukaryotes (see Hackett and Bhattacharya, Chapter 3, this book). On the basis of molecular phylogenies, the predator *Oxyrrhis*, and perhaps syndineans are thought to be sister groups or basal to typical dinoflagellates (Saldarriaga *et al.* 2003).

Apicomplexa: Apicomplexa are obligate parasites of animals and they generally use an apical complex (a collection of secretory organelles that discharge at the apical end of the cell and are associated with a short tubular conoid) to invade host cells. One group, gregarines, are usually large cells that live in gut lumen or coelom. Malaria, the most serious deadly eukaryote parasite of humans, killing more than a million people per year, is caused by *Plasmodium* – an apicomplexan. Most Apicomplexa, including *Plasmodium*, have a remnant non-photosynthetic plastid – the apicoplast (see Carlton, Chapter 2 and Huang and Kissinger, Chapter 8, this book).

Perkinsids (parasites of marine invertebrates or protists, with a flagellated dispersal stage) and Colpodellidae (small flagellated predators) have an apical complex-like structure with "open" conoid (shaped like a curved sheet rather than a tube). Molecular phylogenies often position Colpodellidae as the sister group to true Apicomplexa and perkinsids closer to dinoflagellates, indicating

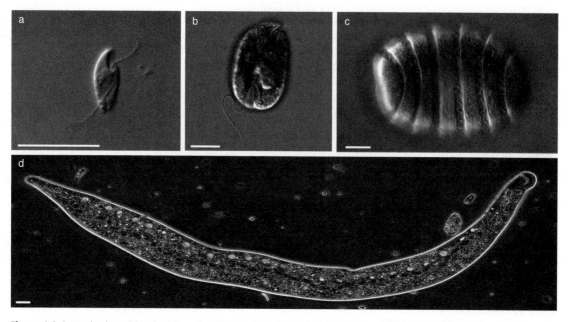

Figure 1.6 Some alveolates. (a) *Colpodella*, a predatory flagellate; (b) *Amphidinium*, a photosynthetic dinoflagellate with two flagella, one looping around the anterior epicone, and the second trailing; (c) *Polykrikos*, a multiflagellated dinoflagellate; (d) *Homalozoon*, a predatory ciliate. All scale bars = 10 μm. All images, D. J. Patterson and provided by micro*scope (http://microscope.mbl.edu), excepting (a) by Alastair Simpson.

an early evolution of open conoids (see Leander and Keeling 2003).

Ciliates: A diverse and ecologically important group of unicellular protists with flagella occurring in rows and beating with a to-and-fro motion (i.e. are cilia). Ciliates are defined by the synapomorphy of nuclear dualism. One kind of nucleus, the micronucleus, contains "germ-line" DNA, and the other, the macronucleus, contains a selectively polyploid genome (McGrath, Zufall, and Katz, Chapter 4, this book). Most ciliates are free-living bacterivores or predators, though some are mixotrophs with algal symbionts or retained prey plastids, and some are commensals or parasites. Several ciliates are model organisms, for example *Tetrahymena*.

"Rhizaria"

This supergroup is used to unite Foraminifera, a revised "Radiolaria", a wide diversity of heterotrophic flagellates and amoeboid organisms, some fungus-like parasites (plasmodiophorids) and the spore-forming Haplosporidia (Fig. 1.7). There is

little morphological similarity across the group. The taxon "Rhizaria" was identified on the basis of well-sampled SSU rRNA and actin phylogenies (Nikolaev *et al.* 2004; and see Fig. 1.2). No morphological synapomorphy is known. The application of the term "Rhizaria" to this group by Nikolaev *et al.* (Nikolaev *et al.* 2004) is the cause of ambiguity as the name was originally introduced for a larger, probably polyphyletic, grouping that also included apusomonads, *Ancyromonas*, and Centroheliozoa (Cavalier-Smith 2002). Nonetheless, the term "Rhizaria" in the sense of Nikolaev *et al.* (Nikolaev *et al.* 2004) seems to be stabilizing and was defined explicitly. They apply the term to the smallest clade containing *Thalassicolla* (polycystinid), *Gromia*, *Chlorarachnion* (in "core Cercozoa"), *Plasmodiophora* (plasmodiophorids), and *Allogromia* (Foraminifera). It would have been desirable to include *Cercomonas* as a specifier.

Cercozoa: This term was introduced by Cavalier-Smith (Cavalier-Smith 1998) for a selection of organisms referred to here as "core Cercozoa", plus the plasmodiophorids. This grouping is indicated by SSU rRNA and actin phylogenies, and, as

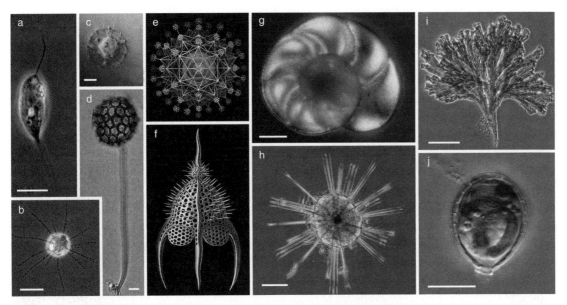

Figure 1.7 Some members of "Rhizaria". (a) *Cercomonas*, a heterotrophic flagellate with two flagella; (b) *Massisteria*, marine bacterivore with radiating arms; (c) *Clathrulina*, a desmothoracid, amoeboid cell within perforated lorica; (d) *Clathrulina*, a desmothoracid, stalked lorica; (e) *Sagenoscena*, a phaeodarean, siliceous shell; (f) *Pterocanium* – skeleton only of this nasellarian polycystinid (Radiolaria); (g) *Ammonia*, foraminifera; (h) *Astrolithium*, one of the acantharean radiolaria; (i) *Rhipidodendron* colony, a spongomonad flagellate (position within "Rhizaria" requires confirmation); (j) *Paulinella*, a shelled amoeba with filose pseudopodia and endosymbiotic cyanobacteria. All scale bars = 10 μm, excepting (h) and (i) which represent 100 μm and (g) which represents 50 μm. All images, D. J. Patterson, except (e) and (f) by Ernst Haeckel, and provided by micro*scope (http://microscope.mbl.edu).

molecular affinities emerge, the group has expanded dramatically in composition (Cavalier-Smith and Chao 2003c). Recent studies indicate that Foraminifera and Haplosporidia might fall within this clade (Nikolaev *et al.* 2004). In the absence of an unambiguous definition of the group, we equate Cercozoa with the smallest clade containing *Cercomonas*, *Chlorarachnion* (core Cercozoa), and *Plasmodiophora* (plasmodiophorid) – and this conforms to current usage.

"Core Cercozoa": refers to a relatively robust clade in SSU rRNA trees that includes the "cercomonads", thaumatomonads, and the amoeboid chlorarachniophyte algae (*inter alia*). It does not include plasmodiophorids, Haplosporidia, or the granuloreticulase (Cavalier-Smith and Chao 2003c; Nikolaev *et al.* 2004). No morphological apomorphy is known. Members of this diverse group are mostly amoebae and small, free-living, gliding flagellates that feed using pseudopodia. Molecular studies have drawn in several types of protists previously regarded as having a distinctive

ultrastructural identity (*Cryothecomonas*, gymnophryids, thaumatomonads, etc.) and some parts of traditional groups of amoebae – for example euglyphids (filose testate amoebae), Phaeodarea (assigned previously to Radiolaria), and desmothoracids (a heliozoan group). The poor state of taxonomic knowledge of small polymorphic flagellates and filopodial amoebae complicates our picture of the diversity of core Cercozoa. Many SSU rRNA sequences are not linked to microscopy records, so there is a risk of mis-assignment of significant taxa to the group because of unverifiable identifications of isolates. For example *"Spongomonas"* has appeared in two separate places within core Cercozoa in SSU rRNA trees – one, possibly both, of these identifications is wrong (Karpov, S.A. pers. comm.; Cavalier-Smith and Chao 2003c). Cavalier-Smith and Chao (Cavalier-Smith and Chao 2003c) used the name Filosa for core Cercozoa. Fortunately, this has not been endorsed as "Filosa" has already had a history of use to refer to amoebae with filose pseudopodia.

• Chlorarachniophytes: This small group is remarkable for having a plastid of secondary origin descended from a green algal group. The green algal nucleus is retained in reduced form as a "nucleomorph" (see Kawach et al., Chapter 13, this book). Typically with rounded cell bodies and elongate microtubule-supported feeding pseudopodia, there is often a flagellated life-cycle stage that in some species is the dominant form.

• Thaumatomonads: This is a small group of a dozen or so genera of heterotrophic flagellates, most of which glide on surfaces and consume prey (bacteria, other eukaryotes) via ventral pseudopodia. Some presumptive members, such as *Protaspis*, are probably important components of benthic microbial communities, but the assemblage has been very little studied.

• Euglyphids: Amoebae with filose pseudopodia and tests of silica scales. Several poorly understood amoebae with organic/agglutinated tests and filose pseudopodia are often considered together with euglyphids as "filose testate amoebae", but few taxa have yet been subject to molecular investigation to confirm this.

• Phaeodarea: Large planktonic marine protists with axopodia, usually with an internal silica skeleton and containing aggregates of waste material – the phaeodium. Flagellated distributive stages may be produced. Phaeodarea were historically united with Acantharea and Polycystinea as the Radiolaria on the basis of light-microscopical similarities, but ultrastructural comparisons and molecular phylogenies refute this placement (Nikolaev et al. 2004).

Plasmodiophorids (also referred to *Phytomyxea*): Endoparasites or phagotrophic ectoparasites of plants (e.g. cabbages) and protists (e.g. oomycetes). A bullet-like extrusome ("stachel") pierces the host cell. The life history includes flagellated cells. There is a distinctive cruciate division profile that is treated as the defining synapomorphy (Patterson 1999).

Haplosporidia: These parasites of marine invertebrates, especially bivalves, produce distinctive lidded spores. The complete life cycle is not known. Recent sophisticated SSU rRNA-based analyses suggest a placement within "Rhizaria" (Cavalier-Smith and Chao 2003c; Nikolaev et al.

2004), and not close to alveolates as suggested by some earlier studies, for example.

Granuloreticulosea: This group is widely regarded as synonymous with the Foraminifera. The Foraminifera are planktonic or benthic amoebae with a reticulate pseudopodial system (see Bowser, Habura, and Pawlowski, Chapter 5, this book). Most foraminifera construct external shells by secreting calcium carbonate, or agglutinating foreign particles. They may be macroscopic and very abundant, with their shells forming massive geological deposits, and are extremely diverse. Traditionally considered marine organisms, recent studies (including environmental PCR approaches) indicate some diversity in freshwater habitats (Holzmann et al. 2003). SSU rRNA phylogenies indicate that *Reticulomyxa* and xenophyophores, previously of uncertain status, are Foraminifera (Pawlowski et al. 1999, 2003). The shelled amoeba *Gromia* is not a member of the group but is probably closely related (see Nikolaev et al. 2004).

"Radiolaria": Historically this term was used to refer to three groups of large planktonic marine cells with internal mineralized skeletons – Acantharea, Polycystinea, and Phaeodarea. This assemblage appears polyphyletic (Nikolaev et al. 2004; Patterson 1999). Unfortunately, the term is re-emerging to embrace the Acantharea, polycystines, and the strange "heliozoon" *Sticholonche* (Taxopodida), which has mobile mineralized "oars" – a grouping indicated by SSU rRNA studies (Nikolaev et al. 2004). Acantharea have radiating spiked skeletons of strontium sulfate. Polycystines usually have silica skeletons of various morphologies. Both groups are significant in marine microbial food webs, and they capture food with radiating microtubule-supported pseudopodia. Many have algal symbionts. Their cell/genome biology is poorly understood and cultures are not available.

Excavata

This supergroup is defined by the "excavate" feeding groove, a conspicuous channel on one side of the cell that is used to capture suspended particles from a feeding current generated by a posteriorly-directed flagellum (Simpson 2003)

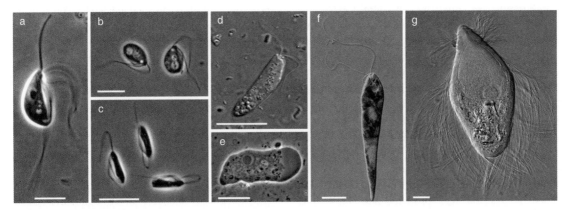

Figure 1.8 Some members of Excavata. (a) *Trimastix*, with four flagella and ventral groove; (b) *Carpediemonas*, each cell with two flagella and ventral groove; (c) *Jakoba*, marine excavate with two flagella and ventral groove; (d) Heterolobosea, swarmer cell with two flagella emerging from an anterior groove (genus not identified); (e) *Vahlkampfia*, a heterolobosean amoeba; (f) *Eutreptiella*, a plastid-bearing marine euglenid with two flagella; (g) *Trichonympha*, endobiotic parabasalid flagellate. All scale bars = 10 μm. All images, D. J. Patterson and provided by micro*scope (http://microscope.mbl.edu), excepting (a) by Alastair Simpson.

(Fig. 1.8). The grouping is contentious because it is neither well supported by molecular phylogenies nor rejected by such analyses. It is best regarded as an unfalsified hypothesis (Simpson 2003). The group is defined by reference to an evolutionary innovation, and so the concept is stable even if membership is not. The validity of the group can be tested by evaluating the homology of the excavate groove. Excavata contains many groups considered at various times to represent "primitive" eukaryotes, for example, diplomonads, parabasalids, Euglenozoa, jakobids and Heterolobosea and several important mitochondrion-lacking lineages.

Preaxostyla are flagellates that lack mitochondria with cristae. The taxon unites oxymonads (intestinal endobionts primarily of termites) and *Trimastix* (free-living organisms from oxygen-poor habitats that consume bacteria using a feeding groove). The grouping is supported by SSU rRNA phylogenies and ultrastructure (see Fig. 1.2 and Simpson 2003).

Malawimonas: Small free-living, mitochondriate, bacterivorous excavate flagellates, of unresolved position in SSU rRNA and tubulin phylogenies (Simpson et al. 2002).

"Excavate clade 1": This grouping reflects growing evidence that four groups of heterotrophic flagellates that lack classical mitochondria are related: parabasalids, diplomonads, retortamonads, and *Carpediemonas* (Embley and Hirt 1998, Simpson *et al.* 2002, Simpson and Roger 2004b). *Carpediemonas* (free-living bacterivores) and retortamonads (mostly endobionts of animals) have feeding grooves and are consecutive sister groups to the better-known diplomonads. This clade lacks a clear synapomorphy. Cavalier-Smith (Cavalier-Smith 2003) proposes the name Trichozoa for this grouping – a confusing re-use of a name created for the polyphyletic assemblage of parabasalids plus *Trimastix* (Cavalier-Smith 1997).

Diplomonads: Heterotrophic flagellates, usually with two sets of nuclei and flagella arranged with rotational symmetry. The doubled-cell condition is apomorphic – molecular phylogenies indicate that non-doubled diplomonads (enteromonads) are not basal to doubled forms, as originally thought (Kolisko *et al.* 2005). Diplomonads inhabit oxygen-poor environments. They have tiny mitosomes that are not involved in energy generation but are argued to be mitochondrial homologs (Tovar *et al.* 2003). Many diplomonads are parasites of vertebrates, including the well-studied human pathogen *Giardia lamblia*.

Parabasalia: A diverse group, mostly endoparasitic or endosymbiotic flagellates of animals. *Trichomonas vaginalis* causes the most common sexually transmitted disease in humans. Many parabasalids are small, usually with four to six flagella, and phagocytose bacteria and host material. Several sublineages have given rise to massive multiflagellated cells that live in the hindguts of wood-eating insects. The mitochondria take the form of "hydrogenosomes" that generate energy anaerobically from pyruvate with hydrogen as a waste product. The apomorphy is the "parabasal apparatus" – striated fibers that attach the Golgi apparatus to the flagellar apparatus.

"Excavate clade 2": This branch includes jakobids, Euglenozoa, and Heterolobosea, which combined nuclear-encoded proteins, and some SSU rRNA and CCTa datasets show to be closely related (see Simpson and Roger 2004b). Heterolobosea and Euglenozoa are often considered related and united within the taxon Discicristata (defined by discoidal mitochondrial cristae). However, combined protein trees suggest that jakobids and Heterolobosea may be sister taxa. Discoidal cristae are also present in *Malawimonas* and other groups not (or not yet) included in the excavates, such as *Stephanopogon*. The sister group relationships within this clade and the status of Discicristata are currently unclear.

Jakobids: Small free-living bacterivorous flagellates with a feeding groove. Some inhabit organic loricae. Jakobids (e.g. *Reclinomonas*) have the most bacterial-like mitochondrial genomes examined to date (Lang *et al.* 1997), and attract interest as possible "primitive" eukaryotes.

Heterolobosea: Most Heterolobosea are amoebae with a flagellated dispersal phase. One or other of the phases is absent or overlooked in several taxa, and one subgroup (acrasids) have evolved a slime mold habit. Common and widespread, for example in soils. Some are facultative pathogens.

Euglenozoa: This grouping unites kinetoplastids with euglenids and includes diplonemids, free-living gliding heterotrophs that are specifically related to kinetoplastids, and some minor groups (e.g. *Postgaardi*). Many Euglenozoa actively deform or squirm. The group was identified by ultrastructural comparisons and is supported by several gene phylogenies (see Simpson 1997; Simpson and Roger 2004a). The group is identified by the presence of a distinctive sort of paraxonemal rods

(non-microtubular cytoskeletal structures within the flagellum that lie parallel to and alongside the flagellar axoneme).

• Euglenids: Flagellated cells with a "pellicle" of interlocking proteinaceous strips lying immediately beneath the cell membrane (this feature is regarded as apomorphic). The most familiar euglenids are photosynthetic, with secondary plastids of green algal origin. Many euglenids are heterotrophs that glide on surfaces, and many prey on other large eukaryotic cells. Contrary to recent proposals that the euglenid plastid was already present in the ancestral euglenozoon (Hannaert *et al.* 2003), morphological and rRNA analyses support a late origin of the euglenid plastid within euglenids, as green euglenids form a single clade deeply nested within non-plastid-bearing heterotrophs (Leander 2004).

• Kinetoplastids: Members of this group of flagellates are identified by the kinetoplast – a massive and bizarrely organized mitochondrial DNA structure. The best studied by far are trypanosomatids, a large group of endoparasites that includes the agents of sleeping sickness, Chagas' disease, and leishmaniases in humans (Stuart and Myler, Chapter 10, this book). Trypanosomatids evolved from within a diverse paraphyletic assemblage of free-living heterotrophs and animal parasites traditionally called "bodonids". Free- living kinetoplastids are important bacterivores in benthic habitats.

Protist lineages that have yet to be confidently assigned to supergroups

Ancyromonas: A single genus of flagellates that glide on surfaces and feed on bacteria. Cells have a thecal layer similar to apusomonads (Cavalier-Smith and Chao 2003a). A few SSU rRNA analyses weakly indicate a relationship with apusomonads (Atkins et al. 2000), but relationships are essentially unresolved.

Apusomonads: These are small biflagellated protists with a submembranous theca that extends as a collar around the anterior flagellum. Cells glide and consume bacteria. Neither morphological data nor SSU rRNA analyses resolve their affinities. Early results weakly uniting apusomonads with opisthokonts (Cavalier-Smith and Chao 1995) are typically not supported with improved methods and/or taxon sampling (e.g. Berney *et al.* 2004).

Biomyxa: Amoeboid cells with stiff anastomosing pseudopodia supported by microtubules. Affinities with foraminifera possible, but this awaits confirmation (Patterson 1999).

"Breviatea": This taxon was introduced for culture ATCC50338 (American Type Culture Collection), bearing the name *Mastigamoeba invertens* (Cavalier-Smith 2004). This culture is misidentified – the "real" *Mastigamoeba invertens* Klebs 1892 is not distinguishable from *Mastigamoeba longifilum* Stokes, 1888. "Breviatea" is an empty taxon because the sole species referred to lacks a formal description. SSU rRNA and RNA polymerase B1 trees (Stiller *et al.* 1998) do not resolve the affinities of ATCC50338, but it is not closely related to other *Mastigamoeba* spp., which are pelobionts (Edgcomb *et al.* 2002).

Centroheliozoa: The most species rich of the UI lineages that had been placed in the polyphyletic Heliozoa. Centrohelids are rounded with stiff radiating pseudopods supported by microtubules in hexagonal array that arise from a central multilaminar microtubular organizing center. They lack flagella and are heterotrophic (some have algal symbionts). SSU rRNA phylogenies do not identify their sister group (Cavalier-Smith and Chao 2003b; Nikolaev *et al.* 2004).

Coelosporidium: Amoeboid endoparasites of Cladocera, forming spores with perforated walls. There are currently no molecular data to indicate affinities.

Collodictyonids: Free-swimming flagellates that ingest eukaryotic cells via a symmetrical feeding gutter that is morphologically and functionally dissimilar to the feeding groove of excavates. SSU rRNA analyses do not identify a sister group (Brugerolle *et al.* 2002).

Copromyxids: A group of cellular slime molds, with tubular mitochondrial cristae (like dictyostelids but unlike acrasids). As yet, there are no molecular data.

Cryptomonads (Cryptophytes): Free-swimming flagellates, mostly photosynthetic with a plastid containing chlorophylls *a* and *c*, phycobilins (but not phycobilisomes), and a nucleomorph. Some taxa are osmotrophs with non-photosynthetic plastids, and one basal group, *Goniomonas*, eats bacteria and may lack plastids. Cells have a groove/channel associated with the flagellar apparatus, which we consider the synapomorphy for the group. Coiled-ribbon extrusomes are associated with the flagellar groove and cell surface. Similar extrusomes are found in kathablepharids, a small but common group of biflagellate predators that may be the sister group to cryptomonads (Okamoto and Inouye 2005; and see Fig. 1.2). Many consider cryptomonads to be related to stramenopiles and alveolates (see above for further discussion).

Dimorphids: Flagellates with two or four similar flagella and radiating pseudopodia that traditionally were placed in the polyphyletic "heliozoa". As yet, there are no molecular data from a confirmed member of this group.

Haptophytes (Prymnesiophytes): Marine unicellular photosynthetic flagellates, usually with a "haptonema" – an elongate microtubule-supported structure used in feeding and/or attachment. The chromalveolate plastids contain chlorophylls *a* and *c*. The most conspicuous haptophytes are the coccolithophorids, which produce tests of calcareous scales (e.g. *Emiliania huxleyi*), and are of global significance both as marine primary producers and as a conduit for carbon export to the deep ocean. The haptonema is synapomorphic. Many consider haptophytes to be related to stramenopiles and alveolates (see above for further discussion).

Discocelis: is a small gliding marine heterotrophic flagellate (Vørs 1988). A placement with bicosoecids (stramenopiles) is suggested from limited morphological data (Karpov 2000) but requires independent confirmation.

Ebriids: Biflagellated heterotrophic marine protists, with internal siliceous skeletons. Sometimes allied with dinoflagellates, but without real evidence (Taylor 1990).

Fonticula: is a slime mold traditionally considered to be an acrasid (Heterolobosea) but the amoebae that aggregate to form the sporulating body are unlike those of acrasids (Blanton 1990).

Gymnosphaerida: Marine "heliozoa", with the pseudopodial microtubules arising from a central nucleating site (Jones 1980). Some with adhering siliceous spicules and/or a stalk. No molecular data are available.

Komokiacea: are large marine benthic organisms with a cytoplasmic mass surrounded by agglutinated material, and producing branching pseudopodial strings for feeding. Possibly related to Foraminifera, although this remains unconfirmed (Cedhagen and Mattson 1991).

Luffisphaera: Marine organisms known primarily from distinctive scales seen by electron microscopy of whole mount preparations (e.g. Vørs 1993).

Nephridiophagids: are spore-forming parasites associated with the malpighian tubules of insects. A recent molecular study suggests affinities with zygomycete fungi (Wylezich *et al.* 2004).

Multicilia: a benthic predator with up to 30 flagella that eats small amoebae (Mikrjukov and Mylnikov 1998), and has surface structures similar to the "glycostyles" of vannelid lobose amoebae. Currently no published molecular studies.

Paramyxea: Endoparasites in marine invertebrates and of economic importance in bivalves, paramyxids mostly occur as amoeboid cells but form multicellular spores. SSU rRNA sequences are very divergent, although one recent analysis indicates a position within Haplosporidia (Cavalier-Smith and Chao 2003a).

Phagodinium: is an endoparasite of synurophyceans (stramenopiles) that forms organic-walled cysts with three points (tridentate) and produces biflagellated dispersive cells from a "sporangium" (Kristiansen 1993). No molecular data to indicate relatedness.

Phalansterium: are heterotrophic flagellates that form colonies with cells embedded in an organic matrix. The single flagellum has a tight-fitting cytoplasmic collar. Phalansterium is unstable in SSU rRNA trees – recent studies favor an affinity with Ramicristates / Amoebozoa, but support is weak (Cavalier-Smithet al. 2004).

Pseudospora: Heterotrophic protists that typically attack plant cells. Trophic cells are amoeboid or plasmodial, while the distributive form has two flagella and is produced after encystment of the amoeboid stage.

Schizoclades: are a small group of marine amoebae not yet subject to molecular analysis.

Spironemidae (Hemimastigophora): Multiflagellated heterotrophic protists with flagella in two longitudinal rows. Thin plates support the cell membrane between the rows. Observed in diverse environments (Foissner and Foissner 1993). No molecular studies.

Stephanopogon: Ciliate-like predatory marine protists. Unlike ciliates there is no nuclear dualism and the cytoskeleton is dissimilar (Patterson and Brugerolle 1988). Affinities with Euglenozoa and Heterolobosea are suggested by the presence of discoidal mitochondrial cristae, but the inference is weak, as other distantly related taxa have similar cristae (e.g. nucleariids). No published molecular data.

Telonema: Predatory marine cells with two flagella that are directed behind the moving cell. One flagellum has stiff hairs. There is a discrete ingestion area adjacent to the basal bodies. The cell is supported by a distinctive and complex submembranous lamina (Klaveness et al. 2005). A recent preliminary ultrastructural study and SSU rRNA phylogeny do not reveal a sister-group (Klaveness et al. 2005).

Vampyrellids: Amoebae with filose/tapering pseudopodia that typically feed on algal and fungal cytoplasm after perforating the cell walls. No molecular data.

Concluding remarks

The current understanding of protistan phylogeny improves remarkably on that available at the end of last century, but still retains many uncertainties. A more complete understanding would be a better framework for comparative biology, including comparative genomics. Further work will aim to: (i) determine the placement of the many "unassigned" lineages; (ii) determine or refute the monophyly of the chromalveolates and the excavates; and (iii) determine the deepest-level phylogeny of eukaryotes – that is, the relationships amongst supergroups and the position of the root of the tree. Some progress on the relationships amongst supergroups is beginning to emerge with evidence that Opisthokonts and Ramicristates/ Amoebozoa share gene duplications (Richards and Cavalier-Smith 2005). Ongoing and future genome projects across a broad taxonomic base of protists will themselves be invaluable in resolving further the eukaryotic evolutionary tree and the relationships among protists.

Acknowledgements

AGBS thanks the Canadian Institute for Advanced Research (CIAR), Program in Evolutionary Biology for salary support as a "Scholar". DJP is supported by the NASA Astrobiology Institute and through the NSF ATOL program. We thank R. A. Andersen for helful comments.

References

Adl, S. M., Simpson, A. G. B., Farmer, M. A., *et al*. (2005). The new higher-level classification of eukaryotes with emphasis on the taxonomy of protists. *J. Eukaryot. Microbiol*., **52**, 399–451.

Andersen, R. A. (2004). Biology and systematics of heterokont and haptophyte algae. *Am. J. Bot*., **91**, 1508–1522.

Atkins, M. S., McArthur, A. G., and Teske, A. P. (2000). Ancyromonadida: a new phylogenetic lineage among the protozoa closely related to the common ancestor of metazoans, fungi, and choanoflagellates (Opisthokonta). *J. Mol. Evol*., **51**, 278–285.

Baldauf, S. L. (2003). The deep roots of eukaryotes. *Science*, **300**, 1703–1706.

Baldauf, S. L. and Doolittle, W. F. (1997). Origin and evolution of the slime molds (Mycetozoa). *Proc. Natl. Acad. Sci. U S A*, **94**, 12007–12012.

Baldauf, S. L., Roger, A. J., Wenk-Siefert, I. and Doolittle, W. F. (2000). A kingdom-level phylogeny of eukaryotes based on combined protein data. *Science*, **290**, 972–977.

Bapteste, E., Brinkmann, H., Lee, J. A., *et al*. (2002). The analysis of 100 genes supports the grouping of three highly divergent amoebae: *Dictyostelium*, *Entamoeba*, and *Mastigamoeba*. *Proc. Natl. Acad. Sci. U S A*, **99**, 1414–1419.

Berney, C., Fahrni, J. F., and Pawlowski, J. (2004). How many novel eukaryotic "kingdoms"? Pitfalls and limitations of environmental DNA surveys. *BMC Biol*., **2**, Art. 13.

Blanton, R. L. (1990). Phylum Acrasea. In L. Margulis, J. O. Corliss, M. Melkonian, and D. J. Chapman, (eds). *Handbook of Protoctista*. Jones and Bartlett, Boston, pp. 75–87.

Bodyl, A. (2005). Do plastid-related characters support the chromalveolate hypothesis? *J. Phycol*., **41**, 712–719.

Brugerolle, G., Bricheux, G., Philippe, H., and Coffe, G. (2002). *Collodictyon triciliatum* and *Diphylleia rotans* (= *Aulacomonas submarina*) form a new family of flagellates (Collodictyonidae) with tubular mitochondrial cristae that is phylogenetically distant from other flagellate groups. *Protist*, **153**, 59–70.

Cavalier-Smith, T. (1989). The kingdom Chromista. In J. C. Green, B. S. C. Leadbeater, and W. L. Diver (eds). *The chromophyte algae: Problems and perspectives*. Clarendon Press, Oxford, pp. 381–407.

Cavalier-Smith, T. (1991). Cell diversification in heterotrophic flagellates. In D.J. Patterson and J. Larsen (eds). *The Biology of Free-Living Heterotrophic Flagellates*. Clarendon Press, Oxford, pp. 113–131

Cavalier-Smith, T. (1997). Amoeboflagellates and mitochondrial cristae in Eukaryote evolution: megasystematics of the new protozoan subkingdoms Eozoa and Neozoa. *Archiv für Protistenkunde*, **147**, 237–258.

Cavalier-Smith, T. (1998). A revised six-kingdom system of life. *Biol. Rev. Camb. Philosoph. Soc*., **73**, 203–266.

Cavalier-Smith, T. (1999). Principles of protein and lipid targeting in secondary symbiogenesis: euglenoid, dinoflagellate, and sporozoan plastid origins and the eukaryotic family tree. *J. Eukaryot. Microbiol*., **46**, 347–366.

Cavalier-Smith, T. (2002). The phagotrophic origin of eukaryotes and phylogenetic classification of Protozoa. *Int. J. System. Evol. Microbiol*., **52**, 297–354.

Cavalier-Smith, T. (2003). The excavate protozoan phyla Metamonada Grassé emend. (Anaeromonadea, Parabasalia, *Carpediemonas*, Eopharyngia) and Loukozoa emend. (Jakobea, *Malawimonas*): their evolutionary affinities and new higher taxa. *Int. J. System. Evol. Microbiol*., **53**, 1741–1758.

Cavalier-Smith, T. (2004). Only six kingdoms of life. *Proc. Roy. Soc. Lond. B Biol. Sci*., **271**, 1251–1262.

Cavalier-Smith, T. and Chao, E. E. (1995). The opalozoan *Apusomonas* is related to the common ancestor of animals, fungi, and choanoflagellates. *Proc. Roy. Soc. Lond. B Biol. Sci*., **261**, 1–6.

Cavalier-Smith, T. and Chao, E. E. (2003a). Phylogeny of choanozoa, apusozoa, and other protozoa and early eukaryote megaevolution. *J. Mol. Evol*., **56**, 540–563.

Cavalier-Smith, T. and Chao, E. E. (2003b). Molecular phylogeny of centrohelid heliozoa, a novel lineage of bikont eukaryotes that arose by ciliary loss. *J. Mol. Evol*., **56**, 387–396.

Cavalier-Smith, T. and Chao, E. E. (2003c). Phylogeny and classification of phylum Cercozoa (Protozoa). *Protist*, **154**, 341–358.

Cavalier-Smith, T., Chao, E. E., and Oates, B. (2004). Molecular phylogeny of Amoebozoa and the evolutionary significance of the unikont *Phalansterium*. *J. Mol. Evol*., **40**, 21–48.

Cedhagen, T. and Mattson, S. (1991). *Globipelorhiza sublittorolis* gen. et sp. n., a komokiaccan (Protozoa: Foraminiferida) from the Scandinavian sublittoral. *Sarsia*, **76**, 209–213.

Edgcomb, V. P., Simpson, A. G. B., Zettler, L. A., *et al.* (2002). Pelobionts are degenerate protists: insights from molecules and morphology. *Mol. Biol. Evol.*, **19**, 978–982.

Embley, T. M. and Hirt, R. P. (1998). Early branching eukaryotes? *Curr. Opin. Genet. Dev.*, **8**, 624–629.

Fahrni, J. F., Bolivar, I., Berney, C., Nassonova, E., Smirnov, A., and Pawlowski, J. (2003). Phylogeny of lobose amoebae based on actin and small-subunit ribosomal RNA genes. *Mol. Biol. Evol.*, **20**, 1881–1886.

Fast, N. M., Kissinger, J. C., Roos, D. S., and Keeling, P. J. (2001). Nuclear-encoded, plastid-targeted genes suggest a single common origin for apicomplexan and dinoflagellate plastids. *Mol. Biol. Evol.*, **18**, 418–426.

Foissner, I. and Foissner, W. (1993). Revision of the family Spironemidae Doflein (Protista, Hemimastigophora), with description of two new specie, *Spironema terricola* n. sp. and *Stereonema geiseri* n.g., n. sp. *J. Eukaryot. Microbiol.*, **40**, 422–438.

Gajadhar, A. A., Marquardt, W. C., Hall, R., Gunderson, J. H., Ariztia-Carmona, E. V., and Sogin, M. L. (1991). Ribosomal RNA sequences of *Sarcocystis muris*, *Theileria annulata* and *Crypthecodinium cohnii* reveal evolutionary relationships among apicomplexans, dinoflagellates and ciliates. *Mol. Biochem. Parasitol.*, **45**, 147–154.

Gilson, P. and McFadden, G. (2002). Jam-packed genomes – a preliminary, comparative analysis of nucleomorphs. *Genetica*, **115**, 13–28.

Graham, L. and Wilcox, L. (2000). *Algae*. Prentice-Hall, Upper Saddle River NJ, p. 640.

Graur, D. and Li, W. (1999). *Fundamentals of molecular evolution*. Sinauer Associates, Sunderland, MA, p. 481.

Hannaert, V., Saavedra, E., Duffieux, F., *et al.* (2003). Plant-like traits associated with metabolism of *Trypanosoma* parasites. *Proc. Natl. Acad. Sci. U S A*, **100**, 1067–1071.

Harper, J. T. and Keeling, P. J. (2003). Nucleus-encoded plastid-targeted glyceraldehyde-3-phosphate dehydrogenase (GAPDH) indicates a single origin for chromalveolate plastids. *Mol. Biol. Evol.*, **20**, 1730–1735.

Harper, J. T., Waanders, E., and Keeling, P. J. (2005). On the monophyly of chromalveolates using a six-protein phylogeny of eukaryotes. *Int. J. System. Evol. Microbiol.*, **55**, 487–496.

Hausmann, K., Hülsmann, N., and Radek, R. (2003). *Protistology*. E. Schweizerbart'sche Verlagsbuchhandlung, Berlin, p. 379.

Holzmann, M., Habura, A., Giles, H., Bowser, S. S., and Pawlowski, J. (2003). Freshwater foraminiferans revealed by analysis of environmental DNA samples. *J. Eukaryot. Microbiol.*, **50**, 135–139.

Jones, W. C. (1980). The ultrastructure of Gymnosphaera albida Sassaki, a marine axopodiate protozoan. *Phil. Trans. Roy. Soc. Lond. B, Biol. Sci.*, **275**, 349–384.

Karpov, S. A. (2000). Ultrastructure of the aloricate *Pseudobodo tremulans*, with revision of the order Bicosoecida. *Protistology*, **1**, 101–109.

Karpov, S. A., Sogin, M. L. and Silberman, J. D. (2001). Rootlet homology, taxonomy, and phylogeny of bicosoecids based on 18S rRNA gene sequences. *Protistology*, **2**, 34–47.

Klaveness, D., Shalchian-Tabrizi, K., Abildhauge Thomsen, H., Eikrem, W., and Jakobsen, K. S. (2005). *Telonema antarcticum* sp. nov. a common marine phagotrophic flagellate. *Int. J. System. Evol. Microbiol.*, **55**, 2595–2604.

Kolisko, M., Cepicka, I., Hampl, V., Kulda, J., and Flegr, J. (2005). The phylogenetic position of enteromonads: a challenge for the present models of diplomonad evolution. *Int. J. System. Evol. Microbiol.*, **55**, 1729–1733.

Kostka, M., Hampl, V., Cepicka, I., and Flegr, J. (2004). Phylogenetic position of *Protopalina intestinalis* based on SSU rRNA gene sequence. *Mol. Phylogenetics Evol.*, **33**, 220–224.

Kristiansen, J. (1993). The "tridentata" parasite of *Mallomonas teilingii* (Synurophyceae) – a new dinophyte? or what? *Archiv für Protistenkunde*, **143**, 195–214.

Lang, B. F., Burger, G., O' Kelly, C. J., *et al.* (1997). An ancestral mitochondrial DNA resembling a eubacterial genome in miniature. *Nature*, **387**, 493–497.

Leander, B. S. (2004). Did trypanosomatid parasites have photosynthetic ancestors. *Trends Microbiol.*, **12**, 251–258.

Leander, B. S. and Keeling, P. J. (2003). Morphostasis in alveolate evolution. *Trends Ecol. Evol.*, **18**, 395–402.

Leipe, D. D., Tong, S. M., Goggin, C. L., Slemenda, S. B., Pieniazek, N. J., and Sogin, M. L. (1996). 16S-like rDNA sequences from *Developayella elegans*, *Labyrinthuloides haliotidis*, and *Proteromonas lacertae* confirm that the stramenopiles are a primarily heterotrophic group. *Eur. J. Protistol.*, **32**, 449–458.

Lewis, L. A. and McCourt, R. M. (2004). Green algae and the origin of land plants. *Am. J. Bot.*, **91**, 1535–1556.

Massana, R., Castresana, J., Balagué, V., *et al.* (2004). Phylogenetic and ecological analysis of novel marine stramenopiles. *Appl. Environ. Microbiol.*, **70**, 3528–3534.

McCarroll, R., Olsen, G. J., Stahl, Y. D., Woese, C. R., and Sogin, M. L. (1983). Nucleotide sequence of the *Dictyostelium discoideum* small-subunit ribosomal ribonucleic acid inferred from the gene sequence: evolutionary implications. *Biochemistry*, **22**, 5858–5868.

McGrath, C. L. and Katz, L. (2004). Genome diversity in microbial eukaryotes. *Trends Ecol. Evol.*, **19**, 32–38.

Mendoza, L., Taylor, J. W., and Ajello, L. (2002). The class Mesomycetozoea: A heterogeneous group of microorganisms at the Animal-Fungal boundary. *Ann. Rev. Microbiol.*, **56**, 315–344.

Mikrjukov, K. A. and Mylnikov, A. P. (1998). The fine structure of a carnivorous multiflagellar protist, *Multicilia marina* Cienkowski, 1881 (Flagellata incertae sedis). *Eur. J. Protistol.*, **34**, 391–401.

Moreira, D., Le Guyader, H., and Philippe, H. (2000). The origin of red algae and the evolution of chloroplasts. *Nature*, **405**, 69–72.

Nikolaev, S. I., Berney, C., Fahrni, J. F., *et al.* (2004). The twilight of Heliozoa and rise of Rhizaria, an emerging supergroup of amoeboid eukaryotes. *Proc. Natl. Acad. Sci. U S A*, **101**, 8066–8071.

Okamoto, N. and Inouye, I. (2005). The katablepharids are a distant sister group of the Cryptophyta: A proposal for Katablepharidophyta divisio nova/Kathablepharida phylum novum based on SSU rDNA and beta-tubulin phylogeny. *Protist*, **156**, 163–179.

O'Kelly, C. J. and Nerad, T. A. (1998). Kinetid architecture and bicosoecid affinities of the marine heterotrophic nanoflagellate *Caecitellus parvulus* (Griessmann, 1913) Patterson *et al.*, 1993. *Eur. J. Protistol.*, **34**, 369–375.

Patron, N. J., Rogers, M. B., and Keeling, P. J. (2004). Gene replacement of fructose-1,6-bisphosphate aldolase supports the hypothesis of a single photosynthetic ancestor of chromalveolates. *Eukaryot. Cell*, **3**, 1169–1175.

Patterson, D. J. (1999). The diversity of eukaryotes. *Am. Naturalist*, **65**, S96–S124.

Patterson, D. J. and Brugerolle, G. (1988). The ultrastructual identity of *Stephanopogon apogon* and the relatedness of the genus to other kinds of protists. *Eur. J. Protistol.*, **23**, 279–290.

Pawlowski, J., Bolivar, I., Fahrni, J. F., de Vargas, C., and Bowser, S. S. (1999). Molecular evidence that *Reticulomyxa filosa* is a freshwater naked foraminifer. *J. Eukaryot. Microbiol.*, **46**, 612–617.

Pawlowski, J., Holzmann, M., Fahrni, J. F., and Richardson, S. L. (2003). Small subunit ribosomal DNA suggests that the Xenophyophorean *Syringammina corbicula* is a foraminiferan. *J. Eukaryot. Microbiol.*, **50**, 483–487.

Philippe, H., Lopez, P., Brinkmann, H., *et al.* (2000). Early-branching or fast-evolving eukaryotes? An answer based on slowly evolving positions. *Proc. Roy. Soc. Lond. B Biol. Sci.*, **267**, 1213–1221.

Richards, T. A. and Cavalier-Smith, T. (2005). Myosin domain evolution and the primary divergence of eukaryotes. *Nature*, **436**, 1113–1118.

Rodriguez-Ezpeleta, N., Brinkmann, H., Burey, S. C., *et al.* (2005). Monophyly of primary photosynthetic eukaryotes: green plants, red algae, and glaucophytes. *Curr. Biol.*, **15**, 1325–1330.

Ruiz-Trillo, I., Inagaki, Y., Davis, L. A., Sperstad, S., Landfald, B., and Roger, A. J. (2004). *Capsaspora owczarzaki* is an independent opisthokont lineage. *Curr. Biol.*, **14**, 946–947.

Saldarriaga, J. F., McEwan, M. L., Fast, N. M., Taylor, F. J. R., and Keeling, P. J. (2003). Multiple protein phylogenies show that *Oxyrrhis marina* and *Perkinsus marinus* are early branches of the dinoflagellate lineage. *Int. J. System. Evol. Microbiol.*, **53**, 355–365.

Silberman, J. D., Clark, C. G., Diamond, L. S., and Sogin, M. S. (1999). Phylogeny of the genera *Entamoeba* and *Endolimax* as deduced from small-subunit ribosomal RNA sequences. *Mol. Biol. Evol.*, **16**, 1740–1751.

Simpson, A. G. B. (1997). The identity and composition of the Euglenozoa. *Archiv für Protistenkunde*, **148**, 318–328.

Simpson, A. G. B. (2003). Cytoskeletal organisation, phylogenetic affinities and systematics in the contentious taxon Excavata (Eukaryota). *Int. J. System. Evol. Microbiol.*, **53**, 1759–1777.

Simpson, A. G. B. and Roger, A. J. (2004a). Protein phylogenies robustly resolve the deep-level relationships within Euglenozoa. *Mol. Phylogenetics Evol.*, **30**, 201–212.

Simpson, A. G. B. and Roger, A. J. (2004b). Excavata and the origin of amitochondriate eukaryotes. In R. P. Hirt and D. S. Horner (eds). *Organelles, Genomes and Eukaryote Phylogeny: an Evolutionary Synthesis in the Age of Genomics*. CRC Press, Boca Raton, pp. 27–53.

Simpson, A. G. B. and Roger, A. J. (2004c). The real "kingdoms" of eukaryotes. *Curr. Biol.*, **14**, R693–696.

Simpson, A. G. B., Roger, A. J., Silberman, J. D., *et al.* (2002). Evolutionary history of "early diverging" eukaryotes: The excavate taxon *Carpediemonas* is closely related to *Giardia*. *Mol. Biol. Evol.*, **19**, 1782–1791.

Smirnov, A., Nassonova, E., Berney, C., Fahrni, J. F., Bolivar, I., and Pawlowski, J. (2005). Molecular phylogeny and classification of the lobose amoebae. *Protist*, **156**, 129–142.

Smith, R. M. and Patterson, D. J. (1986). Analyses of heliozoan interrelationships: an example of the potentials and limitations of ultrastructural approaches to the study of protistan phylogeny. *Proc. Roy. Soc. Lond. B Biol. Sci.*, **227**, 325–366.

Sogin, M. L. (1989). Evolution of eukaryotic microorganisms and their small subunit ribosomal RNAs. *Am. Zool.*, **29**, 487–499.

Sogin, M. L., Gunderson, J. H., Elwood, H. J., Alonso, R. A., and Peattie, D. A. (1989). Phylogenetic significance of the kingdom concept: an unusual eukaryotic 16S-like ribosomal RNA from *Giardia lamblia*. *Science*, **243**, 75–77.

Stechmann, A. and Cavalier-Smith, T. (2002). Rooting the eukaryote tree by using a derived gene fusion. *Science*, **297**, 89–91.

Stiller, J. W., Duffield, E. C. S., and Hall, B. D. (1998). Amitochondriate amoebae and the evolution of DNA-dependent RNA polymerase II. *Proc. Natl. Acad. Sci. U S A*, **95**, 11769–11774.

Taylor, F. J. R. (1990). Incertae sedis Ebridians. In L. Margulis, J.O. Corliss, M. Melkonian, and D.J. Chapman (eds). *Handbook of Protoctista*. Jones and Bartlett, Boston, pp. 720–722.

Tovar, J., Fischer, A., and Clark, C. G. (1999). The mito-some, a novel organelle related to mitochondria in the amitochondrial parasite *Entamoeba histolytica*. *Mol. Microbiol.*, **32**, 1013–1021.

Tovar, J., Leon-Avila, G., Sanchez, L. B. *et al.* (2003). Mitochondrial remnant organelles of *Giardia* function in iron-sulphur protein maturation. *Nature*, **426**, 172–176.

Van de Peer, Y. and De Wachter, R. (1997). Evolutionary relationships among the eukaryotic crown taxa taking into account site-to-site rate variation in 18S rRNA. *J. Mol. Evol.*, **45**, 619–630.

Van de Peer, Y., Ben Ali, A., and Meyer, A. (2000). Microsporidia: accumulating molecular evidence that a group of amitochondriate and suspectedly primitive eukaryotes are just curious fungi. *Gene*, **246**, 1–8.

Vørs, N. (1988). *Discocelis saleuta* gen.nov.et sp.nov. (Protista incertae sedis) – a new heterotrophic marine flagellate. *Eur. J. Protistol.*, **23**, 297–308.

Vørs, N. (1993). Marine heterotrophic amoebae, fla-gellates and heliozoa from Belize (Central America) and Tenerife (Canary Islands), with descriptions of new species, *Luffisphaera bulbochaete* n. sp., *L. longihastis* n. sp., *L. turriformis* n. sp. and *Paulinella intermedia* n. sp. *J. Eukaryot. Microbiol.*, **40**, 272–287.

Wylezich, C., Radek, R., and Schlegel, M. (2004). Phylo-genetische analyse der 18S rRNA identifiziert den parasitischen Protisten *Nephridiophaga blattellae* (Nephridiophagidae) als Vertreter der Zygomycota (Fungi). *Denisia*, **13**, 435–442.

Yoon, H. S., Hackett, J. D., Pinto, G., and Bhattacharya, D. (2002). The single, ancient origin of chromist plastids. *Proc. Natl. Acad. Sci. U S A*, **99**, 15507–15512.

PART II

Evolutionary genomics of eukaryotic microbes

CHAPTER 2

Comparative genomics of *Plasmodium* species

Jane M. Carlton

The Institute for Genomic Research, Rockville, MD, USA

Introduction

Overview

The field of comparative genomics of malaria parasites has recently blossomed with the completion of the whole genome sequence of the human malaria parasite *Plasmodium falciparum* and three species of rodent malaria model. With several more genome sequencing projects of primate, human, and bird *Plasmodium* species close to completion, comparing genomes from multiple *Plasmodium* species promises to provide a wealth of new data that can be used to develop a better understanding of the biology of the parasite. Results from initial comparative analyses have revealed several interesting characteristics of *Plasmodium* genomes: (i) gene synteny is highly conserved in chromosome cores, in contrast to reduced similarity between subtelomeric regions of different malaria parasites; (ii) genes that elicit a host immune response are frequently found to be species-specific; (iii) paralogous gene families exhibit differential expansion in different *Plasmodium* lineages; (iv) comparison of selective constraint between orthologous genes shows orthologs in different species to be under diverse evolutionary pressure; (v) mining of comparative gene expression analysis has identified conserved motifs used in the regulation of gene expression. Genome sequence data from eight different species of malaria parasite provide one of the largest datasets for a parasitic protist and as such has tremendous potential for comparative analysis.

Comparative genomics

The term comparative genomics is used liberally in the post-genomic world, but what does it actually mean? A simple definition is: "a large-scale holistic approach that compares two or more genomes to discover the similarities and difference between the genomes, and to study the biology of the individual genomes" (Wei *et al.* 2002). Comparative genomics can be applied to whole genomes or syntenic regions of different species, subspecies, or different strains of the same species. It also includes both the development of computational tools as well as use of the tools to analyze genomes and further elucidate the biology behind the species under study.

The malaria parasite

Species of parasite that cause the disease malaria are members of the genus *Plasmodium* in the phylum Apicomplexa. The Apicomplexa is a large and complex clade, consisting of more than 5000 described species all of which are parasitic and characterized by a distinctive structure, the apical complex (see Simpson and Patterson, Chapter 1). The taxonomy and phylogeny of the phylum is controversial, involving as it does issues such as the derivation of the *Plasmodium* genus (see for example Escalante and Ayala 1995). The *Plasmodium* genus consists of ~200 species that parasitize birds, reptiles, and mammals (Garnham 1966). Four species infect man: *P. falciparum*, the most severe that causes malignant malaria, *P. vivax*, the most prevalent that causes widespread morbidity but little mortality, *P. ovale*, and *P. malariae* both of which are less common. Recently, probable zoonotic infections of *P. knowlesi* have been found in humans in Indonesia (Singh *et al.* 2004). Studies concerning the phylogeny of the *Plasmodium* genus are beginning to clarify the evolutionary relationship among the species (Fig. 2.1) (Leclerc *et al.* 2004; Perkins and Schall 2002). All the human malaria parasites are phylogenetically very distant (Escalante and Ayala 1994), with the closest relative to *P. falciparum* being the chimpanzee parasite *P. reichenowi* (Ayala 1998), and *P. vivax* forming a monophyletic group with monkey malaria parasites from South East Asia (Escalante *et al.* 2005).

Malaria is a major threat to public health in tropical and subtropical regions of the world. There are an estimated 300 to 500 million cases and 2 to 3 million deaths from *P. falciparum* malaria each year (Breman *et al.* 2001). Approximately 70 to 80 million cases are caused by infection with *P. vivax* (Mendis *et al.* 2001). All malaria parasites are obligate, intracellular parasites that have a complex life cycle in two hosts, the mosquito and man. The life cycle proceeds as

follows: sporozoites are inoculated into the vertebrate host through the bite of a female mosquito. The sporozoites travel to the liver, where successive rounds of mitotic replication in hepatic cells generates the merozoite stage. Merozoites are released from the liver cells and migrate to the blood, where they invade erythrocytes and through successive rounds of mitosis develop into the trophozoite and finally the schizont form. Schizonts rupture at maturity and release merozoites into the blood stream, which invade further erythrocytes, and the asexual cycle is repeated. Some merozoites develop into gametocytes, the sexual stage of the parasite. These are taken up in the next blood meal of a mosquito, and enter the sexual phase. Gametes from the gametocytes fuse to form ookinetes, which cross the wall of the mosquito mid-gut and develop into sporozoite-filled oocysts. When the oocysts burst, sporozoites migrate to the mosquito's salivary glands, and thus the life cycle is repeated.

Monkey, rodent, and bird malaria parasite species adapted for growth in laboratory animals have been developed as model systems to circumvent the problems associated with studies of the human malaria parasites. For example: species of rodent malaria including *P. yoelii*, *P. berghei*, and *P. chabaudi* have been isolated from African thicket rats and adapted to growth in laboratory mice and rats; Old World monkey malaria parasites, for example *P. knowlesi*, have been adapted to growth in New World laboratory monkeys; and species of bird malaria, such as *P. gallinaceum*, have been adapted for growth in domestic fowl. These model malaria parasites have similar life cycles and biology to *P. falciparum* while differing in certain aspects such as virulence, invasion methods, and evasion of the host immune response. All species are transmitted by the *Anopheles* mosquito, except for bird malaria species, which are transmitted by species of *Aedes*.

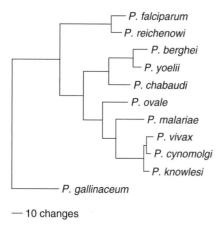

— 10 changes

Figure 2.1 Phylogeny of eleven *Plasmodium* species inferred from partial sequences of the mitochondial gene *cytochrome b*, as discussed by (Perkins and Schall 2002). Branch lengths were reconstructed using parsimony. This group of taxa represents a wide range of evolutionary distances, from the closely related triad that includes *P. vivax*, in which taxa differ from each other by ~1.5%, to the distantly related *P. gallinaceum*, which differs from other taxa by > 15%.

Current status of *Plasmodium* genome sequencing projects

A current list of malaria parasite genome sequencing initiatives is given in Table 2.1. The genomes of four species have been sequenced and published to date: the complete finished sequence of *P. falciparum* (Gardner *et al.* 2002), and the partial sequences of three of the four species of rodent malaria parasites, *P. yoelii yoelii* (Carlton *et al.* 2002), *P. chabaudi chabaudi*, and *P. berghei* (Hall *et al.* 2005). Other sequencing projects close to completion include shotgun coverage of the monkey malaria parasite *P. knowlesi*, and the finished genome sequence of a second human malaria parasite, *P. vivax* (Carlton 2003), with publications describing the annotation and comparative analysis of the genomes expected late 2006. All sequence data are being released by the sequencing centers on their websites or by deposition into PlasmoDB (Bahl *et al.* 2003), a centralized malaria sequence database, so that researchers can use the data for biological experimentation. This has proven highly successful in the case of prior release of the *P. falciparum* genome sequence data, resulting in identification of parasite-specific pathways which may represent unique targets for intervention strategies (see for example Jomaa *et al.* 1999), while acknowledging the entitlement of the sequencing centers to publish a whole genome analysis of the final data.

Due to the novelty of complete *Plasmodium* genomic and expression data, comparative genomics of malaria parasite genomes is still a developing research area. The studies mentioned below focus primarily on comparative genomic studies of the rodent models of malaria and *P. falciparum*, since these are the most advanced. A brief, general background concerning the *Plasmodium* nuclear genome is given (for a recent review see Carlton *et al.* 2005), but greater emphasis is given to more recent developments specifically in the area of comparative whole genome analyses.

The *Plasmodium* nuclear genome, gene content, and chromosome structure

From the completed genomes of four malaria parasite species, *P. falciparum* (Gardner *et al.* 2002), *P. y. yoelii* (Carlton *et al.* 2002), *P. c. chabaudi*, *P. berghei* (Hall *et al.* 2005), and other studies, generalizations concerning the structure and

organization of a typical *Plasmodium* nuclear genome can be made (Table 2.2). The genome is haploid and has a size of 22 to 27 Mb, distributed among 14 linear chromosomes that range in size from 500 kb to over 3 Mb (Carlton *et al.* 1999a; Janse *et al.* 1994a; Kemp *et al.* 1987). Base composition varies from species to species, from 19% GC in *P. falciparum* (Gardner *et al.* 2002) to 35% GC in *P. knowlesi* (Buckee 2002), and does not appear to be host-lineage specific. An additional higher order structuring exists in some species, for example sections of the *P. vivax* genome are compartmentalized into discrete isochores of differing [G + C] content (McCutchan *et al.* 1984). In contrast, *P. falciparum* has a uniform genome composition, with the exception of short regions of > 97% [A + T] on each chromosome which most likely contain the centromeres (Hall *et al.* 2002), in addition to the bias exhibited between coding and non-coding regions (described below).

Several features of chromosome structure are conserved in all *Plasmodium* species. All possess telomeres consisting of degenerate, canonical, tandem repeats (Table 2.2) (Scherf *et al.* 2001), with the mean length of the telomeric array (∼800 to ∼6700 bp) varying among species but remaining constant within species (Figueiredo *et al.* 2002). Subtelomeric regions of many *Plasmodium* species chromosomes consist of a variable number of species-specific repeats that extend 10 to 40 kb towards the center of chromosomes, and which have extensive large-scale similarity between chromosomes, indicative of intrachromosomal exchange (Gardner *et al.* 2002; Carlton *et al.* 2002). Low-resolution restriction maps and YAC (yeast artificial chromosome) contig maps, in conjunction with the completed *Plasmodium* genomes, have now established that species-specific gene families coding for immunodominant antigens and proteins known to be involved in antigenic variation are predominantly found within these regions, whereas conserved housekeeping genes are located within central chromosome regions. *P. falciparum* chromosome ends have been shown to cluster at the periphery of the nucleus, facilitating ectopic recombination among heterologous subtelomeric chromosome regions and thus providing a

Table 2.1 Malaria parasite genome, transcriptome, and proteome projects

Species and strain, line or clone	Genome project goal (current status)	Sequencing center	[a]ESTs in GenBank dbEST	[b]GSSs in GenBank dbGSS	Microarray projects	Proteomic projects
P. falciparum 3D7 clone	Finished genome (published)	TIGR, Wellcome Trust Sanger Institute, Stanford University	20 914 (many isolates)	1766	cDNA arrays gDNA arrays oligo arrays SAGE	asexual stages food vocuoles sporozoite gametocyte gamete
P. falciparum Clinical isolate Ghana	8X (ongoing)	Wellcome Trust Sanger Institute	–	–	–	–
P. falciparum IT clone	8X (ongoing)	Wellcome Trust Sanger Institute	–	–	–	–
P. falciparum HB3 clone	8X (ongoing)	Broad Institute	–	–	–	–
P. falciparum Dd2 clone	8X (ongoing)	Broad Institute	–	–	–	–
P. vivax strain Salvador I	Finished genome (manuscript in prep.)	TIGR	22 238 (many isolates)	10 697	cDNA arrays oligo arrays	asexual stages
P. yoelii yoelii 17XNL clone	Partial to 5X (complete, published)	TIGR	18 930	–	cDNA arrays oligo arrays	asexual stages sporozoite ookinete gametocyte
P. chabaudi chabaudi AS clone	Partial to 4 X (published, additional sequencing in progress)	Wellcome Trust Sanger Institute	–	767	–	–
P. berghei ANKA clone	Partial to 4 X (published, additional sequencing in progress)	Wellcome Trust Sanger Institute	7518	5476	gDNA arrays	asexual stages sporozoite gametocyte ookinete oocyst
P. knowlesi H	Partial to 8X (manuscript in prep.)	Wellcome Trust Sanger Institute	–	–	ongoing	asexual stages
P. reichenowi Oscar	Partial to 3X (1X)	Wellcome Trust Sanger Institute	–	–	–	–
P. gallinaceum A	Partial to 3X (underway)	Wellcome Trust Sanger Institute	–	–	–	–

[a] ESTs: expressed sequence tags from cDNA libraries.
[b] GSSs: genome survey sequences from mung bean nuclease-digested gDNA libraries.

Table 2.2 *Plasmodium* genome characteristics – comparison of general genome characteristics from six *Plasmodium* genome datasets

	P. falciparum	P. vivax [c]	P. knowlesi [d]	P. y. yoelii	P. c. chabaudi	P. berghei
Genome size (Mb)	23.1	26.8	24.3	23.1	~23 [b]	~23 [b]
No. chromosomes	14	14	14	14	14	14
% (A + T)	80.6	67.1	64.2	77.4	75.6	76.1
Isochore structure	Absent	Present	Absent	Absent	Absent	Absent
No. genes	5473	5393	5281	5878	5698 [a]	5864 [a]
Copy no. of largest gene families	149 *rif* 59 *var* 28 *stevo*	~300 *vir* Others ND	194 *SICAvar* 36 *hypothetical* 33 *kir*	838 *yir* 168 *pyst-a* 57 *pyst-b*	138 *cir* 108 *pyst-a* 75 *pcst-g*	180 *bir* 45 *pyst a* 34 *pyst b*
Centromeres	Functionally uncharacterized	Functionally uncharacterized	ND	Functionally uncharacterized	Functionally uncharacterized	Functionally uncharacterized
Telomeric repeat	AACCCTA	AACCCT(G/A)	AACCCTA	AACCCTG	AACCCT(G/A)	AACCCT(G/A)
Complex subtelomeric repeats	Identified	Absent	Identified	Limited	Identified	Identified

[a] Determined for gene predictions where orthologs identified in other *Plasmodium* species.
[b] Determined from karyotype data; ND: not determined.
[c] Unpublished data.
[d] Taken from (Buckee, 2002).

mechanism for the generation of different repertoires of antigen genes (Freitas-Junior *et al.* 2000).

The chromosomes of *Plasmodium* field isolates exhibit size polymorphisms, most likely as a result of unequal recombination between homologous chromosomes of parasite clones during meiosis, but also by deletion and insertion of repeat sequences (Van der Ploeg *et al.* 1985; Kemp *et al.* 1985; Langsley *et al.* 1988). *P. falciparum* chromosomes also vary in size during *in vitro* culture by the process of chromosome breakage followed by healing of the blunt end through the addition of telomeric repeats (Bottius *et al.* 1998). Many of these chromosomal rearrangements affect non-coding repeat sequences in the subtelomeric regions. However, exceptions include the genome rearrangements that occur in parasites under selective pressure, for example the generation of amplicons of tandemly repeated genes during selection for resistance to the antimalarial drug

mefloquine in clones of *P. falciparum* (Peel *et al.* 1994). Thus, chromosomal rearrangements in *Plasmodium* are important in genome evolution, although to what extent this occurs in natural populations remains to be determined.

Each *Plasmodium* genome analyzed to date has a predicted content of 5000 to 6000 genes (Gardner *et al.* 2002; Carlton *et al.* 2002; Buckee 2002; Hall *et al.* 2005). The difference in gene number between species is primarily due to gene family expansion in different lineages, and species-specific antigen and cell-surface genes (Table 2.2; described in more detail below). The number of orthologous genes, that is those which have evolved from a common ancestor by speciation, between different *Plasmodium* species has been variously estimated as ~60% (Carlton *et al.* 2002) to 80% (Hall *et al.* 2005). A comparison of *P. falciparum* and *P. y. yoelii* coding and non-coding regions suggests that different *Plasmodium* species exhibit similar characteristics for these regions (Carlton *et al.* 2002). For example coding regions of the genome have a lower [A + T] content (76%) than non-coding regions (80–87%), and a similar percentage of genes contain introns (54%). The main exception appears to be the mean length of genes, which in *P. falciparum* appears to be almost twice the size of the gene length in *P. y. yoelii*, and also larger than the mean length of genes in the budding yeast *Saccharomyces cerevisiae* and the fission yeast *Schizosaccharomyces pombe* (Gardner *et al.* 2002). Exactly why *P. falciparum* genes are longer is at present not known, but studies on the length of predicted *P. vivax* genes promises to shed light on this phenomenon.

Whole genome analysis of paralogous gene families

The four *Plasmodium* genomes sequenced and analyzed so far have started to provide a significant amount of data regarding gene families that have expanded in different lineages (paralogous gene families). Typically, such families can be defined as groups of proteins that share significant sequence similarity and a common evolutionary history. Proteins within paralogous gene families preserve their structure and thus maintain similar or identical biochemical functions across evolutionary distances. Identification of gene families is therefore an important part of genome annotation since this assists the functional annotation of proteins. There are a number of methods to detect protein families from complete genome sequences, for example the TribeMCL method that uses a Markov clustering algorithm to overcome problems presented by sequence similarity errors and fragmented peptides (Enright *et al.* 2002). The identification of paralogous gene families is somewhat subjective and can depend significantly upon the parameters used.

What is known concerning paralogous gene families in *Plasmodium*? A description of the major families identified in four species is shown in Table 2.3. Of principal interest is the finding that many families have undergone gene family expansion in certain lineages, and that some of the expansions occur in more closely related lineages,

for example the *yir/bir/cir* gene family in the rodent malaria lineage. This family, which is a large variant gene family predicted to be involved in antigenic variation, was first described in *P. vivax* (the *vir* family) (del Portillo *et al.* 2001), and latterly in *P. y. yoelii* (the *yir* family) (Carlton *et al.* 2002), *P. berghei* (the *bir* family) and *P. chabaudi* (the *cir* family) (Janssen *et al.* 2002), and *P. knowlesi* (the *kir* family) (Buckee 2002). As evident from Table 2.3, there are varying copies of this gene family in the three rodent lineages, with massive expansion having occurred in *P. y. yoelii* but less in *P. c. chabaudi*. There are some data to suggest that the *yir/bir/cir* family is part of a superfamily that includes another multigene family of *P. falciparum*, the *rif/stevor* gene family (Janssen *et al.* 2004). However, true homologs of this family have not, so far, been identified in *P. falciparum*, which contains other gene families involved in antigenic variation and evasion of immune responses (the *var* gene families) (Craig and Scherf 2001). In *P. knowlesi*, the *SICAvar* gene family has also been described (al-Khedery *et al.* 1999), which is expressed on the surface of infected erythrocytes and is implicated in antigenic variation in this species. No significant sequence similarity exists between the *var* and *SICAvar* genes.

Other gene families in the *Plasmodium* genome are beginning to be identified through analysis of whole genome data; for example the SURFIN gene family encoded by a family of ten *surf* genes, including three pseudogenes and located within or close to the subtelomeres of five *P. falciparum* chromosomes (Winter *et al.* 2005). SURFINs show structural and sequence similarities with *SICAvar*, *vir*, and *var* genes, and have been implicated in the invasion of erythrocytes by merozoites. It is interesting to note that several of the gene families mentioned above are expressed in different stages of the parasite's life cycle, suggesting that the parasite has adapted certain key genes to perform a variety of quite distinct tasks (pleiotropic genes).

Whole genome alignment and synteny analyses

Genome sequence alignment is one of the most basic tools for comparative genomic analysis. The underlying assumption is that the genome

Table 2.3 *Plasmodium* gene families

	Family name	P. falciparum	P. chabaudi	P. berghei	P. yoelii	Function
Common gene families	Py235	4	4	10	14	Reticulocyte binding protein
	HAD-Hydrolase	5	19	3	5	HAD hydrolase
	pst-a	10	41	6	12	Alpha beta hydrolase
	etramp	11	3	7	11	Early transcribed membrane protein
Rodent malaria-specific gene families	yir/cir/bir	0	138	180	838	Variant antigen
	pyst-a	1	108	45	168	Unknown
	pyst-b	0	10	34	57	Unknown
	pyst-c	0	5	4	21	Unknown
	pyst-d	0	0	1	17	Unknown
	pcst-f	0	10	2	1	Unknown
	pcst-g	0	75	11	7	Unknown
	pcst-h	1	5	6	4	Unknown
P. falciparum-specific gene families	Var	59	0	0	0	Erythrocyte membrane antigen (PfEMP1)
	Rifin/stevor	177	0	0	0	Variant RBC surface antigen
	rhoph1/clag	4	1	1	2	Cytoadherence linked asexual protein
	Pf-fam-a	35	0	0	0	DNAJ domain protein
	Pf-fam-b	11	0	0	0	Unknown
	Pf-fam-c	20	0	1	1	Ser/thr kinase
	Pf-fam-d	17	0	0	0	Unknown
	Pf-fam-e	6	0	0	0	Unknown
	Pf-fam-f	10	0	0	0	Unknown
	Pf-fam-g	13	0	0	0	Unknown
	Pf-fam-h	19	0	0	0	Unknown
	Pf-fam-i	11	2	1	1	Acyl-Co A sythetase

Gene family designations follow previous convention: py (Carlton *et al.* 2002), pc (Fischer *et al.* 2003), and Pf (Gardner *et al.* 2002). Adapted from (Hall *et al.* 2005).

sequences being aligned had a common ancestor, and therefore that every base pair in each sequence can be explained as the combination of this ancestral sequence and the action of evolution. Genome sequence alignment is also one of the more challenging aspects of comparative genomics, since whole genomes can be large and involve large-scale rearrangements which are problematic to align. There are a number of algorithms which have been developed for genome-scale alignment and visualization and which use either local or global alignment techniques; however a discussion of these is not within the scope of this chapter, and further details can be found in two recent reviews (Ureta-Vidal *et al.* 2003; Frazer *et al.* 2003).

Whole genome alignments of *Plasmodium* genomes have been published and yielded much insight into gene synteny (conservation of order of orthologous genes between different species) in the genus. Initial studies involving mapping of conserved genes to separations of *Plasmodium* chromosomes showed that gene location (conserved synteny) and gene order (conserved linkage) are preserved over large regions between all four species of rodent malaria (Janse *et al.* 1994b), between species of rodent malaria and *P. falciparum* (Carlton *et al.* 1998), and between all four human malaria species (Carlton *et al.* 1999b). Further studies also showed that even exon/intron boundaries and the fine-scale organization of genes can be conserved between species (van Lin *et al.* 2001; Tchavtchitch *et al.* 2001; Vinkenoog *et al.* 1995). The degree of conservation of synteny is greatest when comparing genomes of more closely related species; for example the rodent malaria parasites show conservation of whole chromosome synteny (Janse *et al.* 1994b), whereas synteny is reduced to the level of conservation of large chromosomal blocks between *P. falciparum* and the rodent malaria species (Carlton *et al.* 2002).

Initially, using a mixture of computational algorithms and laboratory-based methods, a whole genome synteny map of the complete sequence of *P. falciparum* and the partial sequence of *P. y. yoelii* (Carlton *et al.* 2002) was generated. A local alignment tool MUMmer2 (Delcher *et al.* 2002) was used to identify matches of at least five amino acids long from six-frame translations of both sequences; these seed matches were extended to create a tiling path of *P. y. yoelii* contigs against the *P. falciparum* chromosomes. From a total of 2212 contigs, approximately 70% of the *P. y. yoelii* genome could be aligned to all 14 *P. falciparum* chromosomes. Linkage of the contigs through PCR amplification, scaffolding, and the use of a physical map resulted in the formation of 457 blocks of synteny ranging in length from a few kilobases to more than 800 kb. Long contiguous sections of the *P. y. yoelii* genome with accompanying chromosomal location were thus assigned to each *P. falciparum* chromosome. From a total of 4787 *P. y. yoelii* genes in the tiling path, 3525 (74%) were found to be conserved in order between the two species using the algorithm Position Effect (Carlton *et al.* 2002). This compares with 41/48 (85%) of genes found to be conserved in order in a 200 kb region syntenic between *P. falciparum* and *P. vivax* (Tchavtchitch *et al.* 2001).

More recently, a synteny map between *P. falciparum* and all three species of rodent malaria has been generated (Hall *et al.* 2005). As mentioned above, species of rodent malaria exhibit whole chromosome synteny (Janse *et al.* 1994a) and their sequence similarity ranges from >90% in coding regions to 80 to 90% in non-coding regions.

Capitalizing on this, the partial genome sequence data from *P. berghei* and *P. chabaudi* was used to fill in the gaps in the *P. falciparum*/*P. y. yoelii* map and to create a composite "Frankenstein" rodent malaria genome (Fig. 2.2). Thirty-six synteny blocks of average size 520 kb (range 42–1788 kb) were found to cover the 14 *P. falciparum* chromosomes. Two rodent malaria chromosomes (3 and 9) were found to be completely syntenic with *P. falciparum* chromosomes 2 and 11, respectively. A total of 22 synteny breakpoints were identified, located in regions containing rRNA genes and other repetitive elements, and the maximum number of breakpoints in any chromosome was determined to be five. Further detailed analysis of the map is ongoing (T. Kooij, J. Carlton, S. Bidwell, N. Hall, J. Ramesar, C. Janse, A. Waters, unpublished data).

The construction of synteny maps between other *Plasmodium* species is also in progress, although this is limited by the nature of the genome data. For example creation of a map using partial genome sequence data requires that one of the genomes be finished or at least in megabase scaffolds and preferably with some karyotype and chromosome mapping data available. A synteny map of two human malaria species, *P. falciparum* and *P. vivax*, is planned, with preliminary tiling paths already suggesting a high degree of conservation of synteny between the two (Carlton 2003).

What do synteny maps tell us about the biology of *Plasmodium* species? They are useful for a number of studies: (i) as a means to infer the evolution of the genus, that is tracking the gross chromosomal rearrangements that gave rise to the karyotype of current species seen today; (ii) as a method of identifying orthologs between species which cannot be identified through sequence similarity alone; (iii) for refinement of gene predictions through simultaneous annotation of multiple *Plasmodium* genomes; (iv) for comparative analysis of gene expression, for example through identification of conserved non-coding regions of the *Plasmodium* genome, and the evaluation of co-ordinated gene expression; and (v) as a means for the classification of genes under different evolutionary pressures.

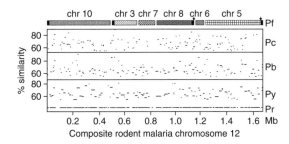

Figure 2.2 Schematic of the composite rodent malaria chromosome 12 compiled from alignment of three rodent malaria parasite genome datasets and syntenic regions in *P. falciparum*. The tiling path of *P. y. yoelii* (Py), *P. berghei* (Pb), and *P. chabaudi* (Pc) contigs are shown, and shown combined as Pr. The contigs for each species are plotted on the y-axis as a function of protein similarity. Six *P. falciparum* syntenic blocks are shown (Pf). Telomeric-proximal regions are shown as black boxes, and rRNA units as black stars.

Comparative gene expression and regulation

Several whole genome microarray datasets are now available for different species of *Plasmodium*, in particular *P. falciparum* asexual stages (Bozdech *et al.* 2003a; Bozdech *et al.* 2003b; Le Roch *et al.* 2002; Le Roch *et al.* 2003; Ben Mamoun *et al.* 2001; Hayward *et al.* 2000) and sexual and asexual stages of species of rodent malaria parasite (Shi *et al.* 2005; Hall *et al.* 2005). These studies have demonstrated that proteins of related function have similar or identical expression profiles, and that genes encoding transcriptional proteins are expressed in the early blood stages, whereas genes encoding proteins involved in invasion are expressed in the late schizont stage, much as would be expected. Moreover, since the transcription profile of many of the 60% hypothetical genes in the genome cluster with those of genes of known function, it may be possible to assign function to the former through "guilt-by-association" (Waters 2003). Whole genome proteomic studies of various life cycle stages have also been completed which have enabled further insights into gene regulation in *Plasmodium* (Lasonder *et al.* 2002; Florens *et al.* 2002; Carlton *et al.* 2002; Hall *et al.* 2005; Khan *et al.* 2005). By comparing the whole genome transcriptome and proteome datasets, a picture is emerging of the malaria parasite controlling gene

expression through regulation at the level of RNA synthesis and/or stability (Hall *et al.* 2005; Le Roch *et al.* 2004). For example Le Roch *et al.* (Le Roch *et al.* 2004) compared transcript and protein abundance for seven different stages of the *P. falciaprum* life cycle, and observed a delay between mRNA generation and protein accumulation. Studies cross-referencing to the proteomic studies of asexual, sexual, and mosquito stages of *P. falciparum* have also determined that large proportions of the genome encode proteins used in multiple stages of the life cycle (Lasonder *et al.* 2002; Florens *et al.* 2002).

The power of comparative genomics has recently become even more apparent in the identification of regulatory motifs involved in gene expression. *Plasmodium* nuclear genes are monocistronically transcribed (a single transcription unit produces a single mRNA that encodes only one protein), which suggests the presence of regulatory sequence elements flanking the coding regions. Functional conservation of promoter elements between different *Plasmodium* species has been shown (Crabb and Cowman 1996), and the few *P. falciparum* promoters that have been identified appear to conform to the structure of other eukaryotic promoters, that is a basal promoter regulated by upstream enhancer elements. However, prior to the generation of whole genome expression datasets, few DNA elements had been identified that direct the transcription of *Plasmodium* genes (van Lin *et al.* 2000; Horrocks *et al.* 1998), for example TATA and CAAT box sequences in *P. knowlesi* (Ruiz i Altaba *et al.* 1987) and *P. falciparum* (Su and Wellems 1994), a 5′ flanking *var* gene regulatory element (Vazquez-Macias *et al.* 2002), and a highly conserved GC-rich element of ∼202 bp in each intergenic region of the internal *var* genes (Hall *et al.* 2002).

Recently, capitalizing on the large whole genome and expression datasets and using bioinformatics approaches, several motifs involved in gene regulation have been identified. For example mining the complete genome sequence of *P. falciparum*, DNA sequences 2 kb upstream from the predicted initiation codon of 18 *P. falciparum* heat shock protein (*hsp*) genes were extracted, the sequences analyzed for over-represented motifs, and a G-rich regulatory element, the G-box, identified as a conserved motif found to be preserved across several *Plasmodium* species (Militello *et al.* 2004). In another study, *P. berghei* microarray and proteomic data were analyzed and a motif in the 3' UTR (untranslated region) of genes that are up-regulated in gametocyte stages but whose protein products appear in the ookinete stage, was identified (Hall *et al.* 2005). This identified the motif as a putative control element involved in the translational repression of transcripts produced during the sexual stages. Orthologous *P. falciparum* genes did not contain the same motif, an indicator that the regulation of gene expression could be a major contributor to *Plasmodium* host specificity and parasite diversification. Finally, two recent studies used a combination of whole genome analysis with computational techniques to identify a conserved motif (called the Pexel motif) in the N-terminal regions of *P. falciparum* proteins involved in protein translocation into the infected erythrocyte cytoplasm (Hiller *et al.* 2004; Marti *et al.* 2004). Interestingly, this motif was found to be conserved across other *Plasmodium* species, in particular *P. vivax* and *P. gallinaceum*.

Whole genome SNP maps and molecular evolution studies

Several SNP (single nucleotide polymorphism) maps of different *Plasmodium* species and isolates have been generated since the availability of whole genome sequence data. They have been used to try to provide answers for a number of primarily evolution-related questions, and have initiated a move towards the generation of a haplotype map or HapMap for the parasite. A HapMap would enable genome-wide association studies to be performed, in which a dense set of SNPs across the parasite's genome is genotyped to survey the most common genetic variation for a role in disease or to identify the heritable quantitative traits that are risk factors for disease (Su and Wootton 2004). One of the first SNP maps resequenced 1.2 to 1.5 kb fragments from each of the 204 predicted genes of *P. falciparum* chromosome 3, and identified 238 SNPs from five parasite clones (Mu *et al.* 2002). Analysis of the synonymous and non-coding SNPs was used to estimate the time to the most recent common

ancestor (MRCA) of *P. falciparum* to be ∼100 000 to 180 000, significantly older than had been previously estimated using variation among a handful of genes (Rich *et al.* 1998; Volkman *et al.* 2001), but in line with other studies (reviewed in Hughes and Verra 2002). Subsequent SNP maps focused on subsets of genes, for example putative transporter genes and transmembrane proteins, and showed that SNPs from multiple transporters are associated with reduced susceptibility to several antimalarial drugs (Mu *et al.* 2003), and that a disproportionate number of polymorphisms are found in genes encoding proteins associated with the cell membrane (Volkman *et al.* 2002), respectively.

Resequencing the mitochondrial genome of isolates of different *Plasmodium* species has also identified SNP haplotypes which can be used to map the genetic relationships among a species. For example, examination of nucleotide diversity in 106 mitochondrial genomes of *P. vivax* has shown the parasite to be ancient with an estimated MRCA of 200 000 to 300 000 years ago (Jongwutiwes *et al.* 2005), in line with other studies (Escalante *et al.* 2005). A similar study that sequenced the mitochondrial genomes of 176 *P. vivax* isolates used the data to infer the genetic and coevolutionary history of the parasite, determining that the parasite is of Asian origin via a host switch from macaque monkeys (Mu *et al.* 2005b).

With several large-scale SNP maps showing the power of studying genetic variation, a recent publication has paved the way for a large-scale HapMap to be generated for *P. falciparum* (Mu *et al.* 2005a). SNP haplotype and population recombination maps for one chromosome sequenced from 99 isolates were generated, and analyzed for linkage disequilibrium and population structure. Geographical isolation and highly variable multiple infection rates were found to be the major factors affecting haplotype structure. However, the study showed the feasibility of genome-wide association studies, at least in some parasite populations.

Plasmodium comparative genomics: what next?

The sequencing of the first *P. falciparum* isolate took 6 years and many millions of dollars to complete, but fortunately due to advances in sequencing technology and expertise, the second generation of *P. falciparum* genomes are proving easier and cheaper to produce. In addition, having available the complete, finished "reference" genome for eight species should allow the generation of partial genome sequence data of other isolates of the same species to be just as informative. For those species for which no reference genome is available, the potential to generate genome-wide synteny maps may enable targeted cloning and sequencing of specified regions through the generation of large insert libraries, similar to what has been proposed for species of primate (Eichler and DeJong 2002). Now that the genome sequencing band wagon is on a roll, what is the future likely to hold for comparative genomics of *Plasmodium* species? This chapter has given a taste of the sorts of comparative genomics studies that whole genome sequence data can support, from inference of parasite evolution and age, to determination of motifs involved in gene expression, identification of low-similarity orthologs, and generation of synteny maps for gene family studies. Such analyses will no doubt continue and evolve as a means to determine new ways to control the parasite and the disease. It is to be hoped that, along the way, the ability to manipulate large datasets and analyze multiple life cycle stages will spread to more than the few core labs that currently hold the necessary cross-disciplinary expertise of *Plasmodium* biology and bioinformatics.

References

al-Khedery, B., Barnwell, J. W., and Galinski, M. R. (1999). Antigenic variation in malaria: a 3′ genomic alteration associated with the expression of a *P. knowlesi* variant antigen. *Mol. Cell*, **3**, 131–141.

Ayala, F. J. (1998). Evolutionary relationships of human malaria parasites. In Sherman, I. W., ed. *Malaria: Parasite Biology, Pathogeneis and Protection* ASM Press, Washington, D.C.

Bahl, A., Brunk, B., Crabtree, J., Fraunholz, M. J., Gajria, B., Grant, G. R., Ginsburg, H., Gupta, D., Kissinger, J. C., Labo, P., Li, L., Mailman, M. D., Milgram, A. J., Pearson, D. S., Roos, D. S., Schug, J., Stoeckert, C. J., Jr., and Whetzel, P. (2003). PlasmoDB: the *Plasmodium* genome resource. A database integrating experimental and computational data. *Nucleic Acids Res.*, **31**, 212–215.

Ben Mamoun, C., Gluzman, I. Y., Hott, C., MacMillan, S. K., Amarakone, A. S., Anderson, D. L., Carlton, J. M., Dame, J. B., Chakrabarti, D., Martin, R. K., Brownstein, B. H., and Goldberg, D. E. (2001). Co-ordinated programme of gene expression during asexual intraerythrocytic development of the human malaria parasite *Plasmodium falciparum* revealed by microarray analysis. *Mol. Microbiol.*, **39**, 26–36.

Bottius, E., Bakhsis, N., and Scherf, A. (1998). *Plasmodium falciparum* telomerase: de novo telomere addition to telomeric and nontelomeric sequences and role in chromosome healing. *Mol. Cell Biol.*, **18**, 919–925.

Bozdech, Z., Llinas, M., Pulliam, B. L., Wong, E. D., Zhu, J., and DeRisi, J. L. (2003a). The transcriptome of the intraerythrocytic developmental cycle of *Plasmodium falciparum*. *PLoS. Biol.*, **1**, E5.

Bozdech, Z., Zhu, J., Joachimiak, M. P., Cohen, F. E., Pulliam, B., and DeRisi, J. L. (2003b). Expression profiling of the schizont and trophozoite stages of *Plasmodium falciparum* with a long-oligonucleotide microarray. *Genome Biol.*, **4**, R9.

Breman, J. G., Egan, A., and Keusch, G. T. (2001). The intolerable burden of malaria: a new look at the numbers. *Am. J. Trop. Med. Hyg.*, **64**, 4–7.

Buckee, C. (2002). Plasmodium Kharlesi *versus* Plasmodium falciparum: *A comparative analysis*. MSc Thesis. University of York, York.

Carlton, J. (2003). The *Plasmodium vivax* genome sequencing project. *Trends Parasitol.*, **19**, 227–231.

Carlton, J., Silva, J., and Hall, N. (2005). The genome of model malaria parasites, and comparative genomics. *Curr. Issues Mol. Biol.*, **7**, 23–37.

Carlton, J. M., Angiuoli, S. V., Suh, B. B., Kooij, T. W., Pertea, M., Silva, J. C., Ermolaeva, M. D., Allen, J. E., Selengut, J. D., Koo, H. L., Peterson, J. D., Pop, M., Kosack, D. S., Shumway, M. F., Bidwell, S. L., Shallom, S. J., van Aken, S. E., Riedmuller, S. B., Feldblyum, T. V., Cho, J. K., Quackenbush, J., Sedegah, M., Shoaibi, A., Cummings, L. M., Florens, L., Yates, J. R., Raine, J. D., Sinden, R. E., Harris, M. A., Cunningham, D. A., Preiser, P. R., Bergman, L. W., Vaidya, A. B., van Lin, L. H., Janse, C. J., Waters, A. P., Smith, H. O., White, O. R., Salzberg, S. L., Venter, J. C., Fraser, C. M., Hoffman, S. L., Gardner, M. J., and Carucci, D. J. (2002). Genome sequence and comparative analysis of the model rodent malaria parasite *Plasmodium yoelii yoelii*. *Nature*, **419**, 512–519.

Carlton, J. M., Galinski, M. R., Barnwell, J. W., and Dame, J. B. (1999a). Karyotype and synteny among the chromosomes of all four species of human malaria parasite. *Mol. Biochem. Parasitol.*, **101**, 23–32.

Carlton, J. M.-R., Galinski, M. R., Barnwell, J. W., and Dame, J. B. (1999b). Karyotype and synteny among the chromosomes of all four species of human malaria parasite. *Mol. Biochem. Parasitol.*, **101**, 23–32.

Carlton, J. M. R., Vinkenoog, R., Waters, A. P., and Walliker, D. (1998). Gene synteny in species of *Plasmodium. Mol. Biochem. Parasitol.*, **93**, 285–294.

Crabb, B. S. and Cowman, A. F. (1996). Characterization of promoters and stable transfection by homologous and nonhomologous recombination in *Plasmodium falciparum. Proc. Natl. Acad. Sci. U S A*, **93**, 7289–7294.

Craig, A. and Scherf, A. (2001). Molecules on the surface of the *Plasmodium falciparum* infected erythrocyte and their role in malaria pathogenesis and immune evasion. *Mol. Biochem. Parasitol.*, **115**, 129–143.

Delcher, A. L., Phillippy, A., Carlton, J., and Salzberg, S. L. (2002). Fast algorithms for large-scale genome alignment and comparison. *Nucleic Acids Res.*, **30**, 2478–2483.

del Portillo, H. A., Fernandez-Becerra, C., Bowman, S., Oliver, K., Preuss, M., Sanchez, C. P., Schneider, N. K., Villalobos, J. M., Rajandream, M. A., Harris, D., Pereira da Silva, L. H., Barrell, B., and Lanzer, M. (2001). A superfamily of variant genes encoded in the subtelomeric region of *Plasmodium vivax. Nature*, **410**, 839–842.

Eichler, E. E. and DeJong, P. J. (2002). Biomedical applications and studies of molecular evolution: a proposal for a primate genomic library resource. *Genome Res.*, **12**, 673–678.

Enright, A. J., Van Dongen, S., and Ouzounis, C. A. (2002). An efficient algorithm for large-scale detection of protein families. *Nucleic Acids Res.*, **30**, 1575–1584.

Escalante, A. A. and Ayala, F. J. (1994). Phylogeny of the malarial genus *Plasmodium*, derived from rRNA gene sequences. *Proc. Natl. Acad. Sci. U S A*, **91**, 11373–11377.

Escalante, A. A. and Ayala, F. J. (1995). Evolutionary origin of *Plasmodium* and other Apicomplexa based on rRNA genes. *Proc. Natl. Acad. Sci. U S A*, **92**, 5793–5797.

Escalante, A. A., Cornejo, O. E., Freeland, D. E., Poe, A. C., Durrego, E., Collins, W. E., and Lal, A. A. (2005). A monkey's tale: the origin of *Plasmodium vivax* as a human malaria parasite. *Proc. Natl. Acad. Sci. U S A*, **102**, 1980–1985.

Figueiredo, L. M., Freitas-Junior, L. H., Bottius, E., Olivo-Marin, J. C., and Scherf, A. (2002). A central role for *Plasmodium falciparum* subtelomeric regions in spatial positioning and telomere length regulation. *EMBO J.*, **21**, 815–824.

Florens, L., Washburn, M. P., Raine, J. D., Anthony, R. M., Grainger, M., Haynes, J. D., Moch, J. K., Muster, N., Sacci, J. B., Tabb, D. L., Witney, A. A., Wolters, D., Wu, Y., Gardner, M. J., Holder, A. A., Sinden, R. E., Yates, J. R., and Carucci, D. J. (2002). A proteomic view of the *Plasmodium falciparum* life cycle. *Nature*, **419**, 520–526.

Frazer, K. A., Elnitski, L., Church, D. M., Dubchak, I., and Hardison, R. C. (2003). Cross-species sequence comparisons: a review of methods and available resources. *Genome Res.*, **13**, 1–12.

Freitas-Junior, L. H., Bottius, E., Pirrit, L. A., Deitsch, K. W., Scheidig, C., Guinet, F., Nehrbass, U., Wellems, T. E., and Scherf, A. (2000). Frequent ectopic recombination of virulence factor genes in telomeric chromosome clusters of *P. falciparum. Nature*, **407**, 1018–1022.

Gardner, M. J., Hall, N., Fung, E., White, O., Berriman, M., Hyman, R. W., Carlton, J. M., Pain, A., Nelson, K. E., Bowman, S., Paulsen, I. T., James, K., Eisen, J. A., Rutherford, K., Salzberg, S. L., Craig, A., Kyes, S., Chan, M. S., Nene, V., Shallom, S. J., Suh, B., Peterson, J., Angiuoli, S., Pertea, M., Allen, J., Selengut, J., Haft, D., Mather, M. W., Vaidya, A. B., Martin, D. M., Fairlamb, A. H., Fraunholz, M. J., Roos, D. S., Ralph, S. A., McFadden, G. I., Cummings, L. M., Subramanian, G. M., Mungall, C., Venter, J. C., Carucci, D. J., Hoffman, S. L., Newbold, C., Davis, R. W., Fraser, C. M., and Barrell, B. (2002). Genome sequence of the human malaria parasite *Plasmodium falciparum. Nature*, **419**, 498–511.

Garnham, P. C. C. (1966). *Malaria parasites and other Haemosporidia*. Blackwell Scientific, Oxford.

Hall, N., Karras, M., Raine, J. D., Carlton, J. M., Kooij, T. W., Berriman, M., Florens, L., Janssen, C. S., Pain, A., Christophides, G. K., James, K., Rutherford, K., Harris, B., Harris, D., Churcher, C., Quail, M. A., Ormond, D., Doggett, J., Trueman, H. E., Mendoza, J., Bidwell, S. L., Rajandream, M. A., Carucci, D. J., Yates, J. R., 3rd, Kafatos, F. C., Janse, C. J., Barrell, B., Turner, C. M., Waters, A. P., and Sinden, R. E. (2005). A comprehensive survey of the *Plasmodium* life cycle by genomic, transcriptomic, and proteomic analyses. *Science*, **307**, 82–86.

Hall, N., Pain, A., Berriman, M., Churcher, C., Harris, B., Harris, D., Mungall, K., Bowman, S., Atkin, R., Baker, S., Barron, A., Brooks, K., Buckee, C. O., Burrows, C., Cherevach, I., Chillingworth, C., Chillingworth, T., Christodoulou, Z., Clark, L., Clark, R., Corton, C., Cronin, A., Davies, R., Davis, P., Dear, P., Dearden, F., Doggett, J., Feltwell, T., Goble, A., Goodhead, I., Gwilliam, R., Hamlin, N., Hance, Z., Harper, D., Hauser, H., Hornsby, T., Holroyd, S., Horrocks, P., Humphray, S., Jagels, K., James, K. D., Johnson, D.,

Kerhornou, A., Knights, A., Konfortov, B., Kyes, S., Larke, N., Lawson, D., Lennard, N., Line, A., Maddison, M., McLean, J., Mooney, P., Moule, S., Murphy, L., Oliver, K., Ormond, D., Price, C., Quail, M. A., Rabbinowitsch, E., Rajandream, M. A., Rutter, S., Rutherford, K. M., Sanders, M., Simmonds, M., Seeger, K., Sharp, S., Smith, R., Squares, R., Squares, S., Stevens, K., Taylor, K., Tivey, A., Unwin, L., Whitehead, S., Woodward, J., Sulston, J. E., Craig, A., Newbold, C., and Barrell, B. G. (2002). Sequence of *Plasmodium falciparum* chromosomes 1, 3–9 and 13. *Nature*, **419**, 527–531.

Hayward, R. E., Derisi, J. L., Alfadhli, S., Kaslow, D. C., Brown, P. O., and Rathod, P. K. (2000). Shotgun DNA microarrays and stage-specific gene expression in *Plasmodium falciparum* malaria. *Mol. Microbiol.*, **35**, 6–14.

Hiller, N. L., Bhattacharjee, S., van Ooij, C., Liolios, K., Harrison, T., Lopez-Estrano, C., and Haldar, K. (2004). A host-targeting signal in virulence proteins reveals a secretome in malarial infection. *Science*, **306**, 1934–1937.

Horrocks, P., Dechering, K., and Lanzer, M. (1998). Control of gene expression in *Plasmodium falciparum*. *Mol. Biochem. Parasitol.*, **95**, 171–181.

Hughes, A. L. and Verra, F. (2002). Extensive polymorphism and ancient origin of *Plasmodium falciparum*. *Trends Parasitol.*, **18**, 348–351.

Janse, C. J., Carlton, J. M., Walliker, D., and Waters, A. P. (1994a). Conserved location of genes on polymorphic chromosomes of four species of malaria parasites. *Mol. Biochem. Parasitol.*, **68**, 285–296.

Janse, C. J., Carlton, J. M.-R., Walliker, D., and Waters, A. P. (1994b). Conserved location of genes on polymorphic chromosomes of four species of malaria parasites. *Mol. Biochem. Parasitol.*, **68**, 285–296.

Janssen, C. S., Barrett, M. P., Turner, C. M., and Phillips, R. S. (2002). A large gene family for putative variant antigens shared by human and rodent malaria parasites. *Proc. R. Soc. Lond. B Biol. Sci.*, **269**, 431–436.

Janssen, C. S., Phillips, R. S., Turner, C. M., and Barrett, M. P. (2004). *Plasmodium* interspersed repeats: the major multigene superfamily of malaria parasites. *Nucleic Acids Res.*, **32**, 5712–5720.

Jomaa, H., Wiesner, J., Sanderbrand, S., Altincicek, B., Weidemeyer, C., Hintz, M., Turbachova, I., Eberl, M., Zeidler, J., Lichtenthaler, H. K., Soldati, D., and Beck, E. (1999). Inhibitors of the nonmevalonate pathway of isoprenoid biosynthesis as antimalarial drugs. *Science*, **285**, 1573–1576.

Jongwutiwes, S., Putaporntip, C., Iwasaki, T., Ferreira, M. U., Kanbara, H., and Hughes, A. L. (2005). Mitochondrial genome sequences support ancient population expansion in *Plasmodium vivax*. *Mol. Biol. Evol.*, **22**, 1733–1739.

Kemp, D. J., Corcoran, L. M., Coppel, R. L., Stahl, H. D., Bianco, A. E., Brown, G. V., and Anders, R. F. (1985). Size variation in chromosomes from independent cultured isolates of *Plasmodium falciparum*. *Nature*, **315**, 347–350.

Kemp, D. J., Thompson, J. K., Walliker, D., and Corcoran, L. M. (1987). Molecular karyotype of *Plasmodium falciparum*: conserved linkage groups and expendable histidine-rich protein genes. *Proc. Natl. Acad. Sci. U S A*, **84**, 7672–7676.

Khan, S. M., Franke-Fayard, B., Mair, G. R., Lasonder, E., Janse, C. J., Mann, M., and Waters, A. P. (2005). Proteome analysis of separated male and female gametocytes reveals novel sex-specific *Plasmodium* biology. *Cell*, **121**, 675–687.

Langsley, G., Patarapotikul, J., Handunnetti, S., Khouri, E., Mendis, K. N., and David, P. H. (1988). *Plasmodium vivax*: karyotype polymorphism of field isolates. *Exp. Parasitol.*, **67**, 301–306.

Lasonder, E., Ishihama, Y., Andersen, J. S., Vermunt, A. M., Pain, A., Sauerwein, R. W., Eling, W. M., Hall, N., Waters, A. P., Stunnenberg, H. G., and Mann, M. (2002). Analysis of the *Plasmodium falciparum* proteome by high-accuracy mass spectrometry. *Nature*, **419**, 537–542.

Leclerc, M. C., Hugot, J. P., Durand, P., and Renaud, F. (2004). Evolutionary relationships between 15 *Plasmodium* species from new and old world primates (including humans): an 18S rDNA cladistic analysis. *Parasitology*, **129**, 677–684.

Le Roch, K. G., Johnson, J. R., Florens, L., Zhou, Y., Santrosyan, A., Grainger, M., Yan, S. F., Williamson, K. C., Holder, A. A., Carucci, D. J., Yates, J. R., 3rd, and Winzeler, E. A. (2004). Global analysis of transcript and protein levels across the *Plasmodium falciparum* life cycle. *Genome Res.*, **14**, 2308–2318.

Le Roch, K. G., Zhou, Y., Batalov, S., and Winzeler, E. A. (2002). Monitoring the chromosome 2 intraerythrocytic transcriptome of *Plasmodium falciparum* using oligonucleotide arrays. *Am. J. Trop. Med. Hyg.*, **67**, 233–243.

Le Roch, K. G., Zhou, Y., Blair, P. L., Grainger, M., Moch, J. K., Haynes, J. D., De La Vega, P., Holder, A. A., Batalov, S., Carucci, D. J., and Winzeler, E. A. (2003). Discovery of gene function by expression profiling of the malaria parasite life cycle. *Science*, **301**, 1503–1508.

Marti, M., Good, R. T., Rug, M., Knuepfer, E., and Cowman, A. F. (2004). Targeting malaria virulence and remodeling proteins to the host erythrocyte. *Science*, **306**, 1930–1933.

McCutchan, T. F., Dame, J. B., Miller, L. H., and Barnwell, J. (1984). Evolutionary relatedness of *Plasmodium* species as determined by the structure of DNA. *Science*, **225**, 808–811.

Mendis, K., Sina, B. J., Marchesini, P., and Carter, R. (2001). The neglected burden of *Plasmodium vivax* malaria. *Am. J. Trop. Med. Hyg.*, **64**, 97–106.

Militello, K. T., Dodge, M., Bethke, L., and Wirth, D. F. (2004). Identification of regulatory elements in the *Plasmodium falciparum* genome. *Mol. Biochem. Parasitol.*, **134**, 75–88.

Mu, J., Awadalla, P., Duan, J., McGee, K. M., Joy, D. A., McVean, G. A., and Su, X. Z. (2005a). Recombination hotspots and population structure in *Plasmodium falciparum*. *PLoS. Biol.*, **3**, e335.

Mu, J., Duan, J., Makova, K. D., Joy, D. A., Huynh, C. Q., Branch, O. H., Li, W. H., and Su, X. Z. (2002). Chromosome-wide SNPs reveal an ancient origin for *Plasmodium falciparum*. *Nature*, **418**, 323–326.

Mu, J., Ferdig, M. T., Feng, X., Joy, D. A., Duan, J., Furuya, T., Subramanian, G., Aravind, L., Cooper, R. A., Wootton, J. C., Xiong, M., and Su, X. Z. (2003). Multiple transporters associated with malaria parasite responses to chloroquine and quinine. *Mol. Microbiol.*, **49**, 977–989.

Mu, J., Joy, D. A., Duan, J., Huang, Y., Carlton, J., Walker, J., Barnwell, J., Beerli, P., Charleston, M. A., Pybus, O. G., and Su, X. Z. (2005b). Host switch leads to emergence of *Plasmodium vivax* malaria in humans. *Mol. Biol. Evol.*, **22**, 1686–1693.

Peel, S. A., Bright, P., Yount, B., Handy, J., and Baric, R. S. (1994). A strong association between mefloquine and halofantrine resistance and amplification, over-expression, and mutation in the P-glycoprotein gene homolog (pfmdr) of *Plasmodium falciparum in vitro*. *Am. J. Trop. Med. Hyg.*, **51**, 648–658.

Perkins, S. L. and Schall, J. J. (2002). A molecular phylogeny of malarial parasites recovered from cytochrome b gene sequences. *J. Parasitol.*, **88**, 972–978.

Rich, S. M., Licht, M. C., Hudson, R. R., and Ayala, F. J. (1998). Malaria's eve: evidence of a recent population bottleneck throughout the world populations of *Plasmodium falciparum*. *Proc. Natl. Acad. Sci. U S A*, **95**, 4425–4430.

Ruiz, I., Altaba, A., Ozaki, L. S., Gwadz, R. W., and Godson, G. N. (1987). Organization and expression of the *Plasmodium knowlesi* circumsporozoite antigen gene. *Mol. Biochem. Parasitol.*, **23**, 233–245.

Scherf, A., Figueiredo, L. M., and Freitas-Junior, L. H. (2001). *Plasmodium* telomeres: a pathogen's perspective. *Curr. Opin. Microbiol.*, **4**, 409–414.

Shi, Q., Cernetich, A., Daly, T. M., Galvan, G., Vaidya, A. B., Bergman, L. W., and Burns, J. M., Jr. (2005). Alteration in host cell tropism limits the efficacy of immunization with a surface protein of malaria merozoites. *Infect. Immun.*, **73**, 6363–6371.

Singh, B., Kim Sung, L., Matusop, A., Radhakrishnan, A., Shamsul, S. S., Cox-Singh, J., Thomas, A., and Conway, D. J. (2004). A large focus of naturally acquired *Plasmodium knowlesi* infections in human beings. *Lancet*, **363**, 1017–1024.

Su, X. Z. and Wellems, T. E. (1994). Sequence, transcript characterization and polymorphisms of a *Plasmodium falciparum* gene belonging to the heat-shock protein (HSP) 90 family. *Gene*, **151**, 225–230.

Su, X. Z. and Wootton, J. C. (2004). Genetic mapping in the human malaria parasite *Plasmodium falciparum*. *Mol. Microbiol.*, **53**, 1573–1582.

Tchavtchitch, M., Fischer, K., Huestis, R., and Saul, A. (2001). The sequence of a 200 kb portion of a *Plasmodium vivax* chromosome reveals a high degree of conservation with *Plasmodium falciparum* chromosome 3. *Mol. Biochem. Parasitol.*, **118**, 211–222.

Ureta-Vidal, A., Ettwiller, L., and Birney, E. (2003). Comparative genomics: genome-wide analysis in metazoan eukaryotes. *Nat. Rev. Genet.*, **4**, 251–262.

Van der Ploeg, L. H., Smits, M., Ponnudurai, T., Vermeulen, A., Meuwissen, J. H., and Langsley, G. (1985). Chromosome-sized DNA molecules of *Plasmodium falciparum*. *Science*, **229**, 658–661.

van Lin, L. H., Janse, C. J., and Waters, A. P. (2000). The conserved genome organisation of non-falciparum malaria species: the need to know more. *Int. J. Parasitol.*, **30**, 357–370.

van Lin, L. H., Pace, T., Janse, C. J., Birago, C., Ramesar, J., Picci, L., Ponzi, M., and Waters, A. P. (2001). Interspecies conservation of gene order and intron-exon structure in a genomic locus of high gene density and complexity in *Plasmodium*. *Nucleic Acids Res.*, **29**, 2059–2068.

Vazquez-Macias, A., Martinez-Cruz, P., Castaneda-Patlan, M. C., Scheidig, C., Gysin, J., Scherf, A., and Hernandez-Rivas, R. (2002). A distinct 5′ flanking var gene region regulates *Plasmodium falciparum* variant erythrocyte surface antigen expression in placental malaria. *Mol. Microbiol.*, **45**, 155–167.

Vinkenoog, R., Veldhuisen, B., Speranca, M. A., del Portillo, H. A., Janse, C., and Waters, A. P. (1995). Comparison of introns in a cdc2-homologous gene within a number of *Plasmodium* species. *Mol. Biochem. Parasitol.*, **71**, 233–241.

Volkman, S. K., Barry, A. E., Lyons, E. J., Nielsen, K. M., Thomas, S. M., Choi, M., Thakore, S. S., Day, K. P., Wirth, D. F., and Hartl, D. L. (2001). Recent origin of *Plasmodium falciparum* from a single progenitor. *Science*, **293**, 482–484.

Volkman, S. K., Hartl, D. L., Wirth, D. F., Nielsen, K. M., Choi, M., Batalov, S., Zhou, Y., Plouffe, D., Le

Roch, K. G., Abagyan, R., and Winzeler, E. A. (2002). Excess polymorphisms in genes for membrane proteins in *Plasmodium falciparum. Science*, **298**, 216–218.

Waters, A. P. (2003). Parasitology. Guilty until proven otherwise. *Science*, **301**, 1487–1488.

Wei, L., Liu, Y., Dubchak, I., Shon, J., and Park, J. (2002). Comparative genomics approaches to study organism similarities and differences. *J. Biomed. Inform.*, **35**, 142–150.

Winter, G., Kawai, S., Haeggstrom, M., Kaneko, O., von Euler, A., Kawazu, S., Palm, D., Fernandez, V., and Wahlgren, M. (2005). SURFIN is a polymorphic antigen expressed on *Plasmodium falciparum* merozoites and infected erythrocytes. *J. Exp. Med.*, **201**,1853–1863.

CHAPTER 3

The genomes of dinoflagellates

Jeremiah D. Hackett[1] and Debashish Bhattacharya[2]

[1] Biology Department, Woods Hole Oceanographic Institution, Woods Hole, MA, USA
[2] Department of Biological Sciences and Roy J. Carver Center for Comparative Genomics, University of Iowa, IA, USA

Introduction

The dinoflagellates are a diverse group of marine and freshwater protists that play a central role in the ecology of aquatic environments and display amazing morphological diversity (Taylor 1987; Fensome *et al.* 1993). Each of their genomes contains a fascinating complement of traits that makes dinoflagellates a model for studying alternative evolutionary pathways. Because of the presence of cellular structures called alveoli, dinoflagellates are included in the Alveolata, together with the ciliates and apicomplexans. Alveoli are flattened vesicles that lie under the plasma membrane and, in many dinoflagellates, contain cellulose armor plates that form the cell covering, or theca. Some species lack armor plates (e.g. *Karenia brevis*, Fig. 3.1a), whereas others have remarkably complex forms, such as *Polykrikos schwartzii* (Fig. 3.1b). The pattern of these plates (termed tabulation) has historically been an important taxonomic character (see Fensome *et al.* 1993). Dinoflagellates have two flagella, one longitudinal flagellum that extends to the posterior of the cell, and one coiled transverse flagellum that runs around the circumference of the cell in a groove (the cingulum) and moves in an undulating motion. These flagella allow the dinoflagellates to be among the fastest moving algae, reaching speeds of 200 to 500 μm s^{-1} (Graham and Wilcox 2000). Many dinoflagellates are also bioluminescent, producing spectacular light displays in the ocean.

Dinoflagellates have an important ecological role as primary producers and can constitute a large proportion of marine phytoplankton (Strickland 1965). They also play a major role in the ocean through bloom formation (e.g. "red tides") and symbiotic associations. Many blooming species also produce potent toxins that have a significant impact on marine ecosystems, fisheries economies, and human health (Shumway 1989). The ability to produce toxins can vary greatly, both within a genus and among different strains of the same species. The mechanisms and control of toxin production in dinoflagellates are not well understood and are the subject of an extensive body of research (e.g. Lehane and Lewis 2000; van Dolah 2000; Ramsdell *et al.* 2005). Different species make a variety of toxins that cause human illness, typically from eating shellfish that have been feeding on the bloom. These illnesses include paralytic, neurotoxic, diarrhetic, and azaspiracid shellfish poisoning. Two other illnesses linked to dinoflagellates, ciguatera fish poisoning and possible estuary-associated syndrome (PEAS), result from consuming fish (ciguatera) or contact with the dinoflagellates (PEAS). These toxins also have a significant impact on marine life, especially fish and marine mammals, and on land animals that feed on marine life (see Landsberg 2002). The toxicity of dinoflagellates has been of increasing interest because of the apparent increase in recent decades of the number and severity of blooms (van Dolah 2000). Dinoflagellates also play a vital ecological role as symbionts with animals. The genus *Symbiodinium* is perhaps the best studied in

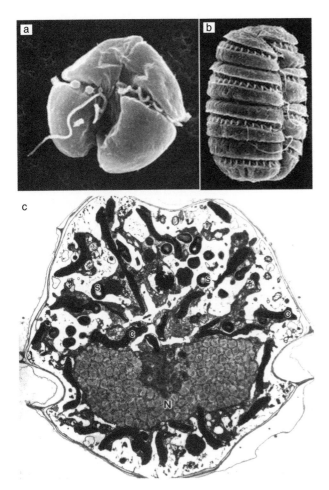

Figure 3.1 Images of dinoflagellates. (a) *Karenia brevis*, cell length 18–40 μm. (b) *Polykrikos schwartzii*, cell length 50–100 μm (images provided by Haruyoshi Takayama). (c) Transverse thin section through an *Alexandrium tamarense* vegetative cell, cell length 27–45 μm. The cell (x 5500) contains numerous chloroplasts (c), starch grains (s), and a large centrally positioned nucleus (N). Reproduced with permission from Fritz (1986).

its symbiosis with corals. This dinoflagellate is responsible for the coral's pigmentation and provides photosynthetic metabolites to the animal. Coral bleaching occurs when the coral loses the dinoflagellate symbiont.

Many dinoflagellates are phototactic, some have "eyespot" light-sensing structures, and migrate through the water column on a rhythmic 24-hour cycle (diel vertical migration). About one-half of dinoflagellate species contain plastids (photosynthetic organelles) but many are mixotrophic (Stoecker 1999; Jeong *et al*. 2005). The most common plastid (see Chapter 6) is surrounded by three membranes and contains the photopigment peridinin. These plastid genomes have undergone drastic evolutionary changes. In addition, four other plastid types are currently known in this lineage that arose through endosymbiosis from other algae. We are just beginning to understand the impact of these endosymbioses on dinoflagellate genomes.

The nuclear genome

The organization of the nuclear genome in dinoflagellates is a distinguishing feature of this group. These taxa contain large amounts of DNA in the nucleus, ranging from 3 to 250 picograms/cell (Spector 1984). The smallest dinoflagellate genomes are about the size of the human genome, whereas the largest are two orders of magnitude greater, which make dinoflagellate genomes among the largest in nature. Recent estimates of haploid cell DNA content, from a diversity of dinoflagellates including many *Symbiodinium* spp., using DAPI stained cells and flow cytometry, have ranged from $1.5\,pg\,cell^{-1}$ to $225\,pg\,cell^{-1}$ (Table 3.1; LaJeunesse *et al.* 2005). Whereas the size estimates are slightly smaller, they are in general agreement with previous estimates (Veldhuis *et al.* 1997). This wide range of DNA content corresponds to approximately 1467 to 220 050 megabases (Mb) of DNA. It is unclear why these single-celled organisms possess such large amounts of nuclear DNA or what accounts for the drastic variation in genome sizes. The genomes do not appear to be comprised primarily of repetitive sequences. DNA reassociation kinetics of nuclear DNA from *Crypthecodinium cohnii* shows that only 55 to 60% of the genome (estimated to be ~6700 Mb) is made of repetitive sequences and the complexity of slowly renaturing unique DNA is about 1.5×10^9 base pairs (bp), an amount typical for eukaryotes (Allen *et al.* 1975; Hinnebusch *et al.* 1980). Other dinoflagellates, such as *Woloszynskia bostoniensis*, have also been shown to have typical eukaryotic reassociation kinetics, with repetitive DNA comprising 45% of the genome (Davies *et al.* 1988). However, the amount of unique DNA was calculated to be 1.32×10^{10} bp, about one order of magnitude larger than mammalian single copy DNA. A 226-bp repetitive sequence (Cc18) from the dinoflagellate *C. cohnii* is present in ca. 25 000 copies (0.06% of the genome) and uniformly distributed on the chromosomes (Moreau *et al.* 1998). Whereas it appears that genome expansion in dinoflagellates is not due to large amounts of repetitive DNA, there is likely not a commensurate increase in the number of genes when compared to other eukaryotes. If this were so, some dinoflagellates with the largest genomes would have over one million genes.

The complete sequence of one of the immense dinoflagellate genomes may not be available for years until sequencing costs come down significantly or an alternative technology for sequence determination is developed (e.g. Rogers and Venter 2005). However, several recent studies using expressed sequence tags (ESTs) suggest that dinoflagellates may possess a large number of genes. An EST gene discovery project in our lab with *Alexandrium tamarense* using normalized and subtracted cDNA libraries has conservatively found 9000 unique transcripts (Hackett *et al.* 2004; Hackett *et al.* 2005; Hackett and Bhattacharya, unpublished). This number is likely to grow because the novelty rate of the subtracted cDNA library is still at 60% after the sequencing of 15 000 ESTs and these libraries were constructed from cultures grown in a single growth condition. An EST project with *K. brevis* produced 5280 unique transcripts (Lidie *et al.* 2005). Another EST project with *Lingulodinium polyedrum* and *Amphidinium carterae* produced 1012 and 2143 unique sequences from these species, respectively (Bachvaroff *et al.* 2004). An independent project produced 2111 unique genes from *L. polyedrum* grown in dark conditions (Tanikawa *et al.* 2004). Future projects using libraries from algae grown under various conditions will likely discover thousands more transcripts. Similarly high novelty rates and cDNA library complexity have been seen with the other dinoflagellate EST projects (Bachvaroff *et al.* 2004; Lidie *et al.* 2005). Ongoing work with *Alexandrium* using Massively Parallel Signature Sequencing (MPSS) may also shed light on the number of genes encoded in dinoflagellate genomes (D. Erdner and D. Anderson, unpublished data).

The large number of dinoflagellate genes may reflect the existence of large gene families, some of which are organized in large tandem arrays. The peridinin chlorophyll *a*-binding protein may be encoded by as many as 5000 copies in *L. polyedrum* and arranged in arrays separated by about 1 kilobase (kb) (Le *et al.* 1997). In *Symbiodinium*, these genes are arranged in arrays with the coding regions separated by spacers of about 900 nucleotides in length that contain a conserved 5′ domain that may function as a promoter (Reichman *et al.* 2003). Tandem arrays have also been reported for genes encoding Form II rubisco and luciferin-binding protein (Lee *et al.* 1993; Rowan *et al.* 1996). Gene families were observed in the ESTs of *A. tamarense* (e.g. *atp*H, 16 copies) in which transcripts with distinct 3′-untranslated regions (UTRs) encoded nearly identical proteins (Hackett *et al.* 2005). In *K. brevis*, clusters of ESTs with 10 or more members had a DNA polymorphism frequency of 1 per 90 bp, also indicating that these transcripts are encoded by unique genes (Lidie *et al.* 2005). Taken together, previous work

Table 3.1 The haploid DNA content of dinoflagellates (reproduced from LaJeunesse *et al.* 2005)

Taxon	pg/cell	Mb (pg978)
Alexandrium andersonii	21.8	21 320.4
Alexandrium insuetum	30.8	30 122.4
Alexandrium lusitanicum	31	30 318
Alexandrium tamarense	103.5	101 223
Amphidinium carterae	5.9	5770.2
Gymnodinium simplex	11.6	11 344.8
Heterocapsa triquetra	24.1	23 569.8
Heterocapsa pygmaea	3.8	3716.4
Polarella glacialis	7	6846
Karenia brevis	57.1	55 843.8
Karenia mikimotoi	100.1	97 897.8
Karlodinium galatheanum	16.9	16 528.2
Katodinium rotundatum	3.6	3520.8
Pfiesteria piscicida	5.5	5379
Pfiesteria shumwayae	19.8	19 364.4
Prorocentrum balticum	8.3	8117.4
Prorocentrum dentatum	6.6	6454.8
Prorocentrum micans	115.2; 225.0[a]	11 2665.6; 22 0050
Prorocentrum minimum	6.9	6748.2
Symbiodinium spp.	1.5–4.8[b]	1467–4694.4

[a] Two different measurements were made for *P. micans*.
[b] A range of *Symbiodinium* spp. were examined, see Lajeunesse *et al.* (2005) for details.

and the recent EST projects have shown that some dinoflagellate genes are present in many, nearly identical copies. The presence of these gene families is likely to contribute, but not fully account for, the immense genome sizes observed in this group.

Analysis of these ESTs has provided several insights into the structure of coding regions in dinoflagellate genomes. The GC-content of coding regions in dinoflagellates investigated thus far has been greater than 50%, but ranges from near 50% (*K. brevis* = 51%, *A. carterae* = 50.44%, *C. cohnii* = 50%) to near 60% (*A. tamarense* = 61%, *L. polyedrum* = 59%; Hackett *et al.* 2004; Lidie *et al.* 2005). Third position GC-content was much less skewed in *K. brevis* (53.5%) compared to *A. tamarense* (77.9%). In *A. tamarense*, it was also observed that the stop codon TGA is heavily favored (81%) over the stop codons TAA and TAG (Hackett *et al.* 2005). In addition, *A. tamarense* ESTs were examined for a potential poly-A signal, a motif within about 30 bp of the polyadenylation site that helps the poly-A polymerase add the poly-A tail in the same location in each transcript from a particular gene. Known poly-A signals include AAUAAA and closely related motifs, as well as some unrelated sequences, but are typically five to six nucleotides in length. No five or six nucleotide sequences or a related set of sequences were found to be over-represented in the 3′ UTR of the *A. tamarense* ESTs. This is consistent with earlier observations of individual dinoflagellate cDNAs that also revealed the absence of a poly-A signal (Lee *et al.* 1993). Polyadenylation clearly occurs in the dinoflagellates, but the mechanism by which this happens apparently does not involve typical poly-A signals. It would be interesting to determine if dinoflagellates have evolved a novel polyadenylation signal system or, perhaps, use signals downstream of the transcribed region, such as the U-rich tracts described for mammalian transcripts (Zarudnaya *et al.* 2003).

Several dinoflagellate genes have been isolated from genomic DNA and these studies shed light on the organization of genes on the genome. The upstream regions of these genes lack recognizable promoter features like TATA boxes and transcription factor binding sites, suggesting the evolution of potentially novel regulatory mechanisms (Lee *et al.* 1993; Le *et al.* 1997; Li and Hastings 1998). Introns are rare in genes studied thus far. Introns have been found in a rubisco gene from *Symbiodinium* spp. and in a luciferase sequence from *Pyrocystis lunula* (Rowan *et al.* 1996; Okamoto *et al.* 2001a). Protists that are closely related to dinoflagellates, such as apicomplexans and *Perkinsus marinus,* often contain genes with multiple introns (Rogozin *et al.* 2003; Schott *et al.* 2003; Robledo *et al.* 2004). Dinoflagellates use both transcriptional and post-transcriptional gene regulation. The peridinin-chlorophyll *a* binding protein, S-adenosyl-homocysteine hydrolase like protein, methionine aminopeptidase like protein, and histone-like protein are transcriptionally regulated (Triplett *et al.* 1993; Taroncher-Oldenburg and Anderson 2000). Okamoto and Hastings (2003) found 246 genes that were differentially expressed in response to redox stress. The extent of transcriptional regulation should become clearer in the near future due to on-going gene expression studies using microarray and other approaches. Luciferin-binding protein and glyceraldehyde-3-phosphate dehydrogenase are post-transcriptionally regulated (Morse *et al.* 1989; Fagan *et al.* 1999). The iron superoxide dismutase of *L. polyedrum* exhibits both modes under different stimuli (Okamoto *et al.* 2001b).

In some dinoflagellates, hydroxymethyluracil (HoMeUra) replaces up to 60% of the thymines in the DNA (Rae 1976; Steele and Rae 1980; Herzog *et al.* 1982). This base analog is found in some phages, such as *Bacillus subtilis* phage SP01, but in eukaryotes HoMeUra is an oxidative DNA damage product and is the target of DNA repair mechanisms in other eukaryotes. This base is non-randomly distributed throughout the DNA and 10% of the DNA in *C. cohnii* has low HoMeUra content (Steele and Rae 1980; Herzog *et al.* 1982; Moreau *et al.* 1998). Restriction enzymes have varying degrees of efficiency when cleaving dinoflagellate DNA, however there does not appear to be a relationship between the presence of HoMeUra and the efficiency of enzyme cleavage (Lee *et al.* 1993; Moreau *et al.* 1998). The DNA of dinoflagellates can also contain 5-methylcytosine and N^6-methyladenine (Rae and Steele 1978).

One of the most striking features of the nucleus is the morphology of the chromosomes. Whereas these chromosomes can vary in shape between species, they are uniform in size and morphology within an organism. Chromosomes numbers vary widely across taxa, from four in *Syndinium borgerti* to 325 in *Endodinium chattoni* (Spector 1984). *A. tamarense,* shown in Fig. 3.1c, has 143 chromosomes in the nucleus. The chromosomes are permanently condensed and attached to the nuclear envelope. During cell division the nuclear envelope does not disappear and the mitotic spindle is extranuclear. Cytoplasmic channels form through the nucleus and microtubules extend from the spindle into the channels (Kubai and Ris 1969; Bhaud *et al.* 2000). The microtubules then attach to the nuclear envelope at a point opposite the chromosomes. The nucleus is then pulled apart and divides by cytokinesis. Dinoflagellate mitosis is unusual in having both a permanent nuclear envelope and an extranuclear mitotic spindle. Hypermastigotes, microbial eukaryotes found in the guts of wood eating insects, also have a closed mitosis with an extranuclear spindle (Ris 1975). Most fungi and many protists have a permanent nuclear envelope; however, the mitotic spindle and mitosis are intranuclear (Heath 1980).

Another striking aspect of the dinoflagellate nuclear genome is the packaging of nuclear DNA. In all other eukaryotes, the nuclear genome is packaged using nucleosomes; that is octamers comprised of histones H2A, H2B, H3, and H4. Histone H1 serves as the "linker" histone between each of the nucleosomes. However, dinoflagellate chromosomes appear to completely lack nucleosomes (Rizzo and Nooden 1976; Herzog and Soyer 1981). Their nuclear DNA is smooth and is associated with very little basic protein. The basic protein:DNA mass ratio in dinoflagellates is about 1:10, rather than the 1:1 ratio found in most eukaryotes (Rizzo and Nooden 1974a). It was long held that dinoflagellates lack histones. However, two recent studies have uncovered transcripts encoding histone proteins. Histone H3 was found among redox-regulated genes from *Pyrocystis lunula* and histone H2A.X was a rare cDNA in an *A. tamarense* library (Okamato and Hastings 2003; Hackett *et al.* 2005). It is still unclear if other histones exist or if they are found throughout the dinoflagellate lineage.

Deciphering the structure of the dinoflagellate chromosomes has been the subject of much research over the last few decades and has resulted in several models (see Spector 1984). Elucidating the true structure has been particularly difficult because the chromosomes begin to disintegrate when the nucleus is ruptured. The most strongly supported model proposes a cholesteric liquid crystal structure for dinoflagellate chromosomes (Livolant and Bouligand 1978; Gautier *et al.* 1986; Bouligand and Norris 2001). Under this model, the chromosomes are constructed of sheets of parallel DNA filaments that are stacked and slightly rotated relative to one another (Fig. 3.2). This structure produces the series of nested arcs of DNA filaments observed with the transmission electron microscope. Loops of DNA extend from the chromosome into the nucleoplasm (Sigee 1984). These may be the sites of active transcription because the DNA within the liquid crystal chromosome would presumably be inaccessible to transcription proteins. The DNA in bacterial nucleoids and sperm heads can also form liquid crystals (Livolant 1984; Bouligand and Norris 2001). Recent work has shown that this structure may also be important for the proper separation of the chromosomes during DNA replication, causing the daughter chromosomes to repel each other and preventing the entanglement of chromosomes (Bouligand and Norris 2001).

Rather than histones, dinoflagellate chromosomes are associated with basic histone-like

proteins (HLPs, Rizzo and Nooden 1974b). The presence of these proteins in dinoflagellate nuclei has been known for some time and recent studies have shown the similarity of these proteins to HU (heat unstable nucleoid protein) and histone-like proteins of bacteria (Wong *et al.* 2003; Hackett *et al.* 2005). These proteins are most closely related to a group of poorly characterized DNA-binding proteins from bacteria that are similar to HU proteins (Fig. 3.3). The primary amino acid sequence similarity between dinoflagellate HLPs and their bacterial homologs is not strong, although there are several highly conserved positions. However, the predicted secondary structures of the dinoflagellate and bacterial HLPs and HU proteins are remarkably similar (Hackett *et al.* 2005). All of them contain two α-helices followed by a series of β-strands. Relative to HU proteins, histone-like proteins from dinoflagellates and bacteria have N-terminal extensions with an α-helix.

The dinoflagellates clearly have evolved a nuclear organization unlike that of any other eukaryotic group. The apicomplexan sisters of dinoflagellates have a typical eukaryotic nuclear organization. The evolution of the nuclear organization in dinoflagellates must therefore have occurred along the branch that unites the dinoflagellates. Other alveolates, such as the Perkinsida (including the oyster pathogen *P. marinus*) are more closely related to dinoflagellates than the apicomplexans (Fong *et al.* 1993; Saldarriaga *et al.* 2003). Morphologically, *P. marinus* has similarities to both apicomplexans and dinoflagellates (Perkins 1996). The nuclear envelope remains intact during nuclear division, cytoplasmic channels form through the nucleus, and an extranuclear spindle forms during mitosis, as in dinoflagellates. However, this organism has a more "typical" genome size for a protist, estimated to be about 28 Mb, that is currently being sequenced at The Institute for Genomic Research. Preliminary sequencing results seem to indicate that the genome is larger than was anticipated. However, it appears that *Perkinsus* diverged after the evolution of a closed mitosis with an external mitotic spindle, but before the development of the immense genomes found in dinoflagellates. The *Perkinsus* genome also contains the genes for canonical histones, but, as yet, no

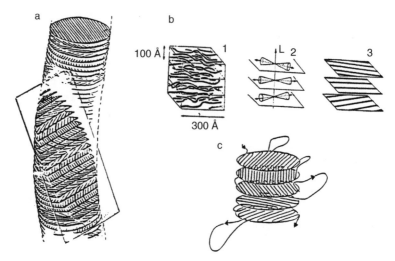

Figure 3.2 Models of dinoflagellate chromosomes. (a) Schematic representation of the chromosome as a stack of disks with parallel DNA filaments in each disk and a continuous rotation along the axis. P1 is a section showing how arcs become apparent. (b) Filaments vary slightly from the mean direction in each disk (Bouligand and Puiseux-Dao 1968). (1) Cube showing the distribution of DNA filaments within three adjacent layers. (2) The mean direction of filaments within each of the layers A, B, and, C are indicated by the double arrow whereas the standard deviation of the directions of the individual filaments from the mean direction is shown by the cone of revolution. (3) Each filament in Fig. 3.2b1 is represented by the mean direction for the layer showing the twist in these directions in successive layers. This yields Fig. 3.2a. (c) DNA can loop out from its liquid crystalline state in each disk. Figure and legend reproduced with permission from Elsevier from Bouligand and Norris (2001).

genes for dinoflagellate histone-like proteins have been detected. This will be the first complete sequenced genome of an organism in the lineage leading to dinoflagellates and may provide insights into the evolutionary leaps leading to the remarkable organization of their nuclear genome.

Oxyrrhis marina is also closely related to dinoflagellates and diverges close to the base of this lineage (Saldarriaga *et al.* 2003). Whereas *Oxyrrhis* has often been considered a dinoflagellate, there are significant differences in the organization of its nuclear genome. The mitotic spindle is intranuclear with microtubules emanating from nuclear plaques that act as microtubule organizing centers (Gao and Li 1986). The chromosomes also lack the periodic banding structure of typical dinoflagellate chromosomes and have weak birefringence (Cachon *et al.* 1989). However, *Oxyrrhis* appears to completely lack nucleosomes. It does have basic histone-like proteins that have been shown to be associated with the nuclear DNA (Saez *et al.* 2001). Whereas the phylogeny of dinoflagellates is presently unclear, *Oxyrrhis* branches near the base of this lineage (Saldarriaga

et al. 2003), suggesting that the loss of nucleosomes and the evolution of histone-like proteins occurred before the divergence of this taxon. It should be noted that several oceanic environmental PCR surveys have revealed a rich diversity of unidentified alveolates that may branch between *Perkinsus* and *Oxyrrhis* in phylogenetic trees (Lopez-Garcia *et al.* 2001; Moon-van der Staay *et al.* 2001). Characterization of these organisms may help clarify the steps in the evolution of the unique dinoflagellate nuclear genome.

The plastid genome

Of the three major lineages within the alveolates, the dinoflagellates are the only group that contains photoautotrophic members. The ciliates do not have plastids and the apicomplexans retain a remnant non-photosynthetic organelle descended from a plastid that is termed the apicoplast (see Chapter 6). Currently, there are five types of plastids known in the dinoflagellates that appear to have independent origins through

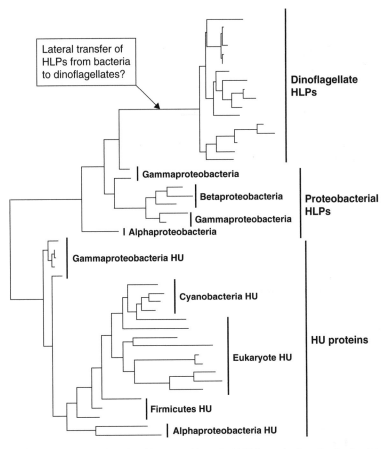

Figure 3.3 Phylogeny of histone-like proteins (HLPs) and heat unstable nucleoid (HU) proteins from the dinoflagellates and bacteria.

endosymbiosis. The most common form is surrounded by three membranes, contains the photopigment peridinin, and is believed to have originated from the red algae. This plastid appears to have arisen early in dinoflagellate evolution because it is the most widely distributed among taxa, although it has been lost many times (Saldarriaga *et al.* 2001). Only dinoflagellates and *Euglena* contain plastids bound by three membranes and each has independently evolved an analogous mechanism to target nuclear-encoded plastid proteins to the organelle (Nassoury *et al.* 2003). The proteins contain a leader sequence that is a signal peptide for targeting to the endoplasmic reticulum (ER), a transit peptide for targeting to the plastid, and a stop-transfer sequence that allows the proteins to be co-translationally inserted into the ER. The protein is then transported to the plastid in membrane-bound vesicles. Another study identified a second class of transit peptides that lack the stop-transfer sequence, suggesting another mechanism for protein trafficking to the plastids (Patron *et al.* 2005).

The genome within peridinin plastids is unlike that of any other eukaryote. Most plastid genomes are single, circular molecules of about 100 to 200 kb in size, and encode around 100 genes. In the peridinin plastids of dinoflagellates, the plastid genome is broken into minicircles of about 2 to 3 kb, most of which encode a single gene. Minicircles have been identified in *Amphidinium* that contain two (*atp*A and *pet*B) or three genes (*psb*D, *psb*E and *psb*I) (Nisbet *et al.* 2004). The minicircles contain a gene coding region and variable non-coding region that contains a highly conserved "core" sequence of about 400 bp. Only

14 plastid proteins have been found to be encoded on these minicircles (*atp*A, *atp*B, *pet*B, *pet*D, *psa*A, *psa*B, *psb*A-E, *psb*I, *ycf*16, *ycf*24), in addition to the plastid large subunit and putative small subunit rDNA (Zhang *et al.* 1999; Barbrook and Howe 2000; Hiller 2001; Zhang *et al.* 2002; Howe *et al.* 2003; Laatsch *et al.* 2004; Nisbet *et al.* 2004). Many minicircles are also comprised entirely of plastid gene fragments, presumably generated through recombination among minicircles via conserved regions in the non-coding portion of these sequences (Zhang *et al.* 2001). Microcircles of 400 to 600 bp have been found in *Amphidinium* that contain a portion of the conserved non-coding region of larger circles, but lack the gene (Nisbet *et al.* 2004). Minicircles that were recently characterized from the dinoflagellate *Adenoides eludens* contain a single gene but these are larger than in other species (about 5 kb) due to the expansion of the non-coding region (Nelson and Green 2005). Minicircles are believed to be located in the plastid, however, few studies have substantiated this hypothesis (Takishita *et al.* 2003). However, a recent study using cyclohexamide and chloramphenicol inhibition of protein synthesis and radiolabeled methionine has shown that the product of the *psb*A gene is synthesized in the plastid, strongly suggesting plastid localization for this gene (Wang *et al.* 2005). One study with *Ceratium horridum* suggests that the minicircles are located in the nucleus of this species (Laatsch *et al.* 2004). In addition, the transcripts from minicircles in *C. horridum* are edited by a substitutional RNA editing mechanism that causes predominantly A to G changes, but also unreported A to C transversions (Zauner *et al.* 2004). The *psb*A transcript was the most extensively edited, with 8.1% of the sites being changed. This activity likely involves a novel RNA editing mechanism, as does the RNA editing seen in the dinoflagellate mitochondrion (see below). Notably absent from the known minicircle genes are critical components of the transcription and translation apparatus, including RNA polymerases, organellar elongation factors, and many ribosomal proteins. This raises the question of how the minicircle genes are transcribed and translated in the organelle. It is possible that the missing transcription and translation machinery is encoded in the nucleus. However, the several EST projects have only identified plastid ribosomal proteins and not yet found the genes encoding the RNA polymerases.

Most of the plastid genes have been moved to the nucleus and two recent studies using ESTs document this remarkable intracellular gene transfer (Hackett *et al.* 2004; Bachvaroff *et al.* 2004). Fifteen of the nuclear-encoded plastid genes are plastid-encoded in every other photosynthetic eukaryote, making dinoflagellates the only organisms that have transferred these genes to the nucleus. What remains in the plastid genome that is encoded on minicircles may be the closest to the minimal set of genes that is required to retain organelle function. These include the core subunits of the photosystem protein complexes (*atp*A, *atp*B, *pet*B, *pet*D, *psa*A, *psa*B, *psb*A-E, *psb*I). One possible explanation for the maintenance of these genes in the plastid may be for the regulation of the redox potential across the plastid membrane. The co-localization for redox regulation hypothesis predicts that the core subunits of the photosystem complexes must remain in the plastid to respond quickly to changes in the redox potential across the plastid membrane to maximize efficiency and minimize the generation of harmful oxygen free-radicals (Allen 2003). Genes encoded in the plastid may be able to respond more quickly to changes in the redox potential than nuclear-encoded genes. For example the photosystem genes *psa*A, *psa*B, *psb*A, and *psb*B have been shown to be under redox regulation (Pfannschmidt 2003). The dramatic reduction of the plastid genome suggests that the dinoflagellates have been under strong pressure to move most of their plastid genome to the nucleus. Whereas it is currently unclear what this pressure might have been, the dinoflagellates clearly have the most reduced plastid genome of any photosynthetic eukaryote and offer a model for understanding the maintenance of organellar genomes.

The dinoflagellates have also acquired several plastids through tertiary endosymbiosis (see Chapter 6). These plastids either replaced the ancestral peridinin plastid or, because plastid loss in dinoflagellates is common (Saldarriaga *et al.* 2001), were acquired by heterotrophic dinoflagellates.

These dinoflagellates with plastid replacements are important to our understanding of genome evolution in this group. The acquisition of a plastid provides the opportunity for the transfer of genes from distant evolutionary lineages, potentially having a significant effect on the genomes of the host. There is the possibility not only for the transfer of plastid-related genes, but also for the transfer of genes involved in other cellular functions. One aspect of genome evolution following tertiary endosymbiosis that has been investigated is the fate of the nuclear-encoded plastid genes after the acquisition of a new plastid. The new endosymbionts would contain typical plastid genomes, rather than the highly reduced genomes of the peridinin plastids, and would have many genes on the plastid genome that are nuclear-encoded in the peridinin-containing dinoflagellates. Have dinoflagellates with tertiary plastids retained the nuclear-encoded genes and lost the plastid-encoded copies? Or have the nuclear-encoded copies been lost in favor of the plastid encoded genes from the new endosymbiont? These questions have been investigated in one group of dinoflagellates, which includes *Karenia*, *Karlodinium*, and *Takayama* and contain a tertiary plastid acquired from a haptophyte. These organelles are also surrounded by three membranes and contain the photopigment 19′ butanoyloxyfucoxanthin and originated from a haptophyte through endosymbiosis (Bjornland and Tangen 1979; Tengs *et al.* 2000; Ishida and Green 2002; Yoon *et al.* 2005). A study by Yoon *et al.* (2005) demonstrated that the plastid genes that are nuclear-encoded only in peridinin-containing dinoflagellates are absent from an extensive EST dataset from the fucoxanthin-containing dinoflagellate *K. brevis*. This suggests that, in this taxon the plastid genes have been lost from the nuclear genome in favor of the plastid-encoded homologs (Yoon *et al.* 2005). It is yet to be determined whether *Karenia* and other fucoxanthin-containing dinoflagellates have a typical plastid genome, although these results suggest that the genome of these tertiary plastids more closely resembles that of other algae.

Plastids of other dinoflagellates have been acquired from prasinophyte green algae (*Lepidodinium viride*) and diatoms (*Peridinium foliaceum*). Each of these lineages has a unique set of genomic characteristics. In *L. viride*, the green algal-like plastid is surrounded by only two membranes, rather than the three membranes seen in most dinoflagellates (Watanabe et al. 1990; Watanabe et al. 1991). It would be interesting to determine if *Lepidodinium* has reverted to the original plastid-protein import pathway found in primary plastids, like those of green algae. Because no remnant of the green algal nucleus has been detected, all the genes necessary for plastid function must have been transferred to the nucleus of the dinoflagellate. It is currently unknown what proportion of nuclear-encoded plastid genes are of green algal origin in this dinoflagellate or if any have been vertically inherited from the peridinin-containing ancestor.

Whereas the species with fucoxanthin and prasinophyte plastids appear to have completely eliminated the endosymbiont genome, the dinoflagellates *P. foliaceum* and *P. balticum* maintain a large endosymbiont genome within the plastid, in addition to other endosymbiont organelles (Tomas and Cox 1973; Kite *et al.* 1988; Schnepf and Elbrachter 1999). The nucleus of the diatom endosymbiont appears to have undergone very little reduction, indicating this plastid may be a relatively recent acquisition. The endosymbiont nucleus has retained genes including HSP90, actin and α-tubulin, strengthening the idea that much of the endosymbiont genome remains (McEwan and Keeling 2004). This is clearly a permanent plastid acquisition because *P. foliaceum* has been autotrophically grown in culture for over a decade. These dinoflagellates also have a three-membrane bound organelle called the eyespot that may be the remnant of the three membrane-bound peridinin plastid (Schnepf and Elbrachter 1999). Under this scenario, *P. foliaceum* potentially provides an example of the simultaneous existence of the tertiary and the secondary plastid within a dinoflagellate. It is still unknown whether gene transfer has occurred between the nucleus of the dinoflagellate host and the diatom endosymbiont.

Finally, some heterotrophic dinoflagellates may acquire temporary plastids from prey called kleptoplasts (Fields and Rhodes 1991; Schnepf and Elbrachter 1992; Skovgaard 1998). Kleptoplasts are maintained within food vacuoles and are often digested within days. *Dinophysis* species have a plastid acquired from a cryptophyte; however, there has been some controversy over whether this organelle is a permanent acquisition or a klepto-plast (Schnepf and Elbrachter 1999; Takishita *et al.* 2002; Hackett *et al.* 2003). A recent study identified a species of cryptophyte with molecular sequences that are virtually identical to previously sequenced genes from *Dinophysis* spp. (Janson 2004). These findings lend more credence to the idea that the plastid of *Dinophysis* spp. is a kleptoplast acquired from a specific cryptophyte species (*Teleaulax amphioxeia*). This may indicate that the relationship between *Dinophysis* and this cryptophyte species represents an early stage of endosymbiosis and plastid acquisition. It is yet unknown if gene transfer has occurred between the prey and host but the specificity of the relationship may indicate a genetic adaptation within *Dinophysis* spp. for this species of cryptophyte. Another species, *Dinophysis mitra*, has recently been shown to acquire a klep-toplast from haptophytes (Koike *et al.* 2005). The *D. mitra* plastid has pigment and ultrastructural features similar to haptophytes and phylogenetic analysis of the plastid SSU rDNA confirm its haptophyte affinity. However, the plastid SSU rDNA sequences are much more variable than those from *Dinophysis* spp. which contains a cryptophyte kleptoplastid.

The mitochondrial genome

The mitochondrial genome of the dinoflagellates is the least well-studied genome in these organisms. There is not yet a complete mitochondrial genome sequence from a dinoflagellate, although sequen-ces from several mitochondrial genes suggest that the mitochondrial genome organization is different from other alveolates and is AT rich (about 65% AT) in contrast to the GC-rich dinoflagellate nuclear genome (see Gray *et al.* 2004). Sequences of *cox*1 from *C. cohnii* and *cob* from *Pfisteria piscicida* are flanked by non-coding DNA and repetitive elements and show a gene organization unlike other alveolates (Zhang and Lin 2002; Norman and Gray 2001; Lin *et al.* 2002). Four different *cox*1 coding regions have been identified in *C. cohnii* that have different upstream and downstream flanking regions (Norman and Gray 2001). The

downstream regions are included in the transcript, resulting in two different cox1 proteins. These flanking regions may also produce complex sec-ondary structures in the mRNA that may be important for gene regulation. Polyadenylated transcripts of *cox*3 have been isolated from *L. polyedrum* that also contain gene fragments from other mitochondrial genes such as *cox*1 and *cob*, suggesting that the mitochondrial genome of dinoflagellates maintains transcriptional units comprised of gene fragments (Chaput *et al.* 2002). The dinoflagellates, like the chlorophycean green algae and apicomplexans, have a *cox*2 gene that has been broken into two subunits and transferred to the nucleus (Funes *et al.* 2002). EST projects with *A. tamarense* and *K. brevis* have found the cox2b subunit in polyA primed cDNA libraries, but have yet to identify the more divergent cox2a subunit (Hackett *et al.* 2004). Phylogenetic analyses unite the split *cox*2 genes, suggesting either convergent, independent evolution of these genes in the chlorophycean green algae and the apicomplex-ans/dinoflagellate lineage or a single origin fol-lowed by lateral gene transfer.

Researchers have also noted extensive and novel RNA editing of mitochondrial transcripts in a diversity of dinoflagellates (Lin *et al.* 2002; Zhang and Lin 2005). RNA editing involves the alteration of the mRNA sequence following transcription of the gene from DNA to RNA. This process results in an mRNA with a different sequence than the DNA sequence that encodes it. RNA editing occurs in the mitochondria of many organisms and can insert/delete nucleotides or substitute one for another, each able to occur by several mechanisms (for review see Gray 2003). In these dinoflagellates, up to 2% of the nucleotides in the mRNA are altered using a novel combination of substitutional editing. Previously known substitutional mRNA editing mechanisms involve either U to C or C to U transitions. Dinoflagellate mitochondrial mRNA editing shows these changes, in addition to A to G transitions and a small number of transversions, which indicates that the dinoflagellates have multiple editing mechanisms or a single novel mechanism that can perform both types of changes (Lin *et al.* 2002; Gray 2003). These RNA editing mechanisms have not yet been observed in the apicomplexans or ciliates, suggesting this mechanism arose independently in the common ancestor of dinoflagellates (Edqvist *et al.* 2000; Rehkopf *et al.* 2000; Gray 2003).

Conclusions

As with many microbial eukaryotes, the dino-flagellates have evolved fascinating alternatives to the "standard" genome biology of eukaryotes. Investigating these aspects has the potential to reveal not only the biology of these organisms, but also fundamental aspects of genome evolution. In some ways, dinoflagellate genomes represent extremes in evolution. The nuclear genomes are among the largest in nature and have a chromosome structure unlike that of any other eukaryote. The plastid genome has undergone drastic reduction and encodes the fewest genes of any photosynthetic plastid, whereas mitochondrial genes show novel organization and RNA editing. It is clear therefore that many more questions exist than answers regarding dinoflagellate genome evolution, which will keep these remarkable taxa at the forefront of evolutionary and genomic studies for years to come.

References

Allen, J. F. (2003). The function of genomes in bioenergetic organelles. *Phil. Trans. Roy. Soc. Lond. B*, *Biol. Sci.*, **358**, 19–37; discussion 37–38.

Allen, J. R., Roberts, M., Loeblich, A. R., 3rd, and Klotz, L. C. (1975). Characterization of the DNA from the dinoflagellate *Crypthecodinium cohnii* and implications for nuclear organization. *Cell*, **6**, 161–169.

Bachvaroff, T. R., Concepcion, G. T., Rogers, C. R., Herman, E. M., and Delwiche, C. F. (2004). Dinoflagellate expressed sequence tag data indicate massive transfer of chloroplast genes to the nuclear genome. *Protist*, **155**, 65–78.

Barbrook, A. C. and Howe, C. J. (2000). Minicircular plastid DNA in the dinoflagellate *Amphidinium operculatum. Mol. Gen. Gen.*, **263**, 152–158.

Bhaud, Y., Guillebault, D., Lennon, J., Defacque, H., Soyer-Gobillard, M. O., and Moreau, H. (2000). Morphology and behavior of dinoflagellate chromosomes during the cell cycle and mitosis. *J. Cell Sci.*, **113**, 1231–1239.

Bjornland, T. and Tangen, K. (1979). Pigmentation and morphology of a marine *Gyrodinium* (Dinophyceae). with a major carotenoid different from peridinin and fucoxanthin. *J. Phycol.*, **15**, 457–463.

Bouligand, Y. and Norris, V. (2001). Chromosome separation and segregation in dinoflagellates and bacteria may depend on liquid crystalline states. *Biochimie*, **83**, 187–192.

Bouligand, Y. and Puiseux-Dao, S. (1968). La structure fibrillaire et l'orientation des chromosomes chez les Dinoflagellates. *Chromosoma*, **24**, 251–287.

Cachon, J., Sato, H., Cachon, M., and Sato, Y. (1989). Analysis by polarizing microscopy of chromosomal structure among dinoflagellates and its phylogenetic involvement. *Biol. Cell*, **65**, 51–60.

Chaput, H., Wang, Y., and Morse, D. (2002). Polyadenylated transcripts containing random gene fragments are expressed in dinoflagellate mitochondria. *Protist*, **153**, 111–122.

Davies, W., Jakobsen, K. S., and Nordby, O. (1988). Characterization of DNA from the dinoflagellate *Woloszynskia bostoniensis. J. Protozool.*, **35**, 418–422.

Edqvist, J., Burger, G., and Gray, M. W. (2000). Expression of mitochondrial protein-coding genes in *Tetrahymena pyriformis. J. Mol. Biol.*, **297**, 381–393.

Fagan, T. F., Morse, D., and Hastings, J. W. (1999). Circadian synthesis of a nuclear-encoded chloroplast glyceraldehyde-3-phosphate dehydrogenase in the dinoflagellate *Gonyaulax polyedra* is translationally controlled. *Biochemistry*, **38**, 7689–7695.

Fensome, R. A., Taylor, F. J. R., Norris, G., Sarjeant, W. A. S., Wharton, D. I., and Williams, G. L. (1993). *A Classification of Fossil and Living Dinoflagellates*. American Museum of Natural History, New York.

Fields, S. D. and Rhodes, R. G. (1991). Ingestion and retention of *Chroomonas* spp. (Cryptophyceae) by *Gymnodinium acidotum* (Dinophyceae). *J. Phycol.*, **27**, 525–529.

Fong, D., Rodriguez, R., Koo, K., Sun, J., Sogin, M. L., Bushek, D., Littlewood, D. T., and Ford, S. E. (1993). Small subunit ribosomal RNA gene sequence of the oyster parasite *Perkinsus marinus. Mol. Marine Biol. Biotechnol.*, **2**, 346–350.

Fritz, L. M. (1986). Ultrastructure if the life cycle of the toxic dinoflagellate *Gonyaulax tamarensis. Botany and Plant Physiology.* Doctoral thesis. New Brunswick, NJ, Rutgers.

Funes, S., Davidson, E., Reyes-Prieto, A., Magallon, S., Herion, P., King, M. P., and Gonzalez-Halphen, D. (2002). A green algal apicoplast ancestor. *Science*, **298**, 2155.

Gao, X. P. and Li, J. Y. (1986). Nuclear division in the marine dinoflagellate *Oxyrrhis marina*. *J. Cell Sci.*, **85**, 161–175.

Gautier, A., Michel-Salamin, L., Tosi-Couture, E., McDowall, A. W., and Dubochet, J. (1986). Electron microscopy of the chromosomes of dinoflagellates in situ: confirmation of Bouligand's liquid crystal hypothesis. *J. Ultra. Mol. Struct. R.*, **97**, 10–30.

Graham, L. E. and Wilcox, L. W. (2000). *Algae*. Prentice-Hall, Upper Saddle River, NJ.

Gray, M. W. (2003). Diversity and evolution of mitochondrial RNA editing systems. *IUBMB Life*, **55**, 227–233.

Gray, M. W., Lang, F., and Burger, G. (2004). Mitochondria of protists. *Ann. Rev. Gen.*, **38**, 477–524.

Hackett, J. D., Maranda, L., Yoon, H. S., and Bhattacharya, D. (2003). Phylogenetic evidence for the cryptophyte origin of the plastid of *Dinophysis* (Dinophysiales, Dinophyceae). *J. Phycol.*, **39**, 440–448.

Hackett, J. D., Scheetz, T. E., Yoon, H. S., Soares, M. B., Bonaldo, M. F., Casavant, T. L., and Bhattacharya, D. (2005). Insights into a dinoflagellate genome through expressed sequence tag analysis. *BMC Genomics*, **6**, 80.

Hackett, J. D., Yoon, H. S., Soares, M. B., Bonaldo, M. F., Casavant, T. L., Scheetz, T. E., Nosenko, T., and Bhattacharya, D. (2004). Migration of the plastid genome to the nucleus in a peridinin dinoflagellate. *Curr. Biol.*, **14**, 213–218.

Heath, I. B. (1980). Variant mitoses in lower eukaryotes: indicators of the evolution of mitosis? *Int. Rev. Cytol.*, **64**, 1–80.

Herzog, M. and Soyer, M. O. (1981). Distinctive features of dinoflagellate chromatin. Absence of nucleosomes in a primitive species *Prorocentrum micans* E. *Eur. J. Cell Biol.*, **23**, 295–302.

Herzog, M., Soyer, M. O., and Daney de Marcillac, G. (1982). A high level of thymine replacement by 5-hydroxymethyluracil in nuclear DNA of the primitive dinoflagellate *Prorocentrum micans* E. *Eur. J. Cell Biol.*, **27**, 151–155.

Hiller, R. G. (2001). "Empty" minicircles and *pet*B/*atp*A and *psb*D/*psb*E (cytb559 alpha) genes in tandem in *Amphidinium carterae* plastid DNA. *FEBS Letters*, **505**, 449–452.

Hinnebusch, A. G., Klotz, L. C., Immergut, E., and Loeblich, A. R., 3rd (1980). Deoxyribonucleic acid sequence organization in the genome of the dinoflagellate *Crypthecodinium cohnii*. *Biochemistry*, **19**, 1744–1755.

Howe, C. J., Barbrook, A. C., Koumandou, V. L., Nisbet, R. E., Symington, H. A., and Wightman, T. F. (2003). Evolution of the chloroplast genome. *Phil. Trans. Roy. Soc. Lond. B, Biol. Sci.*, **358**, 99–107.

Ishida, K. and Green, B. R. (2002). Second- and third-hand chloroplasts in dinoflagellates: Phylogeny of oxygen-evolving enhancer 1 (PsbO). protein reveals replacement of a nuclear-encoded plastid gene by that of a haptophyte tertiary endosymbiont. *Proc. Natl. Acad. Sci. U S A*, **99**, 9294–9299.

Janson, S. (2004). Molecular evidence that plastids in the toxin-producing dinoflagellate genus *Dinophysis* originate from the free-living cryptophyte *Teleaulax amphioxeia*. *Environ. Microbiol.*, **6**, 1102–1106.

Jeong, H. J., Yeong, D. Y., Park, J. Y., Song, J. Y., Kim, S. T., Lee, H. L., Kim, K. Y., and Yih, W. H. (2005). Feeding by phototrophic red-tide dinoflagellates: five species newly revealed and six species previously known to be mixotrophic. *Aquatic Microb. Ecol.*, **40**, 133–150.

Kite, G. C., Rothschild, L. J., and Dodge, J. D. (1988). Nuclear and plastid DNAs from the binucleate dinoflagellates *Glenodinium (Peridinium) foliaceum* and *Peridinium balticum*. *Biosystems*, **21**, 151–163.

Koike, K., Sekiguchi, H., Kobiyama, A., Takishita, K., Kawachi, M., Koike, K., and Ogata, T. (2005). A novel type of kleptoplastidy in *Dinophysis* (Dinophyceae): Presence of haptophyte-type plastid in *Dinophysis mitra*. *Protist*, **156**, 225–237.

Kubai, D. F. and Ris, H. (1969). Division on the dinoflagellate *Gyrodinium cohnii* (Schiller). A new type of nuclear reproduction. *J. Cell Biol.*, **40**, 508–528.

Laatsch, T., Zauner, S., Stoebe-Maier, B., Kowallik, K. V., and Maier, U. G. (2004). Plastid-derived single gene minicircles of the dinoflagellate *Ceratium horridum* are localized in the nucleus. *Mol. Biol. Evol.*, **21**, 1318–1322.

Lajeunesse, T. C., Lambert, G., Anderson, R. A., Coffroth, M. A., and Galbraith, D. W. (2005). *Symbiodinium* (Pyrrhophyta) genome sizes (DNA content) are smallest among dinoflagellates. *J. Phycol.*, **41**, 880–886.

Landsberg, J. (2002). The effects of harmful algal blooms on aquatic organisms. *Rev. Fish. Sci.*, **10**, 113–390.

Le, Q. H., Markovic, P., Hastings, J. W., Jovine, R. V. M., and Morse, D. (1997). Structure and organization of the peridinin-chlorophyll a-binding protein gene in *Gonyaulax polyedra*. *Mol. Gen. Gen.*, **255**, 595–604.

Lee, D.-H., Mittag, M., Sczekan, S., Morse, D., and Hastings, J. W. (1993). Molecular cloning and genomic organization of a gene for luciferin-binding protein from the dinoflagellate *Gonyaulax polyedra*. *J. Biol. Chem.*, **268**, 8842–8850.

Lehane, L. and Lewis, R. J. (2000). Ciguatera: recent advances but the risk remains. *Int. J. Food Microbiol.*, **61**, 91–125.

Li, L. and Hastings, J. W. (1998). The structure and organization of the luciferase gene in the photosynthetic dinoflagellate *Gonyaulax polyedra*. *Plant Mol. Biol.*, **36**, 275–284.

Lidie, K. B., Ryan, J. C., Barbier, M., and van Dolah, F. M. (2005). Gene expression in Florida red tide dinoflagellate *Karenia brevis*: Analysis of an expressed sequence tag library and development of DNA microarray. *Marine Biotechnol.*, **7**, 481–493.

Lin, S., Zhang, H., Spencer, D. F., Norman, J. E., and Gray, M. W. (2002). Widespread and extensive editing of mitochondrial mRNAs in dinoflagellates. *J. Mol. Biol.*, **320**, 727–739.

Livolant, F. (1984). Cholesteric organization of the DNA in the stallion sperm head. *Tissue Cell*, **16**, 535–555.

Livolant, F. and Bouligand, Y. (1978). New observations on the twisted arrangement of dinoflagellate chromosomes. *Chromosoma*, **68**, 21–44.

Lopez-Garcia, P., Rodriguez-Valera, F., Pedros-Alio, C., and Moreira, D. (2001). Unexpected diversity of small eukaryotes in deep-sea Antarctic plankton. *Nature*, **409**, 603–607.

McEwan, M. L. and Keeling, P. J. (2004). HSP90, tubulin and actin are retained in the tertiary endosymbiont genome of *Kryptoperidinium foliaceum*. *J. Eukaryot. Microbiol.*, **51**, 651–659.

Moon-van der Staay, S. Y., De Wachter, R., and Vaulot, D. (2001). Oceanic 18S rDNA sequences from picoplankton reveal unsuspected eukaryotic diversity. *Nature*, **409**, 607–610.

Moreau, H., Geraud, M. L., Bhaud, Y., and Soyer-Gobillard, M. O. (1998). Cloning, characterization and chromosomal localization of a repeated sequence in *Crypthecodinium cohnii*, a marine dinoflagellate. *Int. Microbiol.*, **1**, 35–43.

Morse, D., Milos, P. M., Roux, E., and Hastings, J. W. (1989). Circadian regulation of bioluminescence in *Gonyaulax* involves translational control. *Proc. Natl. Acad. Sci. U S A*, **86**, 172–176.

Nassoury, N., Cappadocia, M., and Morse, D. (2003). Plastid ultrastructure defines the protein import pathway in dinoflagellates. *J. Cell Sci.*, **116**, 2867–2874.

Nelson, M. J. and Green, B. R. (2005). Double hairpin elements and tandem repeats in the non-coding region of *Adenoides eludens* chloroplast gene minicircles. *Gene*, **358**, 102–110.

Nisbet, R. E., Koumandou, L. V., Barbrook, A. C., and Howe, C. J. (2004). Novel plastid gene minicircles in the dinoflagellate *Amphidinium operculatum*. *Gene*, **331**, 141–147.

Norman, J. E. and Gray, M. W. (2001). A complex organization of the gene encoding cytochrome oxidase subunit 1 in the mitochondrial genome of the dinoflagellate, *Crypthecodinium cohnii*: homologous recombination generates two different *cox*1 open reading frames. *J. Mol. Evol.*, **53**, 351–363.

Okamato, O. K. and Hastings, J. W. (2003). Genome-wide analysis of redox-regulated genes in a dinoflagellate. *Gene*, **321**, 73–81.

Okamoto, O. K., Liu, L., Robertsonn, D. L., and Hastings, J. W. (2001a). Members of the dinoflagellate luciferase gene family differ in synonymous substitution rates. *Biochemistry*, **40**, 15862–15868.

Okamoto, O. K., Robertson, D. L., Fagan, T. F., Hastings, J. W., and Colepicolo, P. (2001b). Different regulatory mechanisms modulate the expression of a dinoflagellate iron-superoxide dismutase. *J. Biol. Chem.*, **276**, 19989–19993.

Patron, N. J., Waller, R. F., Archibald, J. M., and Keeling, P. J. (2005). Complex protein targeting to dinoflagellate plastids. *J. Mol. Biol.*, **348**, 1015–1024.

Perkins, F. O. (1996). The structure of *Perkinsus marinus* (Mackin, Owen and Collier, 1950). Levine, 1978 with comments on taxonomy and phylogeny of *Perkinsus* spp. *J. Shellfish Res.*, **15**, 67–87.

Pfannschmidt, T. (2003). Chloroplast redox signals: how photosynthesis controls its own genes. *Trends Plant Sci.*, **8**, 33–39.

Rae, P. M. (1976). Hydroxymethyluracil in eukaryote DNA: a natural feature of the pyrrophyta (dinoflagellates). *Science*, **194**, 1062–1064.

Rae, P. M. M. and Steele, R. E. (1978). Modified bases in the DNAs of unicellular eukaryotes: And examination of distribution and possible roles, with emphasis on hydroxymethyluracil in dinoflagellates. *BioSystems*, **10**, 37–53.

Ramsdell, J. S., Anderson, D. M., and Gilbert, P. M., eds. (2005). *Harmful algal research and response: A national environmental science strategy 2005–2015*. Ecological Society of America, Washington, D. C.

Rehkopf, D. H., Gillespie, D. E., Harrell, M. I., and Feagin, J. E. (2000). Transcriptional mapping and RNA processing of the *Plasmodium falciparum* mitochondrial mRNAs. *Mol. Biochem. Parasitol.*, **105**, 91–103.

Reichman, J. R., Wilcox, T. P., and Vize, P. D. (2003). PCP gene family in *Symbiodinium* from *Hippopus hippopus*: low levels of concerted evolution, isoform diversity, and spectral tuning of chromophores. *Mol. Biol. Evol.*, **20**, 2143–2154.

Ris, H. (1975). Primative mitotic mechanisms. *Biosystems*, **7**, 298–301.

Rizzo, P. J. and Nooden, L. D. (1974a). Isolation and partial characterization of dinoflagellate chromatin. *Biochim. Biophys. Acta*, **349**, 402–414.

Rizzo, P. J. and Nooden, L. D. (1974b). Partial characterization of dinoflagellate chromosomal proteins. *Biochim. Biophys. Acta*, **349**, 415–427.

Rizzo, P. J. and Nooden, L. D. (1976). Chromosomal proteind in the dinoflagellate alga *Gyrodinium cohnii*. *Science*, **176**, 796–797.

Robledo, J. A., Courville, P., Cellier, M. F., and Vasta, G. R. (2004). Gene organization and expression of the divalent cation transporter Nramp in the protistan parasite *Perkinsus marinus*. *J. Parasitol.*, **90**, 1004–1014.

Rogers, Y. H. and Venter, J. C. (2005). Genomics: massively parallel sequencing. *Nature*, **437**, 326–327.

Rogozin, I. B., Wolf, Y. I., Sorokin, A. V., Mirkin, B. G., and Koonin, E. V. (2003). Remarkable interkingdom conservation of intron positions and masive, lineage-specific intron loss and gain in eukaryotic evolution. *Curr. Biol.*, **13**, 1512–1517.

Rowan, R., Whitney, S. M., Fowler, A., and Yellowlees, D. (1996). Rubisco in marine symbiotic dinoflagellates: form II enzymes in eukaryotic oxygenic phototrophs encoded by a nuclear multigene family. *Plant Cell*, **8**, 539–553.

Saez, A. G., Engel, H., Medlin, L. K., and Huss, V. A. R. (2001). Plastid genome size and heterogeneous base composition of nuclear DNA from *Ochrosphaera neapolitana* (Prymnesiophyta). *Phycologia*, **40**, 147–152.

Saldarriaga, J. F., McEwan, M. L., Fast, N. M., Taylor, F. J., and Keeling, P. J. (2003). Multiple protein phylogenies show that *Oxyrrhis marina* and *Perkinsus marinus* are early branches of the dinoflagellate lineage. *Int. J. System. Evol. Microbiol.*, **53**, 355–365.

Saldarriaga, J. F., Taylor, F. J., Keeling, P. J., and Cavalier-Smith, T. (2001). Dinoflagellate nuclear SSU rRNA phylogeny suggests multiple plastid losses and replacements. *J. Mol. Evol.*, **53**, 204–213.

Schnepf, E. and Elbrachter, M. (1992). Nutritional strategies in dinoflagellates: A review with emphasis on cell biological aspects. *Eur. J. Protistol.*, **28**, 3–24.

Schnepf, E. and Elbrachter, M. (1999). Dinophyte chloroplasts and phylogeny – a review. *Grana*, **38**, 81–97.

Schott, E. J., Robledo, J. A., Wright, A. C., Silva, A. M., and Vasta, G. R. (2003). Gene organization and modeling of two iron superoxide dismutases of the early branching protist *Perkinsus marinus*. *Gene*, **309**, 1–9.

Shumway, S. E. (1989). Toxic algae: a serious threat to shellfish aquaculture. *World Aquacult.*, **20**, 65–74.

Sigee, D. C. (1984). Structural DNA and genetically active DNA in dinoflagellate chromosomes. *Biosystems*, **16**, 203–210.

Skovgaard, A. (1998). Role of chloroplast retention in a marine dinoflagellate. *Aquatic Microb. Ecol.*, **15**, 293–301.

Spector, D. L. (1984). Dinoflagellate nuclei. In D. L. Spector, ed. *Dinoflagellates*. Academic Press, Orlando, Florida.

Steele, R. E. and Rae, P. M. (1980). Ordered distribution of modified bases in the DNA of a dinoflagellate. *Nucleic Acids Res.*, **8**, 4709–4725.

Stoecker, D. K. (1999). Mixotrophy among dinoflagellates. *J. Eukaryot. Microbiol.*, **46**, 397–401.

Strickland, J. D. H. (1965). Phytopankton and marine primary production. *Ann. Rev. Microbiol.*, **19**, 127–162.

Takishita, K., Ishikura, M., Koike, K., and Maruyama, T. (2003). Comparison of phylogenies based on nuclear-encoded SSU rDNA and plastid-encoded *psb*A in the symbiotic dinoflagellate genus *Symbiodinium*. *Phycologia*, **42**, 285–291.

Takishita, K., Koike, K., Maruyama, T., and Ogata, T. (2002). Molecular evidence for plastid robbery (Kleptoplastidy). in *Dinophysis*, a dinoflagellate causing diarrhetic shellfish poisoning. *Protist*, **153**, 293–302.

Tanikawa, N., Akimoto, H., Ogoh, K., Chun, W., and Ohmiya, Y. (2004). Expressed sequence tag analysis of the dinoflagellate *Lingulodinium polyedrum* during dark phase. *Photochem. Photobiol.*, **80**, 31–35.

Taroncher-Oldenburg, G. and Anderson, D. M. (2000). Identification and characterization of three differentially expressed genes, encoding S-adenosylhomocysteine hydrolase, methionine aminopeptidase, and a histone-like protein, in the toxic dinoflagellate *Alexandrium fundyense*. *Appl. Environ. Microbiol.*, **66**, 2105–2112.

Taylor, F. J. R., ed. (1987). *The Biology of Dinoflagellates*. Blackwell Scientific Publications, Oxford.

Tengs, T., Dahlberg, O. J., Shalchian-Tabrizi, K., Klaveness, D., Rudi, K., Delwiche, C. F., and Jakobsen, K. S. (2000). Phylogenetic analyses indicate that the 19′Hexanoyloxy-fucoxanthin-containing dinoflagellates have tertiary plastids of haptophyte origin. *Mol. Biol. Evol.*, **17**, 718–729.

Tomas, R. and Cox, E. R. (1973). Observations on the symbiosis of *Peridinium balticum* and its intracellular alga I: Ultrastructure. *J. Phycol.*, **9**, 304–323.

Triplett, E. L., Jovine, R. V., Govind, N. S., Roman, S. J., Chang, S. S., and Prezelin, B. B. (1993). Characterization of two full-length cDNA sequences encoding for apoproteins of peridinin-chlorophyll a-protein (PCP). complexes. *Mol. Marine Biol. Biotechnol.*, **2**, 246–254.

van Dolah, F. (2000). Marine algal toxins: origins, health effects, and their increased occurrence. *Environ. Health Persp.*, **108**, 133–141.

Veldhuis, M. J. W., Cucci, T. L., and Sieracki, M. E. (1997). Celular DNA content of marine phytoplankton using two new fluorochromes: taxonomic and ecological implications. *J. Phycol.*, **33**, 527–541.

Wang, Y., Jensen, L., Højrup, P., and Morse, D. (2005). Synthesis and degradation of dinoflagellate plastid-encoded psbA proteins are light-regulated, not circadian-regulated. *Proc. Natl. Acad. Sci. U S A*, **102**, 2844–2849.

Watanabe, M. M., Sasa, T., Suda, S., Inouye, I., and Takichi, S. (1991). Major carotenoid composition of an endosymbiont is a green dinoflagellate, *Lepidodinium viride*. *J. Phycol.*, **27**, 75.

Watanabe, M. M., Suda, S., Inouye, I., Sawaguchi, T., and Chihara, M. (1990). *Lepidodinium viride* new-genus new-species Gymnodiniales Dinophyta a green dinoflagellate with a chlorophyll *a*-containing and chlorophyll *b*-containing endosymbiont. *J. Phycol.*, **26**, 741– 751.

Wong, J. T. Y., New, D. C., Wong, J. C. W., and Hung, V. K. L. (2003). Histone-like proteins of the dinoflagellate *Crypthecodinium cohnii* have homologies to bacterial DNA-binding proteins. *Eukaryotic Cell*, **2**, 646–650.

Yoon, H. S., Hackett, J. D., van Dolah, F. M., Nosenko, T., Lidie, K. L., and Bhattacharya, D. (2005). Tertiary endosymbiosis driven genome evolution in dinoflagellate algae. *Mol. Biol. Evol.*, **22**, 1299–1308.

Zarudnaya, M. I., Kolomiets, I. M., Potyahaylo, A. L., and Hovorun, D. M. (2003). Downstream elements of mammalian pre-mRNA polyadenylation signals: primary, secondary and higher-order structures. *Nucleic Acids Res.*, **31**, 1375–1386.

Zauner, S., Greilinger, D., Laatsch, T., Kowallik, K. V., and Maier, U. G. (2004). Substitutional editing of transcripts from genes of cyanobacterial origin in the dinoflagellate *Ceratium horridum*. *FEBS Letters*, **577**, 535–538.

Zhang, H. and Lin, S. (2002). Detection and quantification of *Pfiesteria piscicida* by using the mitochondrial cytochrome *b* gene. *Appl. Environ. Microbiol.*, **68**, 989–994.

Zhang, H. and Lin, S. (2005). Mitochondrial cytochrome *b* mRNA editing in dinoflagellates: Possible ecological and evolutionary associations? *J. Eukaryot. Microbiol.*, **52**, 538–545.

Zhang, Z., Cavalier-Smith, T., and Green, B. R. (2001). A family of selfish minicircular chromosomes with jumbled chloroplast gene fragments from a dinoflagellate. *Mol. Biol. Evol.*, **18**, 1558–1565.

Zhang, Z., Cavalier-Smith, T., and Green, B. R. (2002). Evolution of dinoflagellate unigenic minicircles and the partially concerted divergence of their putative replicon origins. *Mol. Biol. Evol.*, **19**, 489–500.

Zhang, Z., Green, B. R., and Cavalier-Smith, T. (1999). Single gene circles in dinoflagellate chloroplast genomes. *Nature*, **400**, 155–159.

CHAPTER 4

Ciliate genome evolution

C. L. McGrath, R. A. Zufall, and L. A. Katz

Department of Biological Sciences, Smith College, Northampton, MA, USA

Introduction

The study of genome evolution in eukaryotic microbes such as ciliates can elucidate trends across the eukaryotic tree of life as well as unique, lineage-specific genome features. Ciliate genomes are unique among eukaryotes in that: (i) two distinct genomes exist in separate nuclei (the "somatic" macronucleus and "germline" micronucleus) within every cell; and (ii) macronuclear genomes undergo developmentally-regulated rearrangements, which can be extensive and lead to the generation of as many as 25 000 000 "chromosomes" in a single ciliate (reviewed in Klobutcher and Herrick 1997; Prescott 1994; Riley and Katz 2001). Ciliates have also been a rich source for the discovery of features found elsewhere across the tree of life, including the existence of self-splicing RNA (Kruger *et al.* 1982) and telomeres (Blackburn and Gall 1978). This chapter reviews the basics of ciliate genomes and discusses the implications of genome processing on ciliate evolution.

Ciliate life cycles

Ciliates encompass a large and diverse clade of microbial eukaryotes and are defined by both the presence of cilia in at least one life cycle phase and the existence of two distinct genomes within each cell (Fig. 4.1). The germline micronucleus houses a "typical" eukaryotic genome, with a small number of large chromosomes, while the somatic macronucleus, with a greater number of smaller chromosomes, is the site of virtually all vegetative transcription. During the development of a new macronucleus, zygotic chromosomes are fragmented, sequences are eliminated, and genes are amplified (Fig. 4.2). In some diverse lineages (Fig. 4.3), termed "extensive fragmenters," each gene is put on its own macronuclear chromosome, which may then be amplified many times (Lahlafi and Méténier 1991; Steinbrück *et al.* 1981; Riley and Katz 2001). This large-scale remodeling leads to the creation of a streamlined macronuclear genome capable of rapid vegetative cell proliferation, with high copy numbers of coding domains that generally lack repetitive elements (Katz 2001; Yao *et al.* 2002; Prescott 1994).

Details of nuclear events during ciliate life cycles vary among species due in part to differences in the number of nuclei: *Tetrahymena* and *Mono-euplotes* species, for example, have one micronucleus; *Paramecium* may have two or more depending on the species; and *Urostyla grandis* can have up to 20 micronuclei (Prescott 1994). The number of macronuclei in a cell is also variable: *Tetrahymena* and *Paramecium* species typically have a single macronucleus; *Oxytricha* species have two; and *Urostyla grandis* has as many as several hundred macronuclei, though the number varies in this taxon throughout its life cycle (Prescott 1994).

For simplicity, the life cycle presented here has been generalized to highlight nuclear events that are shared by many ciliate species (Fig. 4.2). Rounds of asexual reproduction are interspersed in ciliates by sexual conjugation. During sexual reproduction, the micronucleus undergoes meiosis to generate four haploid nuclei (Fig. 4.2); (reviewed in Jahn and Klobutcher 2002; Prescott 1994). Three of these nuclei degrade, while the fourth then divides mitotically to produce two gametic nuclei. During conjugation, one gametic

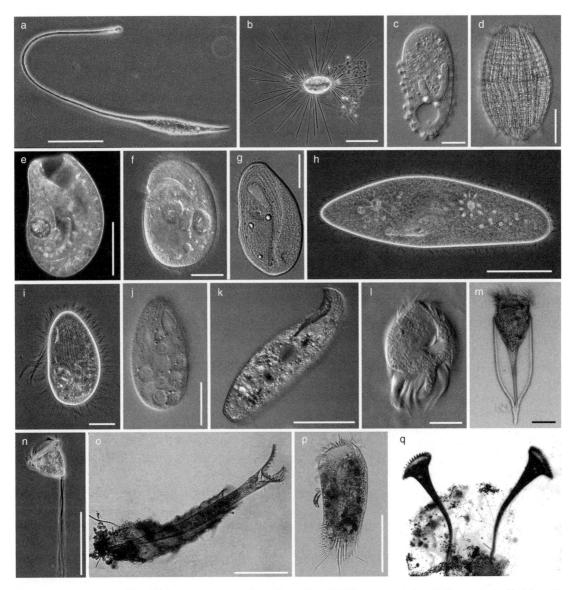

Figure 4.1 Representative ciliates: (a) *Lacrymaria*, a contractile predatory ciliate; (b) *Heliophrya*, a suctorian; (c) *Thysanomorpha*; (d) *Coleps* with calcareous plates; (e) *Bursaria*; (f) *Plagiopyla*; (g) *Chilodonella*; (h) *Paramecium*, (i) *Pleuronema*; (j) *Tetrahymena*; (k) *Loxodes*; (l) *Uronychia*; (m) *Favella*; (n) *Vorticella*; (o) *Folliculina*; (p) *Stylonychia*; and (q) *Stentor*. Scale bars in images (c), (d), (f), (g), (j), (l), and (m) = 20 µm; all others = 100 µm. Images (c) and (j) by W. Bourland; all others by D. J. Patterson, images provided by micro*scope (Http://microscope.mbl.edu).

nucleus from each cell is transferred to the other mating cell, where it fuses with the non-exchanged gametic nucleus. The result is a zygotic nucleus, which divides mitotically to produce identical diploid zygotic nuclei. At least one of these nuclei develops into a macronucleus and one develops into a micronucleus.

During vegetative growth, the macronucleus is transcriptionally active and ciliate cells divide by binary fission (reviewed in Prescott 1994; Yao *et al.* 2002). The diploid micronuclei divide by mitosis, while the polyploid macronuclei divide by amitosis (except in the class Karyorelictea, in which macronuclei are not capable of division).

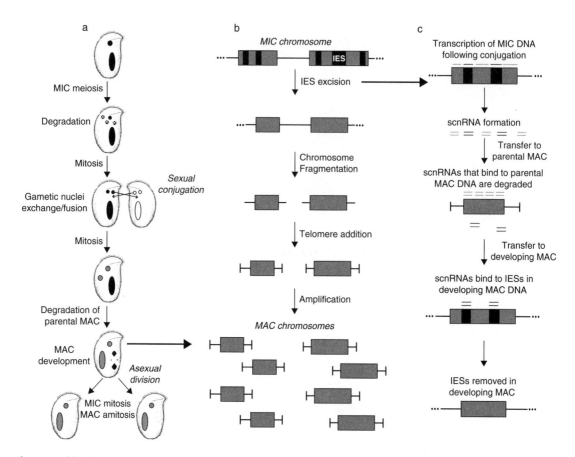

Figure 4.2 (a) Nuclear events in a generalized ciliate life cycle: In preparation for sexual conjugation, a ciliate's micronucleus undergoes meiosis to generate four haploid gametic nuclei, three of which subsequently degrade. The remaining gametic nucleus undergoes mitosis to generate two genetically identical nuclei. During conjugation, one gametic nucleus from each cell is transferred to the other cell, where it fuses with the non-exchanged gametic nucleus, forming a diploid zygotic nucleus. The zygotic nucleus undergoes mitosis. One zygotic nucleus becomes the new micronucleus, and the other undergoes macronuclear development (see b) to become the new macronucleus. The parental macronucleus degrades. During asexual division, the micronucleus divides by mitosis and the macronucleus by amitosis (except in Karyorelictea). (b) Simplified version of Macronuclear development in ciliates with extensive fragmentersion: During macronuclear genome development, internally excised sequences (IESs) are eliminated from the zygotic genome (see c), chromosomes are fragmented, telomeres are added to the ends of new macronuclear chromosomes, and chromosomes are amplified. (c) Simplified model of Epigenetic regulation of DNA elimination: IES removal from the zygotic genome during macronuclear development is regulated by epigenetic processes. Following conjugation, transcription of the micronuclear genome occurs. Transcripts then form short, double-stranded RNA segments called "scan" RNAs (scnRNAs). The scnRNAs are transported from the micronucleus to the parental macronucleus, where scnRNAs homologous to macronuclear DNA are degraded. The remaining scnRNAs (representing micronuclear-limited DNA) are transferred to the developing macronucleus, where they bind to IESs, signaling removal of these DNA segments in new macro nuclear chromosomes (derived from Mochizuki and Gorovsky 2004b).

Amitosis is a poorly understood process whereby chromosomes are relegated to daughter nuclei without the formation of a mitotic spindle (see below) (Raikov 1982; Prescott 1994).

Macronuclear division by amitosis in the Intramacronucleata leads to two similar, but not necessarily identical, nuclei (Raikov 1982). The imprecise mechanism of amitotic division is known to lead to unequal distribution of DNA to daughter nuclei, such that ciliates within a clonal line do not necessarily have identical macronuclear genomes. For example one macronucleus sometimes receives 10 to 15% more DNA than the other in *Tetrahymena pyriformis* and up to 50% more

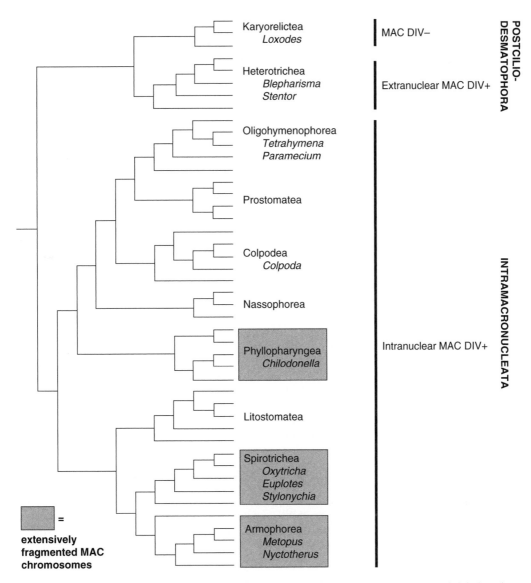

Figure 4.3 Phylogeny of ciliate classes with representative genera: The major division of ciliate classes into two subphyla (Postcilio-desmatophora and Intramacronucleata) based on differences in amitotic division are shown. The three extensively fragmenting lineages are highlighted. Representative genera are listed below class names. Support for many deep nodes is poor (Derived from Riley and Katz, 2001).

DNA than the other in *Paramecium putrinum* (Raikov 1982; Prescott 1994). In *Bursaria truncatella*, unequal macronuclear division is a regular phenomenon, with one daughter macronucleus consistently containing 1.4 to 2.5 times more DNA than the other (Raikov 1982). Similarly, in ciliates such as the Suctoria and Chonotrichia (Cl: Phyllopharyngea), which divide by budding, only

a small portion of the parental macronucleus may be pinched off into a daughter cell (Raikov 1982; Foissner 1996). Finally, in some ciliates, macronuclear genomes can be regenerated following physical disruption; for example a macronucleus of the ciliate *Stentor* can be regenerated from a nodule as small as 1/10 to 1/20 the size of the original macronucleus (Raikov 1982). Presumably,

DNA content is "reset" following sexual conjugation and development of a new macronucleus, which can regenerate a full complement of zygotic chromosomes. However, substantially more data on amitosis and epigenetics (see below) are needed from diverse ciliate lineages to assess the accuracy of macronuclear amitosis.

Differences in macronuclear division form the basis for major phylogenetic divisions within ciliates (Fig. 4.3). The subphylum Intramacronucleata (comprising most ciliate classes) is characterized by the formation of intranuclear microtubules during macronuclear division (Raikov 1982; Lynn and Small 1997). Only in the class Heterotrichea (subphylum Postciliodesmatophora) are the microtubules exclusively extranuclear. Macronuclei of the class Karyorelictea (Postciliodesmatophora) do not undergo amitotic division (Raikov 1996).

Genome processing

During macronuclear development, ciliate genomes undergo a high level of chromosomal processing, the extent of which varies among ciliate groups. For example in *Tetrahymena* (class Oligohymenophorea), the five large zygotic chromosomes (totaling ~110 Mb) are fragmented and DNA is eliminated to produce 200 distinct macronuclear chromosomes between 100 and 1500 kb long. Each of these chromosomes contains several hundred genes and is amplified an average of 45 times (Chalker and Yao 2001; Prescott 1994; Orias and Higashinakagawa 1990). In other species in Oligohymenophorea, such as *Paramecium tetraurelia*, each macronuclear chromosome is amplified an average of 800 times (Jahn and Klobutcher 2002; Klobutcher and Herrick 1997; Yao *et al.* 2002). Some extensive fragmenters of the class Spirotrichea fragment their approximately 120 zygotic chromosomes into as many as 24 000 macronuclear chromosomes, each less than 15 kb in length and most of which contain only a single gene; each chromosome may be amplified 950–15 000 times (Klobutcher and Herrick 1997; Prescott 1994).

At least a portion of the genome, the ribosomal DNA locus, is thought to be highly processed in all ciliates. In all species studied to date, zygotic chromosomes are processed such that rDNA genes are found on macronuclear chromosomes ranging from 8 to 15 kb and containing either a single or pair of rDNA clusters with no other genes (Prescott 1994). In fact, extrasomal processing and amplification of rDNAs occurs in diverse lineages of eukaryotes (i.e. *Euglena gracilis*, *Dictyostelium discoideum*, *Entamoeba histolytica*, *Xenopus laevis* embryos; reviewed in Zufall *et al.* 2005). Perhaps the effect of this processing is to maintain high levels of homogeneous rDNAs that are necessary to meet the requirements of translation.

Patterns of chromosome processing in ciliates

The precise order of the processes involved in development of the macronuclear genome remains unclear for many ciliates lineages, but a working hypothesis for the order of these events is presented here (Fig. 4.2b). The genome rearrangements that occur during macronuclear development include: (i) DNA deletion, (ii) DNA unscrambling (only known in a few lineages of the class Spirotrichea), (iii) fragmentation, (iv) telomere addition, and (v) amplification.

DNA deletion

Chromosomal processing in ciliates includes extensive and programmed elimination of DNA from the zygotic nucleus. Depending on the ciliate species, 10 to 95% of the zygotic genome may be eliminated during macronuclear development (Prescott 1994). This eliminated DNA includes not only repetitive elements and intervening sequences but also sequences that interrupt coding regions, termed internally excised sequences (IESs). IESs, which must be removed for proper macronuclear gene expression, fall into at least two categories: short IESs and transposon IESs (reviewed in Klobutcher and Herrick 1997). Both types of IESs are efficiently and reproducibly eliminated through DNA excision reactions, although the mechanisms may differ somewhat among ciliate species (Betermier 2004; Klobutcher and Herrick 1997).

Table 4.1 *cis*-acting sequences involved in IES excision

Class	Sequence element	Consensus sequence
Oligohymenophorea	*Paramecium* small IESs	TAYAGYNR
Spirotrichea	*Euplotes* small IESs	TATRGCRN
Spirotrichea	*Euplotes* Tec1/Tec2 transposon IESs	TATAGAGG
	Tc1/mariner transposable elements	TACAGTKS
Oligohymenophorea	*Tetrahymena* IESs	None
Phyllopharyngea	*Chilodonella* IESs	YGATTSWS
Spirotrichea	*Stylonychia* IESs	None
Spirotrichea	*Oxytricha* IESs	None

Data on *Chilodonella* from Katz, Lasek-Nesselquist, and Snoeyenbos-West (2003).
Other taxa reviewed in Jahn and Klobutcher (2002) and Yao *et al.* (2002).

One hypothesis for the origin of IESs is that they are derived from transposable elements (Klobutcher and Herrick 1997). For example "TA" IESs from *Paramecium* and *Monoeuplotes* share the same terminal repeats (Klobutcher and Herrick 1997; Betermier 2004; Jacobs and Klobutcher 1996) and are argued to have retained features of ancestral transposons. In *Paramecium*, these unique sequences range from 26 to 882 bp in size, although most (75%) are shorter than 100 bp (Betermier 2004). A degenerate consensus sequence was identified among 20 *Paramecium* IESs: an 8-bp sequence that includes a TA dinucleotide conserved at the IES boundaries (Table 4.1) (Betermier 2004). This TA dinucleotide is observed at the boundaries of *Monoeuplotes* IESs as well, and is very similar to the ends of Tc1/mariner-like transposons, which duplicate a target TA upon insertion and are also eliminated during macronuclear development (Betermier 2004; Jacobs and Klobutcher 1996).

The similarity between "TA" IESs and transposons may reflect a secondary invasion by transposable elements following the evolution of a precise elimination process during macronuclear development in ciliates. Alternatively, IESs could have evolved within ciliates via the developmentally-regulated rearrangements that generate macronuclear chromosomes from germline chromosomes. Under this latter hypothesis, the similarity between *Paramecium* and *Monoeuplotes* IESs could be due to constraints on the *cis*-acting signals required to remove germline sequences (Dubrana *et al.* 1997).

The conservation in IESs between *Paramecium* and *Monoeuplotes* is particularly surprising, given data from other ciliates: *Oxytricha*, *Stylonychia*, *Chilodonella*, and *Tetrahymena* have apparently unique *cis*-acting signals (reviewed in Katz *et al.* 2003; Jahn and Klobutcher 2002; Yao *et al.* 2002). There is no apparent similarity in *cis*-acting sequences in ciliates within the same classes: *Tetrahymena* and *Paramecium* (Cl: Oligohymenophorea) and *Monoeuplotes*, *Oxytricha*, and *Stylonychia* (Cl: Spirotrichea) (Table 4.1). This observation suggests that either *cis*-acting sequences have evolved multiple times in ciliates or that the mechanisms of IES excision are relatively plastic over evolutionary time (see "Ciliate genome evolution" section below).

DNA unscrambling

After IES excision, remaining gene segments must be ligated together; in some ciliate lineages within the class Spirotrichea, however, segments must first be reoriented in relation to each other. DNA unscrambling has been identified among stichotrich ciliates (Cl: Spirotrichea) – for example *Oxytricha* and *Stylonychia* – and involves the reordering of gene segments during macronuclear development (reviewed in Jahn and Klobutcher 2002; Prescott 1999; Ardell *et al.* 2003). The micronuclear copies of some genes from these ciliates are not only interrupted by DNA destined for elimination in the macronucleus, but the segments are also disordered, and in some cases segments can be inverted relative to the others. For example out of

12 genes in members of the Stichotrichia (Cl: Spirotrichea) whose micronuclear organizations have been studied, three genes exhibit scrambled orders (the genes for actin, α-telomere-binding protein, and DNA polymerase α) (Prescott and Greslin 1992; Greslin *et al.* 1989; Prescott *et al.* 1998; Hoffman and Prescott 1996; Landweber *et al.* 2000). The macronuclear-destined segments of the actin gene appear to be randomly scrambled, but the segments of the other two genes are scrambled in a non-random fashion (e.g. 1–3–5–7–9–11–2–4–6–8–10–12–13–14 for the α-telomere-binding protein).

The mechanism of unscrambling appears to involve *cis*-acting sequences. Short sequences (about 11 bp) on the ends of the gene fragments are identical between segments that must be joined during unscrambling (the sequence on the 3′ end of segment 1 is identical to the sequence on the 5′ end of segment 2, etc.). Across gene segments, however, the sequences vary, with no obvious similarity: the sequence used to join segments 1 and 2 is different from that used to join segments 2 and 3, segments 3 and 4, etc. (Prescott 2000). Scrambled genes could have originated in genomes already containing conventional IESs due to germline recombination between different IESs (Prescott 1999).

Fragmentation

In addition to DNA elimination, the zygotic genome is further processed by fragmentation of remaining DNA. There is as yet little evidence of phylogenetic conservation among fragmentation signals. The extent to which zygotic chromosomes are fragmented during macronuclear development varies considerably among ciliates. Chromosomes of some ciliates, extensive fragmenters, are fragmented at up to 40 000 sites across the genome, while those in class Oligohymenophorea are fragmented far less frequently, at only 50 to 200 sites (Yao *et al.* 2002; Riley and Katz 2001; Klobutcher and Herrick 1997).

As with IES excision, different genera of ciliates appear to use different mechanisms for signaling chromosome breakage sites. A conserved chromosome breakage sequence (Cbs) was identified in several *Tetrahymena* species (Yao *et al.* 1987; King and Yao 1982). This 15-bp sequence is found at

multiple chromosome fragmentation sites in the *Tetrahymena* micronuclear genome. Experiments in which this sequence is added or deleted demonstrate that the sequence is both necessary and sufficient for chromosome breakage in these ciliates (Yao *et al.* 1990; Yao and Yao 1989). A chromosome breakage sequence has also been identified in *Monoeuplotes crassus* (E-Cbs) (Klobutcher *et al.* 1998; Baird and Klobutcher 1989). This 10-bp sequence, however, shares little similarity with the *Tetrahymena* Cbs and the fragmentation mechanism appears to be different, suggesting that fragmentation in these two genera may not share a common origin (Yao *et al.* 2002). The E-Cbs does not appear to be sufficient to define the fragmentation sites in the *Monoeuplotes* genome, as strong matches to the consensus sequence can be found within non-fragmented regions of macronuclear chromosomes.

Unlike *Tetrahymena* and *Monoeuplotes* fragmentation sites, no consensus sequence has been identified adjacent to chromosome breakage sites for any other ciliate species studied (*Oxytricha nova* and *O. fallax*, *Paramecium primaurelia* and *P. tetraurelia*; reviewed in Yao *et al.* 2002). However, in *Stylonychia lemnae*, sequences within approximately 300 bp of fragmentation sites resemble the *Monoeuplotes* Cbs and are essential for correct fragmentation (Jonsson *et al.* 1999; Jonsson *et al.* 2001). This suggests that chromosome breakage sequences may go unidentified in other ciliates because they are not located directly at fragmentation sites. Alternatively, the epigenetic phenomena known to be involved in macronuclear development (see below) may play a broader role in fragmentation.

Telomere addition

The chromosome fragmentation process in ciliates involves the *de novo* addition of telomeric repeats to the ends of newly created chromosomes (Table 4.2). Because some ciliates have as many as 25×10^6 macronuclear chromosomes, which means 50×10^6 telomeres, ciliates are particularly tractable for studies of telomere structure and telomerase activity. In some ciliates, the telomeric repeats are always the same number of nucleotides

Table 4.2 Telomere repeats in ciliates

Ciliate class	Genus	Repeat sequence
Oligohymenophorea	*Tetrahymena, Glaucoma*	CCCCAA
	Paramecium	(A/C)CCCAA
Phyllopharyngea	*Chilodonella*	CCCCAAA
Spirotrichea	*Oxytricha, Stylonichia, Euplotes*	CCCCAAAA
Armophorea	*Metopus, Nyctotherus*	CCCCAAAA

Data on Phyllopharyngea and Armophorea from McGrath *et al.* (unpublished).
Other taxa reviewed in Yao *et al.* (2002).

upstream of the coding region, while in others there is a limited degree of variability in sequences (Yao *et al.* 2002). The telomeres are comprised of simple, tandem repeats that differ among ciliate classes (Prescott 1994; McGrath, Zufall and Katz unpublished; Table 4.2).

Studies indicate that, in at least some ciliates, *de novo* telomere addition may not require a specific sequence signal. For example in *Paramecium*, any DNA molecule introduced into the macronucleus will have telomeres added onto its ends (Gilley *et al.* 1988). In *Monoeuplotes crassus*, there also appears to be no specific sequence requirement for telomere addition (Melek *et al.* 1996).

Amplification

A final step in macronuclear development is the differential amplification of new macronuclear chromosomes. This process occurs after telomere addition and is particularly pronounced in ciliates with extensively fragmented macronuclear chromosomes. In some ciliates in the class Spirotrichea, for example, each chromosome can be amplified up to 15 000 times (Prescott 1994). While replication origins in other organisms are usually associated with *cis*-acting sequences, replication of ciliate macronuclear chromosomes is not well understood. One hypothesis is that macronuclear chromosome replication is dependent on telomeres (Skovorodkin *et al.* 2001); macronuclear chromosomes generated by replacement of coding and non-coding sequences continue to replicate and are maintained at high copy number for at least several months. This finding is consistent with electron microscopic studies that show replication

originating at or near the telomeres and the discovery of primase activity synthesizing RNA primers over 3′ telomeric repeats *in vitro* (Skovorodkin *et al.* 2001).

Variation in ciliate genome content

Ciliate species vary considerably in the size, DNA content, and ploidy of both their macronuclear and micronuclear genomes. While there does not appear to be a correlation between macronuclear DNA content and taxonomy, macronuclear DNA content does seem to depend on the size of the ciliate cell (Raikov 1982). The relative DNA content of the macronucleus to the micronucleus as well as the absolute content of macronuclear DNA appears to be higher in large ciliates than in smaller, closely related species (e.g. *Tetrahymena patula* and *T. thermophila*; *Colpoda magna* and *C. steini*; Fig. 4.4).

Data from whole genomes

Because of their importance for cell biology and evolution, several ciliate genome projects have been undertaken. Data from sequencing projects of three ciliates, *Paramecium tetraurelia* (Sperling *et al.* 2002; Dessen *et al.* 2001), *Tetrahymena thermophila* (Orias 2000; Turkewitz *et al.* 2002), and *Oxytricha trifallax* (Cavalcanti *et al.* 2004b; Doak *et al.* 2003; Cavalcanti *et al.* 2004a) will enable further elucidation of the evolution of ciliate genome features. An entire *Tetrahymena* macronuclear genome has been deposited in GenBank (GB#AAGF01000000), and full annotation is now underway (http://www.ciliate.org/). Initial data describing 3568

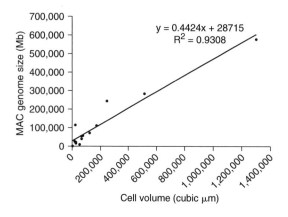

Figure 4.4 Relationship of macronuclear genome size to cell volume: There is a positive correlation between the size of the macronuclear genome and cell volume among ciliates. Cell size and DNA content estimates are from Raikov (1982), Shuter *et al.* (1983), and Lynn and Small (2002).

P. tetraurelia sequences and 1356 *O. trifallax* macronuclear sequences have been published (Sperling *et al.* 2002; Cavalcanti *et al.* 2004b). Analysis of the *Paramecium* genome indicates that there have been four whole genome duplications (Patrick Wincker, pers. comm.).

Preliminary findings from *O. trifallax* indicate that introns range in length from 31 to 137 bp, with most introns near the smaller end of the range (>30% between 31 and 40 bp). These introns are small compared to those of other eukaryotes. *P. tetraurelia* introns are even smaller – between 18 and 35 bp in length. Like other ciliates, *O. trifallax* coding regions are A + T rich (61% A + T) (Cavalcanti *et al.* 2004b), although they are less biased than coding regions in *T. thermophila* (64.3% A + T) (Wuitschick and Karrer 2000) or *P. tetraurelia* (67% A + T) (Sperling *et al.* 2002). In both *Oxytricha* and *Paramecium*, introns are more A + T rich than surrounding exons, and *Oxytricha's* smaller introns are significantly more A + T rich than its larger introns. Clearly, completion and full annotation of all three of these genomes will allow for powerful comparisons of genome features and gene discovery within this clade of eukaryotes.

Epigenetic regulation of genome architecture

The highly variable, "plastic" nature of deletion and fragmentation processes among ciliates is consistent with increasing evidence of a role for epigenetics in regulating genome processing in

eukaryotes (reviewed in Zufall *et al.* 2005). For example the targeting of specific zygotic sequences for removal during macronuclear development is thought to be controlled by epigenetic mechanisms (Fig. 4.2c; Mochizuki and Gorovsky 2004a; Mochizuki and Gorovsky 2004b; Garnier *et al.* 2004; Chalker *et al.* 2005; Betermier 2004; Mochizuki *et al.* 2002; Yao *et al.* 2003). One model for the role of epigenetic regulation in this elimination involves a mechanism resembling RNA interference (RNAi) that enables genome scanning between the zygotic nucleus, the parental macronucleus, and the newly developing macronucleus (Mochizuki *et al.* 2002).

Evidence for the role of epigenetics in ciliate inheritance predates molecular technologies and is consistent with the non-Mendelian inheritance of immobilization antigens in *Paramecium tetraurelia* (reviewed in Jahn and Klobutcher 2002). More recently, when IESs (normally micronuclear-limited) were introduced into the macronucleus of *Tetrahymena* and *Paramecium*, the IESs were maintained in the macronuclei of the cell's offspring (Chalker and Yao 1996; Chalker *et al.* 2005; Duharcourt *et al.* 1995; Duharcourt *et al.* 1998).

The genome scanning hypothesized in these systems involves transcription of the micronuclear genome coinciding with conjugation, followed by subsequent "detection" of sequences maintained in the parental macronucleus through an RNAi-like process (Mochizuki *et al.* 2002). Small fragments of double-stranded RNA, termed "scan" RNAs (scnRNAs), are formed and transported to the parental macronucleus, where the scnRNAs homologous to macronuclear sequences are then degraded (Fig. 4.2). The scnRNAs with no homologous sequence in the macronucleus (micronuclear-limited sequences) are then transported to the newly developing macronucleus, where they direct the elimination of their homologous micronuclear-limited sequences. This model has been supported by data from Yao *et al.* (2003): double-stranded RNAs of sequences normally present in the macronuclei were injected into cells during conjugation, and deletion of the corresponding sequences in the newly developed macronucleus was then observed for all three RNA fragments injected. Further, Mochizuki and Gorovsky (2005) demonstrate that a Dicer-like protein in *Tetrahymena* processes non-genic micronuclear transcripts into scnRNA and that these are required for proper elimination of IESs.

Ciliate genome evolution

Origin of genome duality

Combining data on ciliate genomes with phylogenetic analyses allows for the further development of models for the evolution of nuclear duality, genome processing, and impacts on ciliate protein evolution. Earlier hypotheses posited that dimorphic nuclei evolved in an ancestor of heterokaryotic foraminifera and ciliates, and the inability of somatic nuclei to divide in heterokaryotic foraminifera and karyorelictid ciliates was believed to be a shared ancestral trait (Orias 1991a; Orias 1991b; Herrick 1994). However, there is no evidence that ciliates and foraminifera are sister taxa, indicating that dual genomes have arisen independently in these two lineages (e.g. Katz 2001).

In contrast, Hammerschmidt *et al.* (1996) proposed two more parsimonious models for the evolution of macronuclear division. The first involves an early origin of macronuclear division followed by a loss in the Karyorelictea; the second proposes two independent origins of dividing macronuclei, one in the class Heterotrichea and one in all other ciliates (Hammerschmidt *et al.* 1996). Hammerschmidt *et al.* (1996) favor the first hypothesis, believing the extranuclear microtubules of the Heterotrichea are a derived

character because intranuclear microtubules are present in ciliate micronuclei and other ciliate macronuclei (Oligohymenophorea, Spirotriche and Intramacronucleater).

Emerging data on the phylogenetic distribution of and genomic processes within ciliates are consistent with a single origin of genome duality in ciliates and subsequent evolution of ciliate genome architecture regulated by epigenetic mechanisms. This model suggests that nuclear dimorphism in ciliates evolved along with a single origin of a relatively plastic mechanism that allowed for the diversification of macronuclear structure and chromosomal processing events (Katz 2001). The apparent reliance on epigenetic phenomena guiding development of a new macronucleus from a parental macronucleus provides such a mechanism by allowing for diverse macronuclear genome structures, such as are involved in chromosomal fragmentation and IES excision. This model, instead of arguing for multiple independent origins of macronuclear division *and* of extensive fragmentation *and* of processing mechanisms, provides a more parsimonious explanation for the wide variety of genome structures and mechanisms observed in ciliates.

We have argued that a further by-product of the evolution of genome duality is that ciliates explore protein space in a novel manner, as compared to other eukaryotes (Katz *et al.* 2004; Zufall *et al.* 2005). Because of the differential selection on processed macronuclear genomes vs. the unexpressed micronuclear genomes, ciliates can effectively hide deleterious mutations from selection during rounds of asexual division. Selection acts on processed chromosomes in macronuclei, where differential amplification, recombination, and breakup of linkage groups may mask the effects of deleterious mutations that are carried in the germline micronucleus. These effects are enhanced by the process of amitosis, whereby unequal segregation of macronuclear chromosomes can change the representation of alleles in the macronucleus, or even eliminate specific alleles. As a result, deleterious alleles eliminated from the processed macronucleus are retained in the micronucleus, where they may accumulate compensatory mutations during cycles of asexual divisions. Such processes will be exaggerated in ciliates with extensively-processed genomes, where the fate of each allele is essentially independent of other alleles.

The hypothesis that nuclear dualism enables ciliates to evolve proteins in novel manners is supported by phylogenetic analyses based on protein-coding genes, such as elongation factor 1 alpha, heat shock protein 70, and eukaryotic release factor 1, that are unable to recover a consistent topology of ciliates due to rapid rates of substitution in this lineage (Budin and Philippe 1998; Bhattacharya and Ehlting 1995; Moreira *et al.* 2002; Moreira *et al.* 1999). A striking example of rapid protein evolution in ciliates is seen in histone H4, which is highly conserved in most eukaryotes but shows dramatic patterns of diversification in ciliates (Katz *et al.* 2004). Analyses of histone H4 and five other proteins show that ciliates tend to have more rapid rates of substitution and higher ratios of non-synonymous substitutions per non-synonymous site to synonymous substitutions per synonymous site (d_N/d_S) than other major lineages of eukaryotes. In addition, both substitution rate and subscripts d_N/d_S are elevated in extensively processing ciliates compared to other ciliates (Katz *et al.* 2004; Zufall, McGrath, Muse, and Katz, unpublished). These data suggest that ciliate gene families are experiencing either a relaxation of evolutionary constraints or an increase in adaptive substitutions.

Acknowledgements

This work was support by NSF grants DEB-0092908, DEB-0079325 and DEB 043115 to LAK; and DBI-0301610 to RAZ. Many thanks to David J. Patterson for his help in assembling Fig. 4.1, which captures a small proportion of the beauty of ciliates. We also appreciate the comments of two reviewers on a previous version of this chapter.

References

Ardell, D. H., Lozupone, C. A., and Landweber, L. F. (2003). Polymorphism, recombination, and alternative unscrambling in the DNA polymerase alpha gene of the ciliate *Stylonychia lemnae* (Alveolata; class Spirotrichea). *Genetics*, **165**, 1761–1777.

Baird, S. E. and Klobutcher, L. A. (1989). Characterization of chromosome fragmentation in two protozoans and identification of a candidate fragmentation sequence in *Euplotes crassus. Genes Dev.*, **3**, 585–597.

Betermier, M. (2004). Large-scale genome remodeling by the developmentally programmed elimination of germ line sequences in the ciliate *Paramecium. Res. Microbiol.*, **155**, 399–408.

Bhattacharya, D. and Ehlting, J. (1995). Actin coding regions: gene family evolution and use as a phylogenetic marker. *Arch. Protistenkd.*, **145**, 155–164.

Blackburn, E. H. and Gall, J. G. (1978). A tandemly repeated sequence at the termini of the extrachromosomal ribosomal RNA genes in *Tetrahymena. J. Mol. Biol.*, **120**, 33–53.

Budin, K. and Philippe, H. (1998). New insights into the phylogeny of eukaryotes based on Ciliate Hsp70 sequences. *Mol. Biol. Evol.*, **15**, 943–956.

Cavalcanti, A. R. O., Dunn, D. M., Weiss, R., Herrick, G., Landweber, L. F., and Doak, T. G. (2004a). Sequence features of *Oxytricha trifallax* (class Spirotrichea) macronuclear telomeric and subtelomeric sequences. *Protist*, **155**, 311–322.

Cavalcanti, A. R. O., Stover, N. A., Orecchia, L., Doak, T. G., and Landweber, L. F. (2004b). Coding properties of *Oxytricha trifallax* (*Sterkiella histriomuscorum*) macronuclear chromosomes: analysis of a pilot genome project. *Chromosoma*, **113**, 69–76.

Chalker, D. L., Fuller, P., and Yao, M. C. (2005). Communication between parental and developing genomes during *Tetrahymena* nuclear differentiation is likely mediated by homologous RNAs. *Genetics*, **169**, 149–160.

Chalker, D. L. and Yao, M. C. (1996). Non-Mendelian, heritable blocks to DNA rearrangement are induced by loading the somatic nucleus of *Tetrahymena thermophila* with germ line-limited DNA. *Mol. Cell. Biol.*, **16**, 3658–3667.

Chalker, D. L. and Yao, M. C. (2001). Nongenic, bidirectional transcription precedes and may promote developmental DNA deletion in *Tetrahymena thermophila. Genes Dev.*, **15**, 1287–1298.

Dessen, P., Zagulski, M., Gromadka, R., Plattner, H., Kissmehl, R., Meyer, E., Betermier, M., Schultz, J. E., Linder, J. U., Pearlman, R. E., Kung, C., Forney, J., Satir, B. H., van Houten, J. L., Keller, A. M., Froissard, M., Sperling, L., and Cohen, J. (2001). *Paramecium* genome survey: a pilot project. *Trends Genet.*, **17**, 306–308.

Doak, T. G., Cavalcanti, A. R. O., Stover, N. A., Dunn, D. M., Weiss, R., Herrick, G., and Landweber, L. F. (2003). Sequencing the *Oxytricha trifallax* macronuclear genome: a pilot project. *Trends Genet.*, **19**, 603–607.

Dubrana, K., Le, M. A., and Amar, L. (1997). Deletion endpoint allele-specificity in the developmentally regulated elimination of an internal sequence (IES) in *Paramecium*. *Nucleic Acids Res.*, **25**, 2448–2454.

Duharcourt, S., Bulter, A., and Meyer, E. (1995). Epigenetic self-regulation of developmental excision of an internal eliminated sequence in *Paramecium tetraurelia*. *Genes Dev.*, **9**, 2065–2077.

Duharcourt, S., Keller, A. M., and Meyer, E. (1998). Homology-dependent maternal inhibition of developmental excision of internal eliminated sequences in *Paramecium tetraurelia*. *Mol. Cell. Biol.*, **18**, 7075–7085.

Foissner, W. (1996). Ontogenesis in ciliated protozoa with emphasis on stomatogenesis. In Hausmann, K. and Bradbury, P. C., eds. *Ciliates: Cells as Organisms*. Gustav Fisher, Stuttgart.

Garnier, O., Serrano, V., Duharcourt, S., and Meyer, E. (2004). RNA-mediated programming of developmental genome rearrangements in *Paramecium tetraurelia*. *Mol. Cell. Biol.*, **24**, 7370–7379.

Gilley, D., Preer, J. R., Aufderheide, K. J., and Polisky, B. (1988). Autonomous replication and addition of telomere-like sequences to DNA microinjected into *Paramecium tetraurelia* macronuclei. *Mol. Cell. Biol.*, **8**, 4765–4772.

Greslin, A. F., Prescott, D. M., Oka, Y., Loukin, S. H., and Chappell, J. C. (1989). Reordering of 9 exons is necessary to form a functional actin gene in *Oxytricha nova*. *Proc. Natl. Acad. Sci. U S A*, **86**, 6264–6268.

Hammerschmidt, B., Schlegel, M., Lynn, D. H., Leipe, D. D., Sogin, M. L., and Raikov, I. B. (1996). Insights into the evolution of nuclear dualism in the ciliates revealed by phylogenetic analysis of rRNA sequences. *J. Eukaryot. Microbiol.*, **43**, 225–230.

Herrick, G. (1994). Germline-soma relationships in ciliated protozoa: the inception and evolution of nuclear dimorphism in one-celled animals. *Sem. Dev. Biol.*, **5**, 3–12.

Hoffman, D. C. and Prescott, D. M. (1996). The germline gene encoding DNA polymerase α in the hypotrichous ciliate *Oxytricha nova* is extremely scrambled. *Nucleic Acids Res.*, **24**, 3337–3340.

Jacobs, M. E. and Klobutcher, L. A. (1996). The long and the short of developmental DNA deletion in *Euplotes crassus*. *J. Eukaryot. Microbiol.*, **43**, 442–452.

Jahn, C. L. and Klobutcher, L. A. (2002). Genome remodeling in ciliated protozoa. *Ann. Rev. Microbiol.*, **56**, 489–520.

Jonsson, F., Steinbruck, G., and Lipps, H. J. (2001). Both subtelomeric regions are required and sufficient for specific DNA fragmentation during macronuclear development in *Stylonychia lemnae*. *Genome Biol.*, **2**, 1–11.

Jonsson, F., Wen, J. P., Fetzer, C. P., and Lipps, H. J. (1999). A subtelomeric DNA sequence is required for correct processing of the macronuclear DNA sequences during macronuclear development in the hypotrichous ciliate *Stylonychia lemnae*. *Nucleic Acids Res.*, **27**, 2832–2841.

Katz, L. A. (2001). Evolution of nuclear dualism in ciliates: a reanalysis in light of recent molecular data. *Int. J. Syst. Evol. Microbiol.*, **51**, 1587–1592.

Katz, L. A., Lasek-Nesselquist, E., Bornstein, J., and Muse, S. V. (2004). Dramatic diversity of ciliate histone H4 genes revealed by comparisons of patterns of substitutions and paralog divergences among eukaryotes. *Mol. Biol. Evol.*, **21**, 555–562.

Katz, L. A., Lasek-Nesselquist, E., and Snoeyenbos-West, O. L. O. (2003). Structure of the micronuclear α-tubulin gene in the phyllopharyngean ciliate *Chilodonella uncinata*: implications for the evolution of chromosomal processing. *Gene*, **315**, 15–19.

King, B. O. and Yao, M. C. (1982). Tandemly repeated hexanucleotide at *Tetrahymena* rDNA free end is generated from a single copy during development. *Cell*, **31**, 177–182.

Klobutcher, L. A., Gygax, S. E., Podoloff, J. D., Vermeesch, J. R., Price, C. M., Tebeau, C. M., and Jahn, C. L. (1998). Conserved DNA sequences adjacent to chromosome fragmentation and telomere addition sites in *Euplotes crassus*. *Nucleic Acids Res.*, **26**, 4230–4240.

Klobutcher, L. A. and Herrick, G. (1997). Developmental genome reorganization in ciliated protozoa: the transposon link. *Prog. Nucleic Acid Res. Mol. Biol.*, **56**, 1–62.

Kruger, K., Grabowski, P. J., Zaug, A. J., Sands, J., Gottschling, D. E., and Cech, T. R. (1982). Self-splicing RNA: Autoexcision and autocyclization of the ribosomal RNA intervening sequence of *Tetrahymena*. *Cell*, **31**, 147–157.

Lahlafi, T. and Méténier, G. (1991). Low molecular weight DNA in the heteromeric macronuclei of two cyrtophorid ciliates. *Biol. Cell*, **73**, 79–88.

Landweber, L. F., Kuo, T., and Curtis, E. A. (2000). Evolution and assembly of an extremely scrambled gene. *Proc. Natl. Acad. Sci. U S A*, **97**, 3298–3303.

Lynn, D. H. and Small, E. B. (1997). A revised classification of the phylum Ciliophora Doflein, 1901. *Rev. Soc. Mex. Hist. Nat.*, **47**, 65–78.

Lynn, D. H. and Small, E. B. (2002). Phylum Ciliophora Doflein, 1901. In J. J. Lee, G. F. Leedale, and P. C. Bradbury, eds. *An Illustrated Guide to the Protozoa*, 2nd edn. Society of Protozoologists, Lawrence, Kansas.

Melek, M., Greene, E. C., and Skippen, D. E. (1996). Processing of nontelomeric 3′ ends by telomerase: Default template alignment and endonucleolytic cleavage. *Mol. Cell. Biol.*, **16**, 3437–3445.

Mochizuki, K., Fine, N. A., Fujisawa, T., and Gorovsky, M. A. (2002). Analysis of a piwi-related gene implicates small RNAs in genome rearrangement in *Tetrahymena. Cell*, **110**, 689–699.

Mochizuki, K. and Gorovsky, M. A. (2004a). Conjugation-specific small RNAs in *Tetrahymena* have predicted properties of scan (scn) RNAs involved in genome rearrangement. *Genes Devel.*, **18**, 2068–2073.

Mochizuki, K. and Gorovsky, M. A. (2004b). Small RNAs in genome rearrangement in *Tetrahymena. Curr. Opin. Genet. Dev.*, **14**, 181.

Mochizuki, K. and Gorovsky, M. A. (2005). A Dicer-like protein in *Tetrahymena* has distinct functions in genome rearrangement, chromosome segregation, and meiotic prophase. *Genes Dev.*, **19**, 77–89.

Moreira, D., Kervestin, S., Jean-Jean, O., and Philippe, H. (2002). Evolution of eukaryotic translation elongation and termination factors: Variations of evolutionary rate and genetic code deviations. *Mol. Biol. Evol.*, **19**, 189–200.

Moreira, D., Le Guyader, H., and Philippe, H. (1999). Unusually high evolutionary rate of the elongation factor 1a genes from the Ciliophora and its impact on the phylogeny of eukaryotes. *Mol. Biol. Evol.*, **16**, 234–245.

Orias, E. (1991a). Evolution of amitosis of the ciliate macronucleus: gain of the capacity to divide. *J. Protozool.*, **38**, 217–221.

Orias, E. (1991b). On the evolution of the karyorelict ciliate life cycle: heterophasic ciliates and the origin of ciliate binary fission. *Biosystems*, **25**, 67–73.

Orias, E. (2000). Toward sequencing the *Tetrahymena* genome: exploiting the gift of nuclear dimorphism. *J. Eukaryot. Microbiol.*, **47**, 328–333.

Orias, E. and Higashinakagawa, T. (1990). Genome organization and reorganization in ciliated protozoa. *Zoo. Sci.*, **7**, 59–69.

Prescott, D. M. (1994). The DNA of ciliated protozoa. *Microbiol. Rev.*, **58**, 233–267.

Prescott, D. M. (1999). Evolution of DNA organization in hypotrichous ciliates. *Annal. N.Y. Acad. Sci.*, 301–313.

Prescott, D. M. (2000). Genome gymnastics: Unique modes of DNA evolution and processing in ciliates. *Nat. Rev. Gen.*, **1**, 191–198.

Prescott, D. M. and Greslin, A. F. (1992). Scrambled actin I gene in the micronucleus of *Oxytricha nova. Dev. Genet.*, **13**, 66–74.

Prescott, J. D., Dubois, M. L., and Prescott, D. M. (1998). Evolution of the scrambled germline gene encoding alpha-telomere binding protein in three hypotrichous ciliates. *Chromosoma*, **107**, 293–303.

Raikov, I. B. (1982). *The Protozoan Nucleus: Morphology and Evolution.* Springer-Verlag, Wien.

Raikov, I. B. (1996). Nuclei of ciliates. In K. Hausmann and P. C. Bradbury, eds. *Ciliates: Cells as organisms.* Gustav Fischer, Stuttgart.

Riley, J. L. and Katz, L. A. (2001). Widespread distribution of extensive genome fragmentation in ciliates. *Mol. Biol. Evol.*, **18**, 1372–1377.

Shuter, B. J., Thomas, J. E., Taylor, W. D., and Zimmerman, A. M. (1983). Phenotypic correlates of genomic DNA content in unicellular eukaryotes and other cells. *Am. Naturalist*, **122**, 26–44.

Skovorodkin, I. N., Zassoukhina, I. B., Hojak, S., Ammermann, D., and Gunzl, A. (2001). Minichromosomal DNA replication in the macronucleus the hypotrichous ciliate *Stylonychia lemnae* is independent of chromosome-internal sequences. *Chromosoma*, **110**, 352–359.

Sperling, L., Dessen, P., Zagulski, M., Pearlman, R. E., Migdalski, A., Gromadka, R., Froissard, M., Keller, A. M., and Cohen, J. (2002). Random sequencing of *Paramecium* somatic DNA. *Eukaryot. Cell*, **1**, 341–352.

Steinbrück, G., Haas, I., Hellmer, K.-H., and Ammermann, D. (1981). Characterization of macronuclear DNA in five species of ciliates. *Chromosoma*, **83**, 199–208.

Turkewitz, A. P., Orias, E., and Kapler, G. (2002). Functional genomics: the coming of age for *Tetrahymena thermophila. Trends Gen.*, **18**, 35–40.

Wuitschick, J. D. and Karrer, K. M. (2000). Codon usage in *Tetrahymena thermophila. Methods Cell. Biol.*, **62**, 565–568.

Yao, M. C., Duharcourt, S., and Chalker, D. L. (2002). Genome-wide rearrangements of DNA in ciliates. In N. L.Craig, R. Craigie, M. Gellert, A. and Lambowitz, eds. *Mobile DNA II.* ASM Press, Washington, D.C..

Yao, M. C., Fuller, P., and Xi, X. (2003). Programmed DNA deletion as an RNA-guided system of genome defense. *Science*, **300**, 1581–1584.

Yao, M. C. and Yao, C. H. (1989). Accurate processing and amplification of cloned germ line copies of ribosomal DNA injected into developing nuclei of *Tetrahymena thermophila. Mol. Cell. Biol.*, **9**, 1092–1099.

Yao, M. C., Yao, C. H., and Monks, B. (1990). The controlling sequence for site-specific chromosome breakage in *Tetrahymena. Cell*, **63**, 763–772.

Yao, M. C., Zheng, K. Q., and Yao, C. H. (1987). A conserved nucleotide-sequence at the sites of developmentally regulated chromosomal breakage in *Tetrahymena*. *Cell*, **48**, 779–788.

Zufall, R. A., Robinson, T., and Katz, L. A. (2005). Evolution of developmentally regulated genome rearrangements in eukaryotes. *J. Exp. Zool. Part B*, **304B**, 448–455.

CHAPTER 5

Molecular evolution of Foraminifera

Samuel S. Bowser[1], Andrea Habura[1], Jan Pawlowski[2]

[1] *Laboratory of Cell Regulation, Wadsworth Center, New York State Department of Health, Albany, NY, USA;*
[2] *Department of Zoology and Animal Biology, University of Geneva, Switzerland*

Introduction

The Foraminifera are numerically abundant, widely distributed, and highly diverse protists, comprising approximately 10 000 modern and 40 000 fossil species (Haynes 1981; Sen Gupta 1999; Lee *et al.* 2000). In most species, the protoplasmic mass, or sarcode, is covered by a shell (test). Traditionally, test morphology has served as the main criterion for the classification of Foraminifera (Loeblich and Tappan 1964, 1987; Sen Gupta 1990). Four broad types of test construction are recognized in modern species (Fig. 5.1a–d). In "naked" reticulate species (athalamids), such as *Reticulomyxa* (Pawlowski *et al.* 1999), the cell body is housed in a thickened glycocalyx. Soft-walled, thecate foraminiferans (Order Allogromiida) secrete a proteinaceous cover (Schwab and Plapp 1983). Agglutinated species (Orders Astrorhizida, Trochamminida, Textulariida, Lituolida) assemble a test of particulate matter bound with secreted adhesive. Calcareous foraminiferans, which are morphologically the most diverse group (extant orders Lagenida, Spirillinida, Buliminida, Globigerinida, Robertinida, Miliolida, and Rotalida, as well as the extinct Fusulinida and Involuntinida), produce biomineralized tests, usually composed of calcite. Test shape ranges from single-chambered spheres or tubes (unilocular tests), to geometrically intricate patterns of multiple chambers (multilocular tests). The progressive complexity in test composition and shape has long been thought to mirror the evolution of the group, with the simpler unilocular forms representing relict stem lineages.

Recent structural and behavioral studies (reviewed in Bowser and Travis 2002), together with molecular phylogenetic analyses (e.g. Pawlowski *et al.* 2003), have advanced the concept that the branching and anastomosing network of granular, filose pseudopodia (reticulopodia; Fig. 5.1e) is the defining structural character of the Foraminifera. Reticulopodia display highly exaggerated microtubule (MT) dynamics, involving spring-like tubulin polymorphs called helical filaments, as well as unusual membrane motility and fusion properties (reviewed in Travis and Bowser 1991; Bowser and Travis 2002). Collectively, these characteristics distinguish foraminiferal reticulopodia from the branching filopodia characteristic of sister taxa, for example *Gromia* (Longet *et al.* 2004). The wide areal extent (often >1 cm^2) and rapid deployment of reticulopodia allow foraminiferans to structure and manipulate their immediate surroundings, and thus to be tremendously responsive to the environment. These properties provide obvious selective advantages that, together with the tailored surfaces and architectural features of the test, are thought to have driven the spectacular diversification of the group.

The extensive fossil record of the Foraminifera, extending to the base of the Cambrian and compiled in over a century of micropaleontologic studies, has been used to chart major evolutionary trends within the group. The foraminiferal stratigraphic record

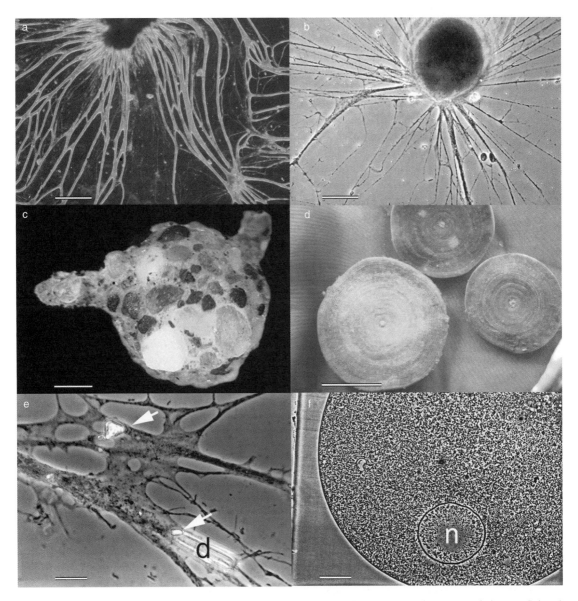

Figure 5.1 Morphological features of Foraminifera. (a) *Reticulomyxa filosa*, a "naked" foram. Thick, cytoplasmic veins, which repeatedly branch and terminate in an extensive pseudopodial network, radiate from a central mass. (b) *Allogromia laticollaris*, a monothalamous "soft walled" species, extends reticulopodia from a single aperture in the organic test. (c) *Astrammina rara*, a larger, monothalamous species, extends pseudopodia from arms on the agglutinated test. (d) Several specimens of *Marginopora vertebralis*, a symbiont-bearing calcareous foram, in the palm of the senior author's hand. (e) Phase-contrast light micrograph of reticulopodia from *Crithionina delacai*, showing surface-bound diatom (d) and mineral particles (arrowheads). These particles are transported bidirectionally, with a net inward bias, along the pseudopodial surface. (f) Phase-contrast light micrograph of a section through *Astrammina triangularis*, a giant monothalamous species from Antarctica, reveals a correspondingly gigantic nucleus (n). Scale bars: (a) 250 mm; (b) 100 mm; (c) 1.0 mm; (d) 5.0 mm; (e) 10 mm; (f) 0.5 mm. Additional images of a wide variety of foraminiferal species are available at the star*sand Web portal (http://www.bowserlab.org/starsand.html).

provides a compelling opportunity to test molecular phylogenetic inferences with fossil data, and vice versa. Because mineralized species are both well preserved as microfossils and extremely common, detailed series of morphological adaptations can often be resolved (e.g. the evolution of *Orbulina universa* and *Globigerinoides trilobus* from a common ancestor in the Miocene; Pearson *et al*. 1997). This level of detail makes it possible to directly test specific hypotheses regarding the evolution of the group. However, phylogenetic relationships at a suprageneric level are not always easily deduced from the morphology of fossilizable parts; they can be obscured by a lack of informative morphological features, an absence of intermediary forms, frequent convergences, and iterative evolution. The results of molecular phylogenetic studies are now forcing us to rethink the classification of several major foraminiferal lineages.

The main purpose of this review is to outline the molecular phylogeny of the Foraminifera, based on analyses of available gene sequence data. We begin with a summary of the molecular genetic features of the group, and conclude with a brief discussion of how the new genetic information is being employed to investigate the molecular ecology of foraminiferans, and speculate on some practical applications that additional genomic information may enable.

Genetics of Foraminifera

Overview: Reproduction, nuclear dimorphism, and endoreplication

Foraminiferal reproductive strategies are complex, with most species examined alternating between distinct sexual and asexual stages (although the details vary considerably among taxa; see Lee *et al*. 1991; Goldstein 1999). Foraminiferan agamonts are produced by the union of two gametes and are multinucleate, although the specific number of nuclei varies widely in different species. It is likely that a large number (probably hundreds) of genome copies is present in the agamont stage (Röttger *et al*. 1989). This stage of the life cycle ends when the body of the parent agamont undergoes

meiosis and multiple fission, to produce a large number of haploid offspring (gamonts). These uninucleate gamonts grow and then in turn produce gametes through a nucleoplasm "cleansing" process known as zerfall, followed by a series of rapid mitotic divisions.

Some foraminiferans have differentiated nuclei in certain stages of the life cycle (reviewed in Grell 1973). Members of the Family Rotaliellidae, and a few other closely related species, contain both generative and somatic nucleus types in the agamont stage. There are usually several of the generative nuclei in the cell, but only one somatic nucleus is observed. The somatic nucleus, which contains more DNA than do the generative nuclei and is the site of DNA synthesis, decays when the generative nuclei enter meiosis at the end of the agamont stage (reviewed in Lee 1991; Goldstein 1999). However, most Foraminifera do not appear to be heterokaryotic. There is no information available on whether the genome, as it exists in the somatic nucleus, is rearranged relative to the generative nucleus in the same cell.

Because foraminiferans are often difficult to culture, usually have translucent or opaque shells, are frequently multinucleate, and may contain endosymbionts, estimates of chromosome number and genome size lag somewhat behind those obtained for other protists. The few reports of chromosome number are summarized in Grell (1973), who describes seven chromosomes in *Myxotheca arenilega*, 18 in *Rotaliella roscoffensis*, and 24 in *Patellina corrugata*. Arnold reported "less than ten" for *Spiroloculina hyalina* (1964) and 10 for *Allogromia laticollaris* (1955), whereas Pawlowski and Lee (1992) illustrated six telocentric chromosomes in *Rotaliella elatiana*.

Endoreplication has also been reported in various foraminiferal taxa. Arnold (1984) reported what appeared to be a series of endoreplication events within the nucleus of *Psammophaga simplora*. The total DNA content of a single *Astrammina rara* nucleus has been measured at 2 ng (Habura, Hayden, and Bowser, unpublished observations), or approximately 4×10^{12} nucleotides. Given that this is approximately 1300 human genome-equivalents, it is highly likely that the giant nuclei in this genus (e.g. Fig. 5.1f) contain hundreds or thousands of genome copies.

Molecular approaches: Sequencing of specimens vs. environmental DNA, and DNA yields

Two main strategies for investigating foraminiferal genomes are currently employed. Conventionally, DNA may be extracted from culturable or easily isolated, morphologically identifiable species. This approach is limited, because only a few foraminiferan species are maintained in culture, and identification of field-collected living specimens is not always straightforward. It is becoming increasingly apparent that many species (especially the allogromiids) present in a given environment may not be recognized as foraminiferans on the basis of morphological criteria (e.g. Habura *et al*. 2004a). Because such cryptic taxa may contain genetic and morphological innovations that are not found in other foraminiferans, sustained efforts must be made to develop new procedures for identifying, isolating, and culturing these species. In such cases, an alternative strategy involves the harvesting of environmental DNA and sorting out the genomic information using, for example, PCR methods (e.g. Bass and Cavalier-Smith 2004) or shotgun sequencing (e.g. Venter *et al*. 2004).

The number of nuclei per organism obviously influences the number of specimens required for adequate yields of DNA; thus, it is important to understand and identify foraminiferal reproductive stages. For example a foraminiferan in the middle phase of gametogenesis can contain several hundred nuclei (Arnold 1955). Unfortunately, the identification of gametogenic specimens is extremely difficult, and only in rare cases is it possible to induce gametogenesis. Fortunately, some giant species, such as *Reticulomyxa filosa* (which can be >1 cm in diameter), routinely harbor thousands of nuclei. Other large species, such as members of the genus *Astrammina*, are typically uninucleate, although the single nucleus may exceed a millimeter in diameter (Fig. 5.1f).

In many cases (especially among benthic species), it is necessary to employ extraction methods optimized for polysaccharide-rich sources. The best yields from certain giant agglutinated species, such as *Astrammina rara*, are obtained using modified sediment-extraction protocols (e.g. Habura *et al*. 2004a).

Characterization of genes sequenced to date

Small subunit and large subunit ribosomal DNA

The ribosomal genes of foraminiferans are highly divergent. The ribosomal small subunit (SSU) genes as a group are the largest reported, ranging from 2300 to over 4000 bp in length. (These genes in other eukaryotes range from 1200 to ~3100 bp, with an average length of 2000 bp; Cole *et al*. 2003.) Most of the additional length is due to a series of sequence expansions and insertions (Pawlowski 2000), which are especially prominent in the SSU rDNA but are also found in the large subunit (LSU) rDNA. These inserts lack the hallmarks of Group I or spliceosomal introns, and at least some of them are retained in the mature SSU rRNA (Habura *et al*. 2004b). Most of the insertions are, surprisingly, concentrated in Domains I and III of the SSU rDNA sequence; in most eukaryotes, Domain II (which contains the highly variable eukaryote-specific helices) is variable, while the sequence in Domains I and III is better conserved.

At least one SSU rDNA insert is a genuine synapomorphy for the Foraminifera (Fig. 5.2). In most other organisms, the affected region is a well-conserved ~30-nt sequence that is predicted to fold into two small helices, 40 and 41. The foraminiferal "Region II" insert expands this region at least four-fold, and the insert is predicted to form a single, large helix (Habura *et al*. 2004b; Ertan *et al*. 2004). The insertion point lies between what is normally the end of Helix 41' and the beginning of Helix 39' (Habura *et al*. 2004b). No insertions have previously been reported at this site, in any organism (Cannone *et al*. 2002). The insert contains two regions of conserved sequence; these contain some phylogenetic information, and are useful targets for clade- and species-specific molecular probes.

The function of the SSU inserts is currently unknown, but structure predictions indicate that they are found primarily on the outer surface of the assembled ribosome. The inserts are a derived

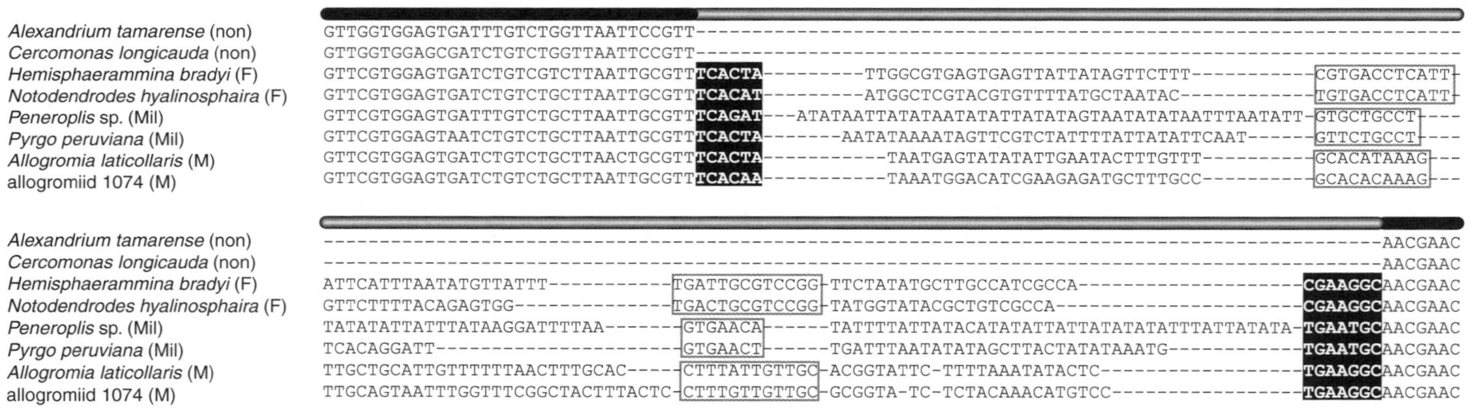
Alexandrium tamarense (non)
Cercomonas longicauda (non)
Hemisphaerammina bradyi (F)
Notodendrodes hyalinosphaira (F)
Peneroplis sp. (Mil)
Pyrgo peruviana (Mil)
Allogromia laticollaris (M)
allogromiid 1074 (M)
GTTGGTGGAGTGATTTGTCTGGTTAATTCCGTT---
GTTGGTGGAGCGATCTGTCTGGTTAATTCCGTT---
GTTCGTGGAGTGATCTGTCGTCTTAATTGCGTTTCACTA---TTGGCGTGAGTGAGTTATTATAGTTCTTT---CGTGACCTCATT-
GTTCGTGGAGTGATCTGTCTGCTTAATTGCGTTTCACAT---ATGGCTCGTACGTGTTTTATGCTAATAC---TGTGACCTCATT-
GTTCGTGGAGTGATTTGTCTGCTTAATTGCGTTTCAGAT---ATATAATTATATAATATATTATATAGTAATATATAATTTAATATT-GTGCTGCCT---
GTTCGTGGAGTAATCTGTCTGCTTAATTGCGTTTCACTA---AATATAAAATAGTTCGTCTATTTTATTATATTCAAT---GTTCTGCCT---
GTTCGTGGAGTGATCTGTCTGCTTAACTGCGTTTCACTA---TAATGAGTATATATTGAATACTTTGTTT---GCACATAAAG---
GTTCGTGGAGTGATCTGTCTGCTTAATTGCGTTTCACAA---TAAATGGACATCGAAGAGATGCTTTGCC---GCACACAAAG---
Alexandrium tamarense (non)
Cercomonas longicauda (non)
Hemisphaerammina bradyi (F)
Notodendrodes hyalinosphaira (F)
Peneroplis sp. (Mil)
Pyrgo peruviana (Mil)
Allogromia laticollaris (M)
allogromiid 1074 (M)
---AACGAAC
---AACGAAC
ATTCATTTAATATGTTATTT---TGATTGCGTCCGG-TTCTATATGCTTGCCATCGCCA---CGAAGGCAACGAAC
GTTCTTTTACAGAGTGG---TGACTGCGTCCGG-TATGGTATACGCTGTCGCCA---CGAAGGCAACGAAC
TATATATTATTTATAAGGATTTTAA---GTGAACA---TATTTTATTATACATATATTATTATATATATTTATTATATA-TGAATGCAACGAAC
TCACAGGATT---GTGAACT---TGATTTAATATATAGCTTACTATATAAATG---TGAATGCAACGAAC
TTGCTGCATTGTTTTTTAACTTTGCAC---CTTTATTGTTGC-ACGGTATTC-TTTTAAATATACTC---TGAAGGCAACGAAC
TTGCAGTAATTTGGTTTCGGCTACTTTACTC-CTTTGTTGTTGC-GCGGTA-TC-TCTACAAACATGTCC---TGAAGGCAACGAAC

Figure 5.2 Group sequence specificity in foraminiferal SSU rDNA. Forams possess a unique insert in Domain III of the SSU rDNA. In this alignment, which is bounded by the two halves of eukaryotic consensus helix 39, non-foram genes (non) contain only the eukaryotic consensus sequence (black bar); all forams contain a long insert (gray bar) predicted to form a single, novel helix. Sequence at the beginning and end of the insert, which forms the base of the helix, is conserved in all forams (black boxes). Individual groups of forams also conserve a region near the distal end of the helix; shown are examples of group-specific conserved regions for the Clade F allogromiids (F), the miliolids (Mil), and the Clade M allogromiids (M). These areas are highly useful targets for the design of group-specific PCR primers and probes.

trait not shared by any of the proposed sister taxa of the Foraminifera (Habura *et al.* 2004b). This fact suggests a period of accelerated evolution at the base of the foraminiferal lineage (see also Pawlowski *et al.* 1997; Pawlowski and Berney 2003), perhaps indicative of selective pressures on the structure of the ribosome early in the evolution of the group.

Protein-coding genes

RNA polymerase. A 5′ fragment of the gene coding for RNA polymerase II (RPB1) was amplified and sequenced for 10 foraminiferan species (five rotaliids, three allogromiids, one miliolid, and one athalamid), as well as two cercozoans and *Gromia oviformis*, a testate protist with branching filose pseudopodia (Longet *et al.* 2003). The lengths of sequences ranged from 1385 to 1554 bp. Most of the size differences observed among species were confined to a highly variable region located near the 5′ end of the molecule. No introns were identified, with the possible exception of an insertion, flanked by canonical GT–AG splicing sites that may represent a putative intron, in the rotaliid *Bulimina marginata*. Important variations in GC content were observed in different foraminiferal groups; GC content ranged from 29.8 to 30.5 % in unilocular species, and from 39.2 to 50.0% in the Rotaliida, and it was 29.4% and 51.3% in the Miliolida and *Reticulomyxa filosa*, respectively. Analysis of these sequence data robustly confirmed the close relationship between the Foraminifera, *Gromia* spp., and the Cercozoa (Longet *et al.* 2004).

Ubiquitin. Ubiquitin is a highly conserved, short (76 amino acid) protein, which plays a major role in numerous cellular processes such as apoptosis, signal transduction, endocytosis, and cell cycle regulation. Most ubiquitin genes are found as polymers whose products are post-translationally processed to ubiquitin monomers. The polyubiquitin genes in forams are peculiar in having an amino acid insertion at the junction between monomers (Archibald *et al.* 2003). In the three examined foraminiferan species, the inserted amino acid was either alanine (*Reticulomyxa*, *Haynesina*) or threonine (*Bathysiphon*). It has been proposed that the extra amino acid is removed

during polyubiquitin processing and therefore it is not under strong evolutionary constraints (Archibald *et al.* 2003).

Foraminiferans share this ubiquitin feature with the Cercozoa, which suggests a common origin for the two groups (Archibald *et al.* 2003). Cercozoa possess either one or two amino acid insertions. An analysis of polyubiquitin genes for a broad taxonomic sample of cercozoans revealed that early-branching genera, such as *Lotharella* and *Bigelowiella*, possess a single amino acid insertion (usually A, T, or S), whereas more derived genera, such as *Cercomonas*, *Gymnophrys*, and *Euglypha*, are characterized by a two amino acid insertion (SG, SA, or AG). Interestingly, the insertion was not found in polycystinean and acantharean radiolarians, which form a sister group to the Cercozoa and the Foraminifera in SSU rDNA trees (Bass *et al.* 2005).

Actin. The actin gene is highly conserved and ubiquitously expressed in eukaryotic cells. The first phylogenetic study of foraminiferal actins revealed the presence of two paralogs, ACT1 and ACT2, which branch together in the eukaryotic tree (Pawlowski *et al.* 1999a, b). Further studies showed the presence of a group of actin-deviating proteins (ADPs) identified within these paralogs (Flakowski *et al.* 2005). The ADPs are neither ARPs (actin-related proteins) nor NAPs (novel actin-like proteins), and they appear in actin phylogenies as long branch sequences that do not cluster with other actins from the same organisms. They also contain indels of varying lengths and lack some canonical actin motifs. They were probably produced during the duplication of the actin gene, followed by rapid postduplication divergence of one of the copies.

Tubulin. The earliest report of tubulin sequences from a foraminiferan, namely *Reticulomyxa filosa*, identified two isoforms each of α- and β-tubulin (Linder *et al.* 1997). While the two foraminiferal α-tubulin isoforms resembled conventional α-tubulins, the β-tubulins were more divergent, with one isoform ("Type 2") being especially heavily substituted. A subsequent study (Habura *et al.* 2005) demonstrated that the Type 2 isoform is the form typically found in Foraminifera, and that foraminiferal β-tubulins (which possess the highest degree of divergence of any β-tubulin gene sequenced to date) represent a novel form of the protein. The observed substitutions appear to have implications for microtubule assembly dynamics, as they are concentrated in areas thought to play a role in lateral contacts between adjacent dimers, and in taxol binding. For example, the M-loop is heavily substituted, and the H1–B2 loop is lengthened by 4 to 11 amino acids relative to other eukaryotes. These changes may serve to strengthen lateral interactions in a manner similar to that induced by taxol binding, allowing the formation of the tubulin-containing helical filaments that have been observed in all foraminiferans examined by electron microscopy (reviewed in Bowser and Travis 2002).

Introns in foraminiferal genes

Intron distribution varies widely among foraminiferal genes. Introns are quite rare in foraminiferal rDNA; they have been identified in the LSU gene of *Rotaliella elatiana* (I. Bolivar, personal communication) and the SSU rDNA of *Cornuspira antarctica* (Habura and Bowser, unpublished), but the vast majority of sequences for these genes lack them. On the other hand, introns are found in a significant percentage of identified sequences of foraminiferal actin and tubulin genes.

In general, foraminiferal introns are small (46–303 nucleotides), dispersed along the whole sequence, and possess the classical splice sites GT–AG of U2-dependent spliceosomal introns. The consensus 5′–3′ splice-site sequence was identified as 5′ GTWW–WYAG 3′ for actins, and 5′ GTWW–WWAG 3′ for tubulins. The phase distribution of extant intron (intron position relative to the codon) of foraminiferal genes (1:1:1) (phases 0, 1, and 2, respectively) is not as biased as is generally observed in eukaryotic introns (5:3:2); instead, it is closer to the approximately uniform phase distribution observed in fungi (Qui *et al.* 2004). Foraminiferal introns exhibit different GC contents in different genes. Actin introns are 72% AT on average, while the exons of actin genes have an average content of 63% AT. Tubulin introns are more AT-rich (85% vs. 64% for exons).

Analysis of 103 actin sequences from 27 foraminiferal species revealed 24 positions at which introns were inserted (Flakowski *et al.* 2005). Among these positions, 20 are specific to the Foraminifera, while four are shared, within the 64 distinct positions previously reported in other eukaryotes (Bhattacharya and Weber 1997; Qui *et al.* 2004). Among 22 introns identified in foraminiferal "conventional" actins, three introns show a clear phylogenetic signal and are lineage-specific: two of these are present in ACT1 and ACT2 of the Rotaliida, and one is characteristic for ACT2 in the Allogromiidae. These two lineages, for which there is a relatively good taxonomic sampling, also contain 10 species-specific introns. Some taxonomic groups, such as the Soritinae, for which four species were examined, appear to lack introns in one of the paralogs (ACT2). Interestingly, among seven species for which two paralogs were sequenced, two species (*Allogromia* sp. A and *Reticulomyxa filosa*) possessed the same introns (C + U and H, respectively) in both paralogs. However, since the data for these two species are derived from long-established laboratory cultures, it is possible that the similarity of these introns is due to homogenization and gene conversion, facilitated by the absence of environmental constraints and continuous clonal reproduction.

In contrast, foraminiferal tubulin introns do not show any particular phylogenetic signal; indeed, introns present in different isoforms of the gene in a single species may vary substantially in sequence.

Molecular phylogeny of Foraminifera

Many of the most important aspects of foraminiferal evolution are likely to be invisible in the fossil record. The physiological innovations that define the Foraminifera as a group, such as the specialized reticulopodia, must have coincided with their inferred Neoproterozoic origins. These structures are essentially unfossilizable, and the morphologically elaborate hard coverings that attest to the physiological dexterity of reticulopodia are not found in the fossil record until the Cambrian. We are therefore forced to rely primarily on comparative ultrastructural and genomic data from modern species in order to infer the evolutionary history of such "soft" parts.

By contrast, hypotheses regarding the evolution of test architecture can be probed using both molecular and morphological data.

Molecular phylogenetic studies may potentially answer many questions regarding foraminiferal evolution, given sufficient availability of molecular data (both in terms of taxon sampling and the number of sequenced genes). Although the available sequence data discussed above are still relatively limited, five general evolutionary trends, in particular concerning the early evolution and the origin of multilocularity, can be deduced for the group.

1. Multiple origins and reversions of agglutination in Foraminifera: Contrary to the traditional morphology-based view, molecular analyses (Fig. 5.3) show that the formation of agglutinated tests arose several times during the evolution of early Foraminifera (Pawlowski *et al.* 2003). Many unilocular lineages contain both thecate and agglutinated species, and there is no compelling evidence for a progressive transformation from one type to the other. Naked genera (*Reticulomyxa*, some species of *Toxisarcon*) also branch in several independent clades within the radiation of the unilocular foraminiferans, and these probably lost their tests secondarily (e.g. as an adaptation to a freshwater environment, in the case of *Reticulomyxa*).

In all phylogenetic trees, the unilocular forms appear as a paraphyletic group from which the multilocular taxa emerged (Pawlowski 2000; Pawlowski *et al.* 2003; Flakowski *et al.* 2005; Habura *et al.* 2005). The radiation of unilocular species comprises a large number of genetically distinctive phylotypes, which can be grouped in higher-level lineages. Thirteen such lineages were distinguished in the first SSU-based phylogeny of allogromiid Foraminifera (Pawlowski *et al.* 2002b). This number is steadily increasing, however, as new sequences are obtained from new isolates and environmental samples (Habura *et al.* 2004a; Pawlowski *et al.* 2005).

It has long been asserted that once calcareous test formation appeared within a lineage, test wall structure did not revert to the supposedly more "primitive" agglutinated pattern. However, a series of recent studies (Fahrni and Pawlowski 1995; Fahrni *et al.* 1997; Flakowski *et al.* 2005; Habura

et al. 2005) has demonstrated that at least one species assigned to the Textulariida based on its agglutinated wall structure (*Miliammina fusca*) is, in fact, descended from calcareous ancestors.

2. Multiple origins of multichambered Foraminifera: Multilocular foraminiferans are commonly divided into a dozen orders (Loeblich and Tappan 1987; 1989), including groups possessing agglutinated walls (Lituolida, Trochamminida, Textulariida) and those with calcareous walls (Miliolida, Spirillinida, Fusulinida, Involutinida, Carterinida, Lagenida, Rotaliida, Robertinida, Globigerinida). These orders are often grouped in two clades, with the miliolids, spirillinids, fusilinids, and involutinids in one clade, and the remaining taxa in the other. These two supergroups are suggested to have evolved separately from within the allogromiids (Tappan and Loeblich 1988), a view supported by biological studies of live foraminiferans (e.g. Arnold 1978b). Molecular data (SSU rDNA sequences) exist for all orders except the extinct Fusilinida and Involutinida, and the modern Carterinida are represented by a single deep-sea genus (*Carterina*). Phylogenetic analyses show that the multichambered test originated at least twice in foraminiferal evolution: once in the lineage leading to the Spiriliniida and Miliolida, and the second time in the lineage leading to the Rotaliida, Textulariida, and other multilocular taxa (Pawlowski *et al.* 2003). The monolamellar Lagenida may have also originated independently, but this point needs to be confirmed by further molecular studies of the group. Molecular phylogenetic analyses also reveal that the calcareous test evolved at least twice, a fact consistent with the distinctive differences in test wall architecture between the two groups.

3. The paraphyly of "Textulariida": Traditionally, all unilocular and multilocular agglutinated genera were grouped together in the Order Textulariida (Loeblich and Tappan 1987). More recently, this group was divided into one unilocular order (Astrorhizida) and three multilocular orders (Lituolida, Trochamminida, and Textulariida), based on the structure of the test and the type of coiling (Loeblich and Tappan 1989; Sen Gupta 1999). This division was proposed so as to reflect the biochemical and structural differences in the formation

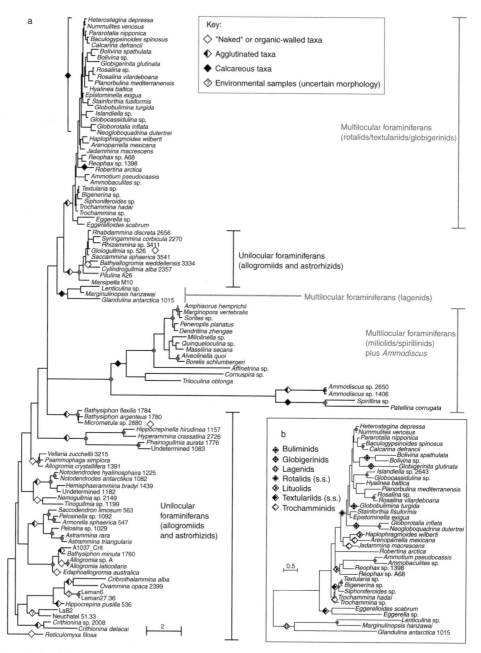

Figure 5.3 Molecular phylogeny of Foraminifera (forarms), based on partial ribosomal SSU gene sequence data. (a) Many of those morpho-logical features that have traditionally been used for classification are distributed in surprising ways on the molecular tree. The basal taxa appear as a "bush" of several lineages of unilocular forams, some of which have agglutinated tests. Multilocularity and calcareous wall formation have clearly arisen multiple times in the forams, with different lineages arising from within the allogromiid bush. (b) Molecular phylogeny of one lineage of multilocular taxa (the lagenids, the textulariids *sensu lato*, and the rotalids *sensu lato*; see text for explanation). Traditional morphological distinctions, such as those among the buliminids and rotalids *sensu stricto*, or between the agglutinated lituolids, textulariids *sensu stricto*, and trochamminids, are not supported. Planktonic taxa (the globigerinids) are shown to have emerged from within the rotalids.

of the agglutinated test, in particular the method of cementing particles: a proteinaceous (Lituolida) or mineralized (Trochamminida) matrix, or a calcitic cement (Tetxulariida). The descriptive term "textulariid" is still in common use for members of all three orders.

The available molecular data do not confirm the division of textulariids *senso lato* into groups based strictly on test composition and architecture. As shown in Fig. 5.3, the representatives of the three orders intermingle in phylogenetic trees (e.g. *Trochammina* sp. and *T. hadai* branch close to *Textularia* sp., but separately from other trochamminids). The textulariids form a series of independent lineages, and they most probably represent a paraphyletic group from which the calcareous Rotaliida and Robertinida evolved. However, the taxon sampling for molecular data in the textulariids is relatively limited, and the resolution of relationships between various agglutinated taxa is not very strong. The paraphyly of the Textulariida needs to be confirmed by the analysis of complete SSU rDNA sequences and corroborated by the analysis of protein-coding gene sequence data.

4. The common origin of Rotaliida, Buliminida, and Globigerinida: The Rotaliida *sensu lato* are the most diverse group of modern benthic Foraminifera. Rotaliids construct a test with a perforate, low-magnesium calcite, bilamellar wall. Some authors (Loeblich and Tappan 1987) consider them to be a single, morphologically diverse group, while others separate rotaliids into two distinct groups: (i) the Buliminida, which usually exhibit triserial, biserial or uniserial chamber arrangements and apertures with an internal toothplate, and (ii) the Rotaliida *sensu stricto*, with mainly trochospiral or planispiral chamber arrangements (Haynes 1981; Sen Gupta 1999). In the past, the Rotaliida were also held to comprise all of the planktonic foraminiferans grouped together in the superfamily Globigerinacea (Loeblich and Tappan 1964). Later, however, the same authors separated the planktonic species from their benthic relatives, and considered the former to be a separate Order, the Globigerinida (Loeblich and Tappan 1987).

The first molecular phylogenies of rotaliids, based on phylogenetic analyses of partial SSU rDNA sequences, supported the separation of the Buliminida and Rotaliida *sensu stricto* (Ertan *et al.* 2004). However, these results were not confirmed by a more taxon-rich analysis of *Uvigerina* and closely related species (Schweizer *et al.* 2005), nor by our present analysis (see the positions of two genera of Buliminida – *Bolivina* and *Globobulimina* – in Fig. 5.3). The representatives of the two groups intermingle in the SSU trees, and there is no evidence to separate the Buliminida and Rotaliida, at least not on the basis of available data.

The phylogenetic position of the Globigerinida was more difficult to establish because their ribosomal genes evolved 50 to 100 times faster than those of benthic foraminiferans (Pawlowski *et al.* 1997). This acceleration may have been driven by the significant changes in reproductive strategies and life habits exhibited by these species relative to their benthic counterparts. In earlier molecular studies, the Globigerinida branched in two or three independent clusters dispersed among rotaliids and textulariids (Darling *et al.* 1997; de Vargas *et al.* 1997). The polyphyly of planktonic foraminiferans was never specifically tested, for example through the use of phylogenetic methods less sensitive to high heterogeneity of substitution rates. However, it is now generally accepted that the Globigerinida emerged from within the Rotaliida *sensu lato* (Fig. 5.3). An important recent confirmation appeared in analyses of actin sequences from the planktonic species *Globigerinella siphonifera*; these analyses showed it to branch within the Rotaliida and to possess an intron specific for that group (Flakowski *et al.* 2005).

5. Monophyly of the Miliolida and Spirillinida: The origin of the Miliolida has been a longstanding question in foraminiferal phylogenetics (e.g. Arnold 1978a, 1979). In the first published SSU-based molecular phylogenies, the Order Miliolida was placed at the base of foraminiferal trees (Pawlowski *et al.* 1997; Pawlowski 2000). However, later studies showed this basal position to be an artifact caused probably by a strong bias in base composition of miliolid SSU rDNA (Pawlowski and Holzmann 2002; Pawlowski *et al.* 2003). Indeed, all analyses of foraminiferal rDNA sequences, which used probabilistic methods that are less susceptible to base composition bias

and rate heterogeneity, show the miliolids branching among unilocular lineages (Fig. 5.3). This position is confirmed by actin (Flakowski *et al.* 2005) and tubulin (Habura *et al.* 2005) sequence analyses.

An important result of molecular analyses was the finding that the Miliolida branch together with the calcareous Spirillinida and the agglutinated genus *Ammodiscus* (Pawlowski *et al.* 2003). The close relationship between the Spirillinida and *Ammodiscus* has been suggested based on the similarly coiled tubular test (Mikhalevich 1998), although conventional taxonomy places the ammodiscids with astrorhizid taxa (Tappan and Loeblich 1987). *Spirillina* and *Ammodiscus* have been considered as possible ancestors of the Miliolida, by Galloway (1933) and Cushman (1948), respectively. The current molecular evidence is consistent with one proposed taxonomy in which the Spirillinida and Miliolida derive from extinct Fusulinida, together with the Lagenida and Involutinida (Tappan and Loeblich 1988). As several important groups in this proposed lineage (namely, the fusulinids and involuntinids) are extinct, the full resolution of this relationship using molecular data will depend on very careful taxon sampling, guided by the fossil record.

Although the relationships between the miliolids and the clade Spirillinida + Ammodiscus are relatively well supported (Fig. 5.3), the SSU genes of these organisms have evolved very rapidly, and analysis of protein sequences will be necessary to ensure that this grouping is not due to a long branch attraction artifact.

Molecular-clock analyses of foraminiferal evolution and its paleoecologic significance

The continuous and abundant fossil record of the Foraminifera offers a powerful tool with which to calibrate phylogenetic trees and to study rates of evolution in different lineages. Through the use of the multiple and accurate stratigraphic data to calibrate the molecular phylogenies of Foraminifera, the absolute rates of nucleotide substitution in foraminiferal ribosomal genes have been estimated. Such studies have revealed an extraordinary heterogeneity of substitution rates, both within

and between foraminiferal taxonomic groups (Pawlowski *et al.* 1997). For example, 50-fold differences in substitution rates were observed between certain planktonic and benthic species. Within the planktonic Foraminifera, substitution rates were relatively constant among spinose globigerinids, but they varied seven-fold among non-spinose globorotalids (de Vargas and Pawlowski 1998). The higher rates observed for some species are thought to result from independent lineage accelerations (de Vargas *et al.* 2001). The stem-branch leading to the Foraminifera in the eukaryotic tree also exhibits a 30-fold acceleration of rates relative to other organisms (Berney and Pawlowski 2003).

Because of the heterogeneity of substitution rates, the use of ribosomal genes as chronometers of foraminiferal phylogeny seems relatively limited. Molecular clock analyses were used to show the role of Quaternary climate changes in diversification of the polar planktonic foraminiferan *Neogloboquadrina pachyderma* (Darling *et al.* 2004). Local molecular clocks were used to calibrate the molecular phylogeny of early-evolving Foraminifera, and to date their origins to the Neoproterozoic (Pawlowski *et al.* 2003). Newer methods for dating molecular phylogenies, based on relaxed molecular clocks, allow us to compensate for some of the bias introduced by heterogeneity of evolutionary rates. More general use of such methods will probably improve our ability to infer divergence times from ribosomal gene sequences; however, protein-coding genes certainly offer corroborative (and perhaps more reliable) data for timing foraminiferal phylogeny.

Some basal foraminiferans use reticulopodia to capture and rend small metazoan prey (Bowser *et al.* 1992), while others feed by rending and consuming biofilms (Bernhard and Bowser 1992). The emergence of early foraminiferans with true reticulopodia in the Neoproterozoic (as suggested by molecular clock data) would have caused significant predation pressure on relatively non-motile benthic organisms. Indeed, the emergence of mobile predators, as exemplified here by modern basal foraminiferans, during the late Proterozoic has been proposed as one of the important factors influencing the emergence of larger and more complex life forms in the Cambrian (Stanley 1973).

Molecular ecology of Foraminifera

Despite the limited genomic information available for the group, molecular approaches are now commonly used to address fundamental aspects of foraminiferal ecology such as diversity estimates and gene flow between populations. This refinement of foraminiferal taxon sampling is illustrated by work at Explorers Cove, an intensively studied site in McMurdo Sound, Antarctica. Initial surveys (Ward *et al.* 1987; Bernhard 1987), using micropaleontological approaches, revealed a low-diversity foraminiferal assemblage dominated by small calcareous taxa; no soft-bodied allogromiids were reported. Subsequent work, using refined "wet-picking" methods, identified numerous allogromiid morphotypes (Gooday *et al.* 1996). Many of these newly identified foraminiferans comprised a bewildering and morphologically unclassifiable array of "quartz spheres" and "mudballs". The introduction of molecular methods subsequently uncovered a strikingly rich diversity among these problematic unilocular morphotypes (Pawlowski *et al.* 2002a, b; Bowser *et al.* 2002), and environmental DNA screens revealed an even larger repertoire of phylotypes that branch within the basal radiation (Habura *et al.* 2004a). From these newer data we estimate that at least 75% of the allogromiid diversity at this site remains uncharacterized. Our ongoing studies (e.g. Habura and Bowser 2005) show that this severe underestimation of allogromiid diversity is likely to be a feature of near-shore foraminiferal assemblages in lower latitudes as well. New foraminiferal morpho- and phylotypes continue to be found in unexpected places, ranging from terrestrial (Meisterfeld *et al.* 2001), to freshwater (Holzmann *et al.* 2003), and to anoxic environments (Bernhard *et al.* 2004).

Investigations are currently underway to determine the degree of gene flow between foraminiferal populations and the physical parameters that control such flow. In planktonic species, which are distributed in latitudinal "provinces" bracketing the equator (Arnold and Parker 1999), there is evidence that subpolar North and South populations exchange genetic information, despite being separated by a wide zone of inhospitable environmental conditions (Darling *et al.* 2000). However, at least one species (*Neogloboquadrina pachyderma*) has allopatrically differentiated into two isolated polar populations (Darling et al. 2004), suggesting that there is at least some barrier to gene flow even in these pelagic forms. Preliminary comparisons of Arctic and Antarctic species indicate that certain benthic populations may also be genetically differentiated (Pawlowski et al. 2004), but this question requires further study.

Ongoing genomic studies and expected outcomes

Foraminiferal expressed sequence tag (EST) projects

An EST project for the athalamid *Reticulomyxa filosa* is in progress. Thus far, 700 clones have been sequenced, producing 579 good-quality sequences (about 600 bp long, on average) after base calling, vector masking, and poly-A trimming. One hundred and fifty-four sequences were grouped into 32 clusters, with the remaining 425 sequences deemed "unique." To distinguish between protein coding sequences and untranslated regions, we identified 557 open reading frames, among which 466 are longer than 100 amino acids. A BLASTX search against the Uniprot database (Altschul *et al.* 1997) revealed 144 different proteins (E-value threshold, 1e-20). A preliminary analysis of these hits shows that actin is the most common protein (9%), followed by polyubiquitin (3%), tubulin (2%), and glyceraldehyde-3-phosphate dehydrogenase (1%). Thus, 85% of the total proteins comprised those present at the lowest abundance (F. Burki, personal communication).

The *Reticulomyxa* EST project is also an important step in understanding the origins and evolution of Foraminifera. A comparison of foraminiferal and *Gromia* ESTs may help to identify the key structural proteins responsible for the development of foraminiferal reticulopodia from filose pseudopodia. A comparison of ESTs within the Foraminifera may provide new insights into the genetic changes that accompanied the transformation of single-chambered tests into multichambered ones, and the development of the calcareous wall. Finally, access to a multigenic database will also make possible the identification of those genes that have evolved in a relatively constant manner, and thus can be used for calibrating foraminiferal phylogeny more accurately than can the ribosomal genes.

Another objective of this EST project, in providing a multigenic database, is to help in better resolving the position of the Foraminifera within the phylogeny of eukaryotes. The available molecular database on Rhizaria, the supergroup to which foraminiferans are currently assigned (Nikolaev *et al.* 2004), is very sparse. The *Reticulomyxa* cDNA library is the first foraminiferal cDNA library sequenced and, together with the cDNA library of *Chlorarachnion*, it constitutes the only available EST data for the Rhizaria. In the near future, we and others are planning to obtain EST sequences for another cercozoan (*Lecythium*) and *Gromia oviformis*, as well as other foraminiferans (probably *Astrammina* and *Ammonia*). These projects should provide enough data to establish the position of the Rhizaria within the eukaryotic tree. They should also help to resolve relationships within the Rhizaria, and they should in particular provide an answer to the question of whether the Foraminifera derived from Cercozoa, or whether the two are sister taxa. One caveat is that, in order to establish a solid phylogeny of the Rhizaria, an EST project for at least one representative each of the Polycystinea and the Acantharea will be necessary.

Biomedical implications and future prospects

The >500 million-year-old evolutionary history of the Foraminifera represents a compelling "natural experiment" in molecular innovation, and several of the observed alterations to foraminiferal structural genes may have potential for biotechnological or medical applications. For example, the Region II insert in the SSU rRNA might be tailored to tether functional ribosomes to a solid substrate, enabling polypeptide nanoproduction. The unique properties of foraminiferal MTs offer another possibility. As the major building blocks of mitotic spindles,

MT proteins are prime targets for anticancer drug research. The extensive foraminiferal β-tubulin modifications, some of which appear to mimic the effects of taxol binding in mammalian MTs, offer insight into the effects of altered lateral contact between tubulin subunits in MT assembly and dynamics. At a more basic level, the complex morphology of foraminiferan tests provides an avenue for the investigation of cellular morphogenetic processes. Foraminifera undergo an elaborate series of steps (the details of which vary by taxon) to construct a new test or test chamber. Preliminary investigations (Habura, unpublished observations) have identified differences in RNA expression profile between resting cells and those that are constructing tests. It seems likely that, as EST and genome sequencing projects on the group proceed, genes that control test construction and pattern formation will be uncovered. Comparison of these genes with their metazoan counterparts – if the counterparts exist – will broadly expand our understanding of life's developmental programs.

Acknowledgments

Portions of our work reviewed here was supported by grants from the Swiss National Science Foundation (3100A0–100415 to J.P.) and the US National Science Foundation (DEB-0445181, OPP-0216043, and ANT-0440769 to S.S.B.).

References

Altschul, S. F., Madden, T. L., Schaffer, A. A., Zhang, J., Zhang, Z., Miller, W., and Lipman, D. J. (1997). Gapped BLAST and PSI-BLAST: A new generation of protein database search programs. *Nucleic Acids Res.*, 25, 3389–3402.

Archibald, J. M., Longet D., Pawlowski J., and Keeling, P. J. (2003). A novel polyubiquitin structure in Cercozoa and Foraminifera: evidence for a new eukaryotic supergroup. *Mol. Biol. Evol.*, 20, 62–66.

Arnold, A. J. and Parker, W. C. (1999). Biogeography of planktonic foraminifera. In B.K. Sen Gupta (ed). *Modern Foraminifera*. Kluwer Academic Publishers, Dordrecht, The Netherlands, pp. 103–122.

Arnold, Z. M. (1955). An unusual feature of miliolid reproduction. *Contributions to the Cushman Foundation for Foraminiferal Research*, 6, 94–96.

Arnold, Z. M. (1964). Biological observations on the foraminifer *Spiroloculina hyalina* Schulze. *University of California Publications in Zoology*, **72**, 1–93.

Arnold, Z.M. (1978a). An allogromiid ancestor of the miliolidean foraminifera. *J. Foramin. Res.*, **8**, 83–96.

Arnold, Z.M. (1978b). Biological evidence for the origin of polythalamy in foraminifera. *J. Foramin. Res.*, **8**, 147–166.

Arnold, Z. M. (1979). Biological clues to the origin of miliodean foraminifera. *J. Foramin. Res.*, **9**, 302–321.

Arnold, Z. M. (1984). The gamontic karyology of the saccamminid foraminifer *Psammophaga simplora* Arnold. *J. Foramin. Res.*, **14**, 171–186.

Bass, D. and Cavalier-Smith, T. (2004). Phylum-specific environmental DNA analysis reveals remarkably high global biodiversity of Cercozoa (Protozoa). *Int. J. System. Evol. Microbiol.*, **54**, 2393–2404.

Bass, D., Moreira, D., Lopez-Garcia, P., Polet, S., Chao, E. E., Herden S., Pawlowski, J., and Cavalier-Smith, T. (2005). Polyubiquitin insertions and the phylogeny of Cercozoa and Rhizaria. *Protist* **156**, 149–161.

Berney, C. and Pawlowski, J. (2003). Revised small subunit rRNA analysis provides further evidence that Foraminifera are related to Cercozoa. *J. Mol. Evol.*, **57** (suppl. 1), 120–127.

Bernhard, J. M. (1987). Foraminiferal biotopes in Explorers Cove, McMurdo Sound, Antarctica. *J. Foramin. Res.*, **17**, 286–297.

Bernhard, J. M. and Bowser, S. S. (1992). Bacterial biofilms as a trophic resource for certain benthic foraminifera. *Mar. Ecol. Prog. Ser.*, **83**, 263–272.

Bernhard, J. M., Habura, A., and Bowser, S. S. (2004). Saccamminid foraminifers from anoxic sediments: Implications for Proterozoic ecosystem dynamics. *J. Eukaryot. Microbiol.*, **52**, 9S.

Bhattacharya, D. and Welser, K. (1997). The actin gene of the glauco cystophyte *Cyanophora paradoxa*: analysis of the coding region and introns and an actin phylogeny of eukaryotes. *Curr. Genet.*, **31**, 439–446.

Bowser, S. S., Alexander, S. P., Stockton, W., and DeLaca, T. E. (1992). Extracellular matrix augments mechanical properties of pseudopodia in the carnivorous foraminiferan *Astrammina rara*: Role in prey capture. *J. Protozool.*, **39**, 724–732.

Bowser, S. S., Bernhard, J. M., Habura, A., and Gooday, A. J. (2002). Structure, taxonomy and ecology of *Astrammina triangularis* (Earland), an allogromiid-like agglutinated foraminifer from Explorers Cove, Antarctica. *J. Foramin. Res.*, **32**, 364–374.

Bowser, S. S. and Travis, J. L. (2002). Reticulopodia: Structural and behavioral basis for the suprageneric placement of granuloreticulosan protists. *J. Foramin. Res.*, **32**, 440–447.

Cannone, J. J., Subramanian, S., Schnare, M. N., Collett, J. R., D Souza, L. M., Du, Y., Feng, B., Lin, N., Madabusi, L. V., Müller, K. M., Pande, N., Shang, Z., Yu, N., and Gutell, R. R. (2002). The Comparative RNA Web (CRW) Site: an online database of comparative sequence and structure information for ribosomal, intron, and other RNAs. *BMC Bioinformatics*, **3**, 2.

Cole, J. R., Chai, B., Marsh, T. L., Farris, R. J., Wang, W., Kulam, S. A., Chandra, S., McGarrell, D. M., Schmidt, T. M., Garrity, G. M., and Tiedje, J. M. (2003). The Ribosomal Database Project (RDP-II): previewing a new autoaligner that allows regular updates and the new prokaryotic taxonomy. *Nucleic Acids Res.*, **31**, 442–443.

Cushman, J. A. (1948). *Foraminifera–Their Classification and Economic Use*, 4th edn. Harvard University Press, Cambridge, MA.

Darling, K. F., Wade, C. M., Kroon D., and Leigh Brown, A. J. (1997). Planktonic foraminiferal molecular evolution and their polyphyletic origins from benthic taxa. *Marine Micropaleontol.*, **30**, 251–266.

Darling, K. F., Wade, C. M., Stewart, I. A., Kroon, D., Dinglek, R., and Brown, A. J. L. (2000). Molecular evidence for genetic mixing of Arctic and Antarctic subpolar populations of planktonic foraminifers. *Nature*, **405**, 43–47.

Darling, K. F., Kucera, M., Prudsey, C. J., and Wade, C. M. (2004). Molecular evidence links cryptic diversification in polar planktonic protists to Quaternary climate dynamics. *Proc. Natl. Acad. Sci. U S A*, **101**, 7657–7662.

de Vargas, C., Zaninetti, L., Hilbreud H., and Pawlowski, J. (1997). Phylogeny and rates of molecular evolution of planktonic forminifera: SSU rDNA sequences compared to the fossil record. *J. Mol. Evol.*, **45**, 285–294.

de Vargas, C. and Pawlowski, J. (1998). Molecular versus taxonomic rates of evolution in planktonic foraminifera. *Mol. Phylogenet. Evol.*, **9**, 463–469.

de Vargas, C., Renaud, S., Hilbrecht, H., and Pawlowski, J. (2001). Pleistocene adaptive radiation in *Globorotalia truncatulinoides*: genetic, morphologic, and environmental evidence. *Paleobiology*, **27**, 104–125.

Ertan, K. T., Hemleben, V., and Hemleben, C. (2004). Molecular evolution of some selected benthic foraminifera as inferred from sequences of the small subunit ribosomal DNA. *Marine Micropaleontol.*, **53**, 367–388.

Fahrni, J. F. and Pawlowski, J. (1995). Identification of actins in foraminfera: Phylogenetic perspectives. *Eur. J. Protistol.*, **31**, 161–166.

Fahrni, J. F., Pawlowski J., Richardson, S., Debenay, J. P., and Zaninetti, L. (1997). Actin suggests *Miliammina fusca* (Brady) is related to porcellanous rather than to agglutinated foraminifera. *Micropaleontology*, **42**, 211–214.

Flakowski, J., Bolivar, I., Fahrni, J., and Pawlowski, J. (2005). Actin phylogeny of Foraminifera. *J. Foramin. Res.*, **35**, 93–102.

Galloway, J. J. (1933). *A Manual of Foraminifera*. Principia Press, Bloomington, Indiana.

Goldstein, S. T. (1999). Foraminifera: A biological overview. In B. K. Sen Gupta (ed). *Modern Foraminifera*. Kluwer Academic Publishers, Dordrecht, The Netherlands, pp. 37–56.

Gooday, A. J., Bernhard, J. M., and Bowser, S. S. (1995). The taxonomy and ecology of *Crithionina delacai* sp. nov., an abundant large foraminifer from Explorers Cove, Antarctica. *J. Foramin. Res.*, **25**, 290–298.

Gooday, A. J., Bowser, S. S., and Bernhard, J. M. (1996). Benthic foraminiferal assemblages in Explorers Cove, Antarctica: A shallow-water site with deep-sea characteristics. *Prog. Oceanography*, **37**, 117–166.

Grell, K. G. (1973). *Protozoology*. Springer-Velag, Berlin.

Habura, A. and Bowser, S. S. (2005). How the other half lives: Cryptic diversity in low-latitude foraminiferal populations. *East Coast Conference on Protozoology, Virginia Commonwealth University School of Medicine*, June 3–5, abstract 36.

Habura, A., Pawlowski, J., Hanes, S. D., and Bowser, S. S. (2004a). Unexpected foraminiferal diversity revealed by small-subunit rDNA analysis of Antarctic sediment. *J. Eukaryot. Microbiol.*, **51**, 173–179.

Habura, A., Rosen, D. R., and Bowser, S. S. (2004b). Predicted secondary structure of the foraminiferal SSU 3′ major domain reveals a molecular synapomorphy for granuloreticulosean protists. *J. Eukaryot. Microbiol.*, **51**, 464–471.

Habura, A., Wegener, L., Travis, J. L., and Bowser, S. S. (2005). Structural and functional implications of an unusual foraminiferal β-tubulin. *Mol. Biol. Evol.*, **22**, 2000–2009.

Haynes, J. R. (1981). *Foraminifera*. Macmillan Publishers, London.

Holzmann, M., Habura, A., Giles, H., Bowser, S. S., and Pawlowski, J. (2003). Freshwater foraminiferans revealed by analysis of environmental DNA samples. *J. Eukaryot. Microbiol.*, **50**, 135–139.

Lee, J. J., Faber, W. W., Anderson, O. R., and Pawlowski, J. (1991). Life-cycles of foraminifera. In O. R. Anderson and J. J. Lee (eds). *Biology of Foraminifera*. Academic Press, London, pp. 285–334.

Lee, J. J., Pawlowski, J., Debeney, J.-P., Whittaker, J., Banner, F., Gooday, A. J., Tendal, O., Haynes, J., and Faber, W. W. (2000). Phylum Granuloreticulosa. In J. J. Lee, G. F. Leedale, and P. Bradbury (eds). *Illustrated Guide to the Protozoa*, vol. 2. Allen Press, Lawrence, Kansas, pp. 872–951.

Linder, S., Schliwa, M., and Kibe-Granderath, E. (1997). Sequence analysis and immuno flourescence study of x- and ρ-tublins in *Retucykibgxa Filosa*: implications of the high degree of βz-tublin divergence. *Cell. Mobil.-Cytoskel.*, **36**, 164–178

Loeblich, A. R. and Tappan, H. (1964). Sarcodina, chiefly "Thecamoebians" and Foraminiferida. In R.C. Moore (ed). *Treatise on Invertebrate Paleontology*, Part C, vol. 1–2. Geological Society of America, Boulder.

Loeblich, A. R. and Tappan, H. (1987; 1988). *Foraminiferal Genera and their Classification*, vol. 1–2. Van Nostrand Reinhold, NY.

Loeblich, A. R. and Tappan, H. (1989). Implication of wall composition and structure in agglutinated foraminifers. *J. Paleontol.*, **63**, 769–777.

Longet, D., Archibald, J. M., Keeling, P. J., and Pawlowski, J. (2003). Foraminifera and Cercozoa share a common origin according to RNA polymerase II phylogenies. *Int. J. System. Evol. Microbiol.*, **53**, 1735–1739.

Longet, D., Burki, F., Flakowski, J., Berney, C., Polet, S., Fahrni, J., and Pawlowski, J. (2004). Multiple evidence for close relations between *Gromia* and Foraminifera. *Acta Protozoologica*, **43**, 303–311.

Meisterfeld, R., Holzmann, M., and Pawlowski, J. (2001). Morphological and molecular characterization of a new terrestrial allogromiid species: *Edaphoallogromia australica* gen. et spec. nov. (Foraminifera) from northern Queensland (Australia). *Protist*, **152**, 185–192.

Mikhalevich, V. I . (1998). The macrosystem of the foraminifera. *Izvestija of the Russian Academy of Sciences, Ser. Biol.*, **2**, 266–271.

Nikolaev, S. I., Berney, C., Fahrni, J. F., Bolivar, I., Polet, S., Mylnikov, A. P., Aleshin, V. V., Petrov, N. B., and Pawlowski, J. (2004). The twilight of Heliozoa and rise of Rhizaria, an emerging supergroup of amoeboid eukaryotes. *Proc. Natl. Acad. Sci. U S A,.* **101**, 8066–8071.

Pawlowski, J. (2000). Introduction to the molecular systematics of foraminifera. *Micropaleontology*, **46** (suppl. 1), 1–12.

Pawlowski, J. and Berney, C. (2003). Episodic evolution of nuclear small subunit ribosomal RNA gene in the stem lineage of Foraminifera. In P.C. Donoghue and M. P.Smith (eds). *Telling the Evolutionary Time: Molecular Clocks and the Fossil Record*. Systematics Association Special Volume No. 66. Taylor and Francis, London, pp. 107–118.

Pawlowski, J., Bolivar, I., Fahrni, J., de Vargas, C., and Bowser, S. S. (1999). Molecular evidence that *Reticulomyxa filosa* is a freshwater naked foraminifer. *J. Eukaryot. Microbiol.*, **46**, 612–617.

Pawlowski., J., Bolivar, I., Fahrni, J. F., de Vargas, C., Gouy, M., and Zaninetti, L. (1997). Extreme differences in rates of molecular evolution of Foraminifera revealed by comparison of ribosomal DNA sequences and the fossil record. *Mol. Biol. Evol.*, **14**, 498–505.

Pawlowski, J., Fahrni, J., Brykczynska, U., Habura, A., and Bowser, S. S. (2002a). Molecular data reveal high taxonomic diversity of allogromid foraminifera in Explorers Cove (McMurdo Sound, Antarctica). *Polar Biology*, **25**, 96–105.

Pawlowski, J., Fahrni, J. F., Guiard, J., Conlan, K., Hardecker, J., Habura, A., and Bowser, S. S. (2005). Allogromiid foraminifera and gromiids from under the Ross Ice Shelf: morphological and molecular diversity. *Polar Biology*, **28**, 514–522.

Pawlowski, J., Gooday, A., Korsun, S., Cedhagen, T., Habura, A., and Bowser, S. S. (2004). Molecular analysis of polar benthic foraminifera. *XXVIII Meeting of the Scientific Committee for Antarctic Research*, Bremen, Germany, July 25–31. Abstract S4/011, p. 165.

Pawlowski, J., Holzmann, M., Berney, C., Fahrni, J., Gooday, A. J., Cedhagen, T., Habura, A., and Bowser, S. S. (2003). Evolution of early Foraminifera. *Proc. Natl. Acad. Sci. U S A*, **100**, 11494–11498.

Pawlowski, J., Holzmann, M., Fahrni, J., Cedhagen, T., and Bowser, S. S. (2002b). Phylogenetic position and diversity of allogromiid foraminifera inferred from rRNA gene sequences. *J. Foramin. Res.*, **32**, 334–343.

Pawlowski, J. and Lee, J. J. (1992). The life cycle of *Rotaliella elatiana* n.sp.: a tiny macroalavorous foraminifer from the Gulf of Elat. *J. Protozool.*, **39**, 131–143.

Pearson, P. N., Shackleton, N. J., and Hall, M A. (1997). Stable isotopic evidence for the sympatric divergence of *Globigerinoides trilobus* and *Orbulina universa* (planktonic foraminifera). *J. Geological Soc.*, **154**, 295–302.

Qui, W.-G., Schisler, N., and Stoltzfus, A. (2004). The evolutionary gain of spliceosomal introns: Sequence and phase preferences. *Mol. Biol. Evol.*, **21**, 1252–1263.

Röttger, R., Schmaljohann, R., and Zacharias, H. (1989). Endoreplication of zygotic nuclei in the larger foraminifer Heterostegina depressa. *Eur. J. Protistol.*, **25**, 60–66.

Schwab, D. and Plapp, R. (1983). Quantitative chemical analysis of the shell of the monothalamous foraminifer *Allogromia laticollaris* Arnold. *J. Foramin. Res.*, **13**, 69–71.

Schweizer, M., Pawlowski, J., Duijnstee, I. A. P., Kouwenhoven, T. J., and van der Zwaan, G. J. (2005). Molecular phylogeny of the foraminiferan genus *Uvigerina* based on ribosomal DNA sequences. *Marine Micropaleontol.*, **57**, 51–67.

Sen Gupta, B. K. (1999). Systematics of modern foraminifera. In B. K. Sen Gupta (ed). *Modern Foraminifera*. Kluwer Academic Publishers, Dordrecht, The Netherlands, pp. 7–36.

Stanley, S. M. (1973). An ecological theory for the sudden origin of multicellular life in the Late Precambrian. *Proc. Natl. Acad. Sci. U S A*, **70**, 1486–1489.

Tappan, H. and Loeblich, A. R. (1988). Foraminiferal evolution, diversification, and extinction. *J. Paleontol.*, **62**, 695–714.

Travis, J. L. and Bowser, S. S. (1991). The motility of Foraminifera. In O. R. Anderson and J. J. Lee (eds). *Biology of Foraminifera*. Academic Press, London, pp. 91–155.

Venter, J. C., Remington, K., Heidelberg, J. F., Halpern, A. L., Rusch, D., Eisen, J. A., Wu, D. Y., Paulsen, I., Nelson, K. E., Nelson, W., Fouts, D. E., Levy, S., Knap, A. H., Lomas, M. W., Nealson, K., White, O., Peterson, J., Hoffman, J., Parsons, R., Baden-Tillson, H., Pfannkoch, C., Rogers, Y. H., and Smith, H. O. (2004). Environmental genome shotgun sequencing of the Sargasso Sea. *Science*, **304**, 66–74.

Ward, B. L., Barrett, P. J., and Vella, P. (1987). Distribution and ecology of benthic foraminifera in McMurdo Sound, Antarctica. *Paleogeogr. Paleocl.*, **58**, 139–153.

CHAPTER 6

Photosynthetic organelles and endosymbiosis

M. S. Sommer, S. B. Gould, O. Kawach, C. Klemme, C. Voß, U.-G. Maier, and S. Zauner

Philipps University of Marburg, Cell Biology, Marburg, Germany

Introduction

Current hypotheses regarding the origin of photosynthetic eukaryotes argue for a single acquisition of plastids in the common ancestor of red and green algae, and the little-known glaucophytes. All remaining photosynthetic eukaryotes, including stramenopiles (heterokonts), dinoflagellates, euglenophytes, chlorarachniophytes, and cryptophytes, acquired the capability to photosynthesize by the engulfment of a phototrophic eukaryote. This chapter summarizes current knowledge about evolution involving these diverse symbiotic events.

There is ongoing discussion about the age of photosynthesis. Whereas the interpretation of early Archaen microfossils and fossilized stromatolithes (Schopf and Packer 1987) has come into question (Brasier *et al.* 2002), a recent report provided footprints of photosynthetic microbes in a 3.4 billion-year-old rock (Tice and Lowe 2004). Because these microbes were bacteria capable of anoxic photosynthesis, cyanobacteria-like cells might not have been the dominant form at this time. However, the dramatic change in the ancient atmosphere to an oxygenated one was induced by photosynthesis that evolved in cyanobacteria. Cyanobacteria changed not only the atmosphere but also strongly influenced the evolution of phototrophic eukaryotes. These originated at least 1.2 billion years ago (Butterfield 2000) and evolved by the cellular amalgamation of two phylogenetically distantly related organisms. In this evolutionary milestone, the so-called primary endosymbiosis, a cyanobacteria-like

cell was engulfed by a eukaryote and, contrary to the normal situation, not metabolized. Instead it was reduced to a DNA-containing organelle, the primary plastid (Margulis 1970).

It is thought that the reduction of an intracellular cyanobacterium into a primary plastid is a step-by-step process, in which the sequence of changes over time is hard to reconstruct. However, at least two processes led to an indissoluble unit of symbiotic interaction: the evolution of new genetic membranes (Cavalier-Smith 2003) and the rearrangement and relocation of the genetic material (Timmis *et al.* 2004). New membranes were created from the membrane systems of the intracellular cyanobacterium, evolving to the double-membrane-bounded primary plastids and the thylakoid membranes (Cavalier-Smith 2003). With respect to rearrangements of the genetic material, it is obvious that in modern plastids only a remnant of the original cyanobacterial genome, the plastid genome, is still resident (Martin *et al.* 1998, 2002).

Genomics has led to abundant sequence data, allowing deep insights into the molecular evolution of plastid genomes, especially by comparing plastid genome sequences with each other and with those of free-living cyanobacteria (Martin *et al.* 1998, 2002). Supposing that the ancestor of modern plastids had a genome similar in size (2–8 Mb) and with the same genetic capacity as extant cyanobacteria, most of the genetic material has been lost from the plastid genome – either deleted, if the gene products were no longer necessary, or transferred into other genetic compartments (Martin 2003). A limited amount of transferred cyanobacterial DNA has been detected in some mitochondrial genomes (e.g. Cummings *et al.* 2003) and to a much larger degree in the nuclear genome. In contrast to the DNA transferred into the mitochondrial genome, the cell nucleus reuses the new genetic material not only for some functions of its own biochemistry, but also to supply the plastid with proteins formerly encoded by its ancestor. For the latter, the proteins gained topogenic signals for their post-translational import (Soll and Schleiff 2004).

Comparisons of homologous genes present in cyanobacteria, plastomes, and in the model plant *Arabidopsis thaliana* showed that approximately 18% of the nuclear-encoded genes of this plant are of cyanobacterial origin, and therefore most likely of plastid origin (Martin *et al.* 2002). This surprisingly high proportion of nuclear genes with a cyanobacterial origin might not be restricted to this model plant but is a more general characteristic. Analysis of the recently completed sequences of the genomes of the red algae *Cyanidioschyzon merolae* (Matsuzaki *et al.* 2004) and the diatom *Thalassiosira pseudonana* (Armrust *et al.* 2004) will show whether a similar high percentage of nuclear genes with a cyanobacterial origin will be common to phototrophs or if it is restricted to the chlorophyll a/b harboring organisms (see below).

The integration of a cyanobacterium into a heterotrophic eukaryote and its conversion to a reduced organelle led, as outlined above, to primary plastids surrounded by two membranes (Fig. 6.1). Such organelles are found in three recent lineages: 1. the chlorophyll a/b harboring organisms (green algae and land plants), 2. the red algae, and 3. the glaucophytes, a small group of unicellular algae with several ancient characteristics (Steiner and Löffelhardt 2002). Phylogenetic analyses of symbiont (plastid) and host genomes indicated that, irrespective of the diversity, primary plastids evolved monophyletically (e.g. Martin *et al.* 1998; Stoebe and Kowallik 1999; Cavalier-Smith 2000). However, endosymbioses may be an ongoing process. In a recent study Marin *et al.* (2005) described an intracellular bacterium providing the thecate, filose amoeba *Paulinella chromatophora* with products of

photosynthesis. Phylogenetic analyses demonstrated that the endosymbiont is of cyanobacterial origin. Thus, nature may be in the process of repeating the evolution of a plastid.

Primary plastid plastomes

At least in terms of cell biology, glaucophytes may possess the most ancient plastids. They are morphologically similar to cyanobacteria and are therefore called cyanelles (Steiner and Löffelhardt 2002). Moreover, these unicellular algae still bear a bacterial-like peptidoglycan layer between the two membranes of the plastid envelope (Pfanzagl *et al.* 1996). The cyanelle genome of the prototype of glaucophytes, *Cyanophora paradoxa*, is 136 kb in size and encodes 182 genes (Table 6.1) (Löffelhard *et al.* 1997). An additional ancient characteristic of cyanelles was revealed by studying the degree of gene loss in different plastomes, indicating that the *Cyanophora paradoxa* plastid genome is enriched in genes not found in other plastids (Martin *et al.* 1998, 2002).

One hallmark of most plastid genomes is that they can be arranged as a circular map, which consists of an inverted repeat region dividing the plastid chromosome into a small and a large single-copy region (quadripartite structure). In *Cyanophora paradoxa* the inverted repeat region (IR) is comparatively small and encodes, in addition to some tRNA genes, an rRNA-operon (Table 6.1).

In red algae, these typical structures are only known from *Porphyra yezoensis* (Shivji *et al.* 1992), but not from plastid genomes of other rhodophytes. In addition, red algal plastid genomes are larger in size than that of the glaucophytes (Table 6.1). In *Porphyra purpurea* about 250 genes are encoded on the 190 kb plastid genome (Reith and Munholland 1995). A significant proportion of these additional genes encode for proteins of metabolic pathways not found in other groups or for ribosomal proteins, which are present in relatively high amounts. Furthermore, some red algae are highly specialized organisms, such as *Cyanidium caldarium* which lives only in hot and acidic environments. The adaptation to this lifestyle is clearly reflected in the plastid genome: not only in gene content, as unusual pathways are at

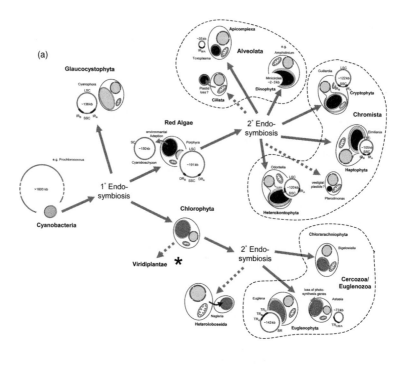

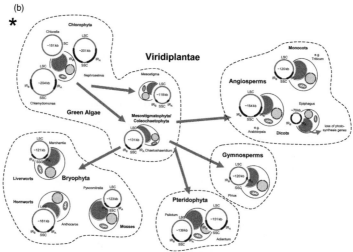

Figure 6.1 The origin of plastids in primary and secondary endosymbiosis. Schematic overview of the acquisition of primary and secondary plastids for different lineages of phototrophic eukaryotic organisms. Order is without phylogenetic or systematic scale. Displayed genomes (as circular maps) reflect their real proportions in size and arrangement (except of *Synechocystis*). (a) The origin of plastids by primary (1°) and secondary (2°) endosymbiosis and the divergence into the red and green line. (b) Display of the divergence plastids and their genomes within the green line (Viridiplantae = green algae and land plants). For more details refer to text. Abbreviations: LSC = large single copy region; SSC = small single copy region; IR = inverted repeat; TR = tandemly arranged direct repeats.

Table 6.1 Summary of sequenced genomes of Cyanobacteria, Primary, and Secondary Plastids

	NCBI taxonomy ID	genome size [kbp]	overall A+T content [%]	unique genes #	protein encoding genes *	unique tRNA + rRNA (+ other small RNA) species	inverted repeat	genes with intron(s) ~		
Prochlorococcus marinus CCMP1986	1218	1657	69.2	1755	1712	37+3 (+3)	-	-		
Synechococcus sp. WH 8102	1129	2434	36.9	2570	2517	44+6 (+3)	-	-	cyanobacteria	
Synechocystis sp. PCC 6803	1142	3573	52.3	3217	3167	43+6 (+1)	-	-		
Anabaena sp. PCC 7120	1177	7212	58.7	5430	5366	48+12 (+4)	-	1		
Cyanophora paradoxon	2761	136	64.0	182	142	36+3 (+1)	1	-	1˚ plastids	1˚ plastids
Cyanidioschyzon merolae 10D	45156	150	62.4	243	207	31+3 (+2)	-	-		
Cyanidium calandarium	2770	165	67.3	232	199	30+3	(1⁻)	-	rhodophyta	1˚ plastids
Gracilaria tenuistipitata var. *liui*	2774	184	70.9	238	205	29+3 (+1)	-	-		
Porphyra purpurea	2784	191	67.0	251	203	35+4	(1⁻)	-		
Odontella sinensis	2839	120	68.2	174	140	24+3	1	-	stramenopiles	
Thalassiosira pseudonana	35127	129	69.0	?	144	33+?	?	?		
Emiliania huxleyi CCMP1516	280463	105	63.2	140	109	28+3	1	-	haptophyta	
Guillardia theta	55528	122	67.0	178	147	28+3	1	-	cryptophyta	2˚ plastids
Toxoplasma gondii	5811	35	78.5	58	32	24+2	1	1	apicomplexa	
Emeria tenella	5802	35	79.4	58	32	24+2	1	-		
Plasmodium falciparum 3D7	5820	35	86.9	57	30	25+2	1	1		
Chlorella vulgaris C-27	3071	151	68,4	115	81	31+1	-	3	Chlorophyta	1˚ plastids
Nephroselmis olivacea	31311	201	57.9	127	91	31+3 (+1)	1	-		
Chlamydomonas reinhardtii	3055	204	65.4	135	99	30+5 (+1)	1	4		
Mesostigma viride	41881	118	69.9	135	99	33+3	1	-	Mesostigmatophyta	
Chaetosphaeridium globosum	96476	131	70.4	124	89	32+3	1	18	Coleochaetophyta	
Marchantia polymorphora	3197	121	71.2	122	86	32+4	1	18	liver worts	
Anthoceros formosae	3233	161	76.1	113	77	32+4	1	18	hornworts	
Physcomitrella patens subsp. *patens*	3217	123	71.5	118	83	31+4	1	18	mosses	
Psilotum nudum	3240	139	64.0	108	72	32+4	1	17	ferns	
Adiantum capillus-veneris	13818	151	63.7	118	85	29+4	1	17		
Pinus koraiensis	3337	117	61.2	?	?	32+4	1	14	gymnospems	
Pinus thunbergii	3337	120	61.5	108	72	32+4	1	14		
Oryza nivara (wild rice)	4536	135	61.0	110	76	30+4	1	16		
Oryza sativa (cultivated rice)	39947	135	61.1	110	76	30+4	1	16		
Triticum aestivum	4565	135	61.7	110	76	30+4	1	16		
Zea mays	4577	140	61.5	110	76	30+4	1	16	angiosperms, monocot	
Saccharum officinarum	4547	141	61.6	108	74	30+4	1	16		
Phalaenopsis aphrodite	308872	149	64.4	110	76	30+4	1	15		
Epifagus virginiana	4176	70	64.1	42	21	17+4	1	6		
Medicago truncatula	3877	124	66.0	88	64	24+0	-	-		
Lotus corniculatus var. *japonicus*	3867	151	64.0	111	77	30+4	1	18		
Spinacia oleracea	3561	151	63.2	113	79	30+4	1	17		
Calycanthus floridus var. *glaucus*	3428	153	60.7	115	81	30+4	1	18		
Arabidopsis thaliana	3701	154	63.7	113	79	30+4	1	18		
Panax ginseng	4053	156	62.2	114	80	30+4	1	18		
Nicotiana tobaccum	4085	156	62.2	113	78	30+4 (+1)	1	18		
Atropa belladonna	24609	157	62.4	113	78	30+4 (+1)	1	18		
Nymphaea alba	4418	160	60.8	118	84	30+4	1	20		
Oenothera elata subsp. *hookeri*	3939	163	60.8	108	74	30+4	1	17		
Amborella trichopoda	13332	163	61.7	114	80	30+4	1	14		
Astasia longa	3037	73	77.6	58	28	27+3	(2⁻)	-	euglenophya	2˚ plastids
Euglena gracilis	3039	143	73.9	97	67	27+3	(2⁻)	44		

Evaluation parameters are genome size, A/T content, number of unique genes and protein-encoding genes, number of tRNAs and rRNAs, existence of inverted repeats and introns. All plastome sequences, maps or the according references can be found under the taxonomy ID of each organism in the NCBI databases.

?= missing data; # = number of unique genes excluding pseudo- and duplicated genes; * = number including all protein-encoding genes, hypothetical ORFs and ycfs (excluding pseudo- and duplicated genes); + = number of intron-containing also non-protein encoding genes excluding pseudo- and duplicated genes (not total number of introns); ~ = tandemly arranged repeats.

least in part encoded within the plastid, but also in its architecture, as it possesses a large intergenic region, which can be folded *in silico* into higher order structures, a characteristic not found in other red algal plastid genomes (Gloeckner *et al.* 2000).

The comparatively large number of genes encoded by red algal plastid genomes was thought to be of major advantage in creating new organisms in secondary endosymbiosis (see later) and is described in the *portable plastid hypothesis* (Grzebyk *et al.* 2003).

In organisms with green (chlorophyll a/b-containing) primary plastids (green algae and land plants), research on plastid genomes provide the unique possibility to study chromosome evolution from unicellular algae to multicellular land plants. Moreover, plastid genome reduction as a consequence of parasitic lifestyle can also be studied. Hence, differences between these genomes were expected and indeed found. Whereas in land plants the plastid genome consists of a basic unit of from 120 up to 160 kb, green algae show a wide range in size (Table 6.1). For example the plastid genome of the charopyhte *Chaetospaeridium globosum* is 131 kb (Turmel *et al.* 2002), whereas in the prasinophytes *Mesostigma viridis* (Lemieux *et al.* 2000) and *Nephroselmis olivaceae* (Turmel *et al.* 1999) it was determined to be 188 and 200 kb, respectively. Surprisingly, other green algae are known to harbor comparatively huge plastid genomes; for example some members of *Acetabularia* possess plastid genomes ranging up to 1.5 Mb, a chromosome size similar to smaller bacterial genomes (Simpson and Stern 2002). On the other hand *Codium fragile* (Manhart *et al.* 1989) and *Nanochlorum eukaryotum* (Schreiner *et al.* 1995) have plastid genomes of less than 100 kb.

The transition from green algae to land plants can be studied in *Chaetosphaeridium globosum*, a green alga whose plastid genome shows an ancient architecture (Turmel *et al.* 2002). Therefore this organism is regarded as a model for the transition from aquatic to sessile, ashore lifestyle. The ancient character of this genome may be seen in the quadripartite structure, which is similar to that in land plants, whereas in other green algae, for example in *Chlorella* or *Chlamydomonas*, no IRs are

present (Simpson and Stern 2002). Together with the differences in plastome architecture in red algae, one can argue that IRs are an ancient feature, which was lost during evolution several times, independently.

The establishment of the plastid as a photosynthetic organelle within a eukaryotic cell led to an interlocked biochemistry in different compartments. As a consequence, higher plant chloroplasts are responsible for more than photosynthesis and carbon fixation. Their total elimination therefore seems to be impossible, even in parasitic, nonphotosynthetic cells. The epiphyte *Epiphagus virginiana*, for example, possesses a plastid genome of only 70 kb (Table 6.1). The extreme reduction of this genome is caused by the loss of nearly all genes of the photosynthesis machinery (Wolfe *et al.* 1992).

The evolution of land plants from a green algae-like progenitor required a lot of new morphological adjustments. The most conspicuous, new acquisition at the genetic level was the development of a plastid editing machinery, which does not exist in algal plastids (Bock 2000). By this editing machinery, mRNA-transcripts of plastid genes are post-transcriptionally modified, thereby generating conserved codon sequences, important for the proper function of the encoded proteins. Whereas in mosses and ferns several hundred editing sites were found in the plastid genome (Kugita *et al.* 2003; Wolf *et al.* 2004), seed plants contain only approximately 30 editing sites in their plastid genomes, at which typically cytidines are converted into uridines (Tsudzuki *et al.* 2001). The origin of editing in land plant plastids, its evolutionary advantage, and the difference in the number of editing sites between the different groups still need to be explained.

Another change at the genomic level was the acquisition of a second RNA polymerase for plastomic transcripts. Whereas in green and other algae plastomic transcripts are transcribed by a plastid-encoded RNA polymerase (PEP), mRNAs of the plastids of land plants are also partly transcribed by a specialized, nuclear- encoded phage-type RNA polymerase (NEP) (Hedtke *et al.* 1997).

Plastome architecture in higher plants: circular or linear?

Nearly all plastid genomes can be arranged as circular maps *in silico*. Electron microscopical studies (Herrmann *et al.* 1975) as well as the separation of plastid genomes by pulse-field gel electrophoresis support the circular structure *in vivo*. In addition, a significant proportion of the chromosomes were detected in a linear form (Deng *et al.* 1989). The latter finding was explained by broken circles. A recent report on the genome size of plastid DNA of maize seedlings showed that in 90% of all cases, the plastid genome was linear with defined ends (Oldenburg and Bendich 2004). Moreover, it was demonstrated that these molecules can be arranged in head-to-tail concatemers within the IR regions. Because these concatemers can invert a single copy region by recombination, previous interpretations of a flipping between the IR (Palmer 1983) can now be explained. In addition, Oldenburg and Bendich (2004) detected multigenomic branched linear chromosomes, leading to a new replication model for plant chloroplast DNA. Under the supposition that the plastid genome organization in maize is similar to that of other organisms, pulse field data of other plastid genomes will have to be revised.

Secondary endosymbiosis

Major groups of photosynthetic organisms do not share the plastid morphology found in primary evolved plastids. The most striking difference is a plastid surrounded by three or four membranes instead of two (Stoebe and Maier 2002; Hjorth *et al.* 2004). In the last decades of the twentieth century it was speculated that these multimembranous plastids evolved by the engulfment of a phototrophic eukaryote by a heterotrophic eukaryotic host cell (Whatley *et al.* 1979; Gibbs 1981). This was found to be correct and was called secondary endosymbiosis. Today, it is generally accepted that the stramenopiles, haptophytes, cryptophytes, chlorarachniophytes, and the apicomplexa, as well as the phototrophic euglenophytes, all contain plastids surrounded by three or four membranes, evolved by secondary endosymbiosis (Douglas

1998; Hjorth *et al.* 2004; Keeling 2004) (Fig. 6.1). However, the origin of peridinin-containing dinoflagellates seems to be still a mater of a debate and is either explained by secondary or tertiary endosymbiosis (see below; for discussion see Bachvaroff *et al.* 2005).

As in secondary evolved organisms, two formerly independent eukaryotes merged to give one new unicellular unit. At least two eukaryotic genomes (two nuclei) and three prokaryotic ones (two mitochondrial and one plastid genome) were initially pooled into one single cell, thereby creating a "super cell" with a, so far, unknown mixture of genetic capacities. This is thought to be one of the reasons for the success of these organisms.

In an international effort, the phylogeny of secondary evolved organisms has been studied. The generally accepted outcome was that at least two independent symbiotic events led to the current groups with secondary, multimembranous plastids (Van de Peer *et al.* 1996; Cavalier-Smith 2003; Keeling 2004) (Fig. 6.1).

The chlorophyll a/b lines

Phylogenetic analyses of plastid and nuclear-encoded plastid genes demonstrated that the phototrophic euglenophytes and chlorarachniophytes evolved by the engulfment and reduction of a green alga to a secondary plastid (McFadden and Gilson 1995; Van de Peer *et al.* 1996; Palmer and Delwiche 1996; Cavalier-Smith 2000). Both euglenophytes and chlorarachniophytes harbor chlorophyll a/b as the main photosynthetic pigments. It has been suggested that they may have originated monophyletically and were therefore grouped together into the phylum Cabozoa (Cavalier-Smith 2000). However, the single origin of euglenophytes and chlorarachniophytes still remains controversial. Indeed, there is strong evidence that the hosts of the chlorarachniophytes and the euglenophytes are not closely related (Archibald *et al.* 2002; Baldauf *et al.* 2000). However, the lack of a sequence of a plastid genome of a chlorarachniophyte prohibits a statement about a possible monophyly of the symbionts.

Euglenophytes are suitable for studying the effects of photosynthetic loss in unicellular organisms. Two of them, the photosynthetically active *Euglena gracilis* (Hallick *et al.* 1993) and the leucoplast bearing and photosynthetically inactive *Astasia longa* (Gockel and Hachtel 2000), share many common features in their plastid genomes; for example two tandemly arranged direct repeats (TR). On the other hand they differ in gene content, especially in genes encoding for components of the photosynthesis machinery, which have been lost in *A. longa*. Therefore, the situation seems very similar to that in parasitic land plants, in which loss of functions led to the loss of the corresponding genes. Thus, their plastid genomes are dynamic structures from which genes are simply eliminated when their encoded functions are no longer necessary. The deletion of all photosynthetic genes led to the shrunken plastome of 73 kb in *Astasia longa*, which is only half the size of the *Euglena gracilis* plastome (143 kb) (Table 6.1). Moreover, *Euglena gracilis* shows the highest number of introns inserted into plastid genes (Table 6.1). Some of these introns were inserted within other introns, resulting in so called twintrons (Hallick *et al.* 1993). It was speculated that mobile genetic elements have invaded this plastid genome and resulted in the unusual quantity and quality of introns.

The Chromalveolata

All other groups with secondary plastids, the chromista stramenopiles, haptophytes, and cryptophytes) and the plastid-containing alveolates (peridinin-containing dinoflagellates and apicomplexa) evolved by the engulfment and intracellular reduction of a red alga (Stoebe and Maier 2000; Keeling 2004). As it appeared that both groups, the chromista (Yoon *et al.* 2002) and the alveolates (Cavalier-Smith 2000), have evolved monophyletically, and the summarizing clade Chromalveolata was proposed by Cavalier-Smith (2000). This view was recently supported in phylogenetic analyses (Fast *et al.* 2001; Harper and Keeling 2003; Patron *et al.* 2004), but some additional data remain in clear conflict with the Chromalveolata hypothesis (discussed in Bachvaroff *et al.* 2005).

If the chromalveolate hypothesis is correct, the individual groups with a red algal symbiont started at the same point, but led in different directions. This may be indicated by the plastid genomes of haptophytes, stramenopiles, and cryptophytes. All of them show most similarity to that of red algae in respect to gene density, gene content, and genome size (Kowallik *et al.* 1995; Douglas and Penny 1999; Sánchez Puerta *et al.* 2005). However, it is still a matter of debate whether the plastids of cryptophytes and stramenopiles evolved from a common ancestor (Yoon *et al.* 2002) or not, as some data from phylogenetic reconstructions do not confirm a monophyletic origin of these plastids (e.g. Martin *et al.* 2002) whereas others do (e.g. Nozaki *et al.* 2003).

Comparisons of plastid genomes of cryptomonads and stramenopiles with those of red algae indicate that once a secondary plastid was established, significant but individual gene losses occurred, thereby streamlining the plastid genome in size (Martin *et al.* 1998, 2002). However, these plastomes show a significant overlap of genes typically located in all plastomes. This picture is different in the plastid-bearing members of the alveolates, described below.

Reduced plastids in apicomplexan parasites

The characterization of the genomes of the malaria parasite *Plasmodium falciparum* resulted in the surprising finding that this organism bears a third genome, in addition to a mitochondrial and a nuclear genome, which was not expected to be present in intracellular parasites (Wilson *et al.* 1997). The additional genome is a 35–kb circle and has characteristics of a minimal plastid genome (Table 6.1). Indeed, this was soon confirmed by electron microscopical investigations with the experimentally more accessible apicomplexa *Toxoplasma gondii* (McFadden *et al.* 1996; Köhler *et al.* 1997). In these studies it was shown that the small circle resides within an organelle-like structure, the apicoplast, which is surrounded by four membranes, implying that these intracellular parasites evolved secondarily from an ancient algal progenitor (Williams and Keeling 2003). As a consequence of the intracellular, non-photosynthetic lifestyle, none of the 57 or 58 genes of the apicoplasts investigated thus far encode components of a photosynthesis machinery (Wilson *et al.* 1997; Foth and McFadden 2003).

The apicoplast shows rapid evolution. This, combined with the relatively small number of informative genes, complicates a definitive statement about the origin of the apicoplast. Moreover, as some of the apicoplast-encoded ribosomal proteins might be derived from mitochondrial copies (Obornik *et al.* 2002), the dataset available for phylogenetic investigations could be much smaller than the total 57 or 58 apicoplast-encoded proteins. However, it can be concluded that the occurrence of genes known only from red algal plastomes indicates a red algal origin (Stoebe and Kowallik 1999).

Results from a genomics study of the nucleus of apicomplaexan parasites have now been published (Gardner *et al.* 2002). Particularly in the case of *Plasmodium falciparum*, the generated data can be used to predict the localization of the nuclear-encoded proteins within the cell. In a recent study, about 500 proteins were predicted to be apicoplast located, allowing a reconstruction of the major parts of the apicoplast metabolism (Ralph *et al.* 2004).

From studies of apicomplexa, it has became clear that once a plastid is established and the biochemistry of host and symbiont is – at least in part – interlocked, plastid losses, even in non-photosynthetic parasites, is unlikely. Nevertheless, plastid loss may be possible, and indeed common, for a number of different organisms (see later).

Peridinin-containing dinoflagellates

Since the 1980s, work on the plastid genomes of some peridinin-containing dinoflagellates showed a plastid genome of around 120 kb in size (Boczar *et al.* 1991). However, it was complicated to isolate plastid-located genes from these organisms. The break through eventually came from the identification of small plasmid-like structures, encoding one or rarely up to three genes of cyanobacterial origin, the so-called minicircles (Zhang *et al.* 1999, 2001; Barbrook and Howe 2000; Barbrook *et al.*

2001; Hiller 2001). The minicircle-encoded genes are known from, and specific for, plastid genomes of other phototrophs. Because the genes do not encode preproteins with a transit peptide, it may be concluded that the minicircles are localized inside the plastid (Koumandou *et al.* 2004). Further support for this theory came from the *in situ* localization of a minicircle-derived RNA within a *Symbiodinium* strain (Takishita *et al.* 2003) and the proof of its translation inside the plastid from experiments with strains of *Amphidinium* and *Gonyaulax* (Wang *et al.* 2005).

As shown in the section "Tertiary endosymbiosis" below, dinoflagellates are a group of algae with unusual traits (Stoebe and Maier 2002; Hackett *et al.* 2004b). This is true for the localization of minicircles, as in the dinoflagellate *Ceratium horridum* in which minicircles were identified in high amounts in the cell nucleus (Laatsch *et al.* 2004). If not a minor fraction of the minicircles is located in the plastid, a transport system of unknown nature has to be postulated for plastid import. *C. horridum* differs from other dinoflagellates in other characters, for example in its morphology; a reinvestigation of the plastid morphology showed differences in the thylakoid structure (stacks of two with an opaque lumen) from that of other dinoflagellates (Laatsch *et al.* 2004).

Additional differences within the peridinin-containing dinoflagellates were revealed from studies of the editing of minicircle-encoded gene transcripts. Whereas editing of mRNAs of minicircles genes from *Amphidinium* was Whereas undetected (Barbrook *et al.* 2001), *Ceratium horridum* shows a high editing rate in a nucleus-specific manner (Zauner *et al.* 2004). Moreover, editing leads to an increased similarity between the corresponding genes of *Ceratium horridum* and other dinoflagellates (Zauner *et al.* 2004).

Until now, only a couple of different minicircle genes have been isolated from the different dinoflagellates (Green 2004). Therefore, a compensatory gene transfer into the cell nucleus has to be assumed. That this is exactly the case, was recently shown in independent EST projects on peridinin-containing dinoflagellates, in which genes originally present in all known plastomes were found as nuclear transcripts (Hackett *et al.* 2004a; Bachvaroff *et al.* 2004).

Tertiary endosymbiosis

Dinoflagellates are an unusual group of protists (Hackett and Bhattacharya, Chapter 3). Approximately 50% of the known taxa are heterotrophs (Stoebe and Maier 2002; Hackett *et al.* 2004b). Most, but not all, of the remaining phototrophs are peridinin-containing dinoflagellates which evolved from secondary endosymbiosis. The few remaining taxa are classified by their pigmentation, which is different from peridinin. Morphological and phylogenetic studies showed that most of these organisms harbor symbionts which trace back to secondary endosymbiosis. Thus, this new step of symbiosis was termed tertiary endosymbiosis. Today, haptophytes, cryptophytes, and stramenopiles are known as tertiary symbionts within dinoflagellates (e.g. Schnepf and Elbrächer 1988; Tengs *et al.* 2000; Ishida and Green 2002; Stoebe and Maier 2002; Yoon *et al.* 2005). Additionally, one dinoflagellate was shown to harbor a green alga symbiont (Watanabe *et al.* 1990). Analyses supporting tertiary endosymbiosis refer to similarities in morphology and phylogeny of plastid genes. In most cases, tertiary endosymbiosis obviously led to plastid replacement, in which the original peridinin plastid was exchanged for the tertiary one. This scenario is thought to be substantially easier than establishing a secondary plastid, due to the pre-existing protein transport systems (Cavalier-Smith 2003). However, at least in *Glenodinium foliaceum* a so-called stigma is also present beside the plastid (Jeffrey and Vesk 1976). This spherical structure is surrounded by three different membranes (Stoebe and Maier 2002). If this stigma originated from a typical peridinin-containing plastid, *Glenodinium foliaceum* would be an organism harboring two different plastids within one cell and thus may represent an intermediate stage in tertiary endosymbiosis.

In particular, work on *Karenia brevis* and *Karenia mikimotoi* has influenced our understanding of tertiary endosymbiosis (Takishita *et al.* 2004; Yoon *et al.* 2005). Both organism harbor a tertiary endosymbiont of haptophytic orgin. Interestingly in these symbioses, not only the symbiont is strongly influenced by the host, but also the host genome. This was shown by analyzing EST data from *K. brevis*, which indicated that nuclear-encoded genes detected in secondary evolved dinoflagellates are lost and genes of the symbiont are used instead (Yoon *et al.* 2005).

Reduced and lost plastid genomes

As outlined above, plastid loss obviously does not occur easily, even in organisms that have lost their photosynthetic capability. Nevertheless, individual plastid losses have been described in stramenopiles, euglenophytes, and in dinoflagellates (Cavalier-Smith *et al.* 1995; Preisfeld *et al.* 2001; Saldarriaga *et al.* 2001).

Phylogenetic analyses indicated that dinoflagellates, apicomplexa, and ciliates are monophyletic and are therefore grouped together as alveolates (e.g. Van de Peer *et al.* 1996; Fast *et al.* 2001; Cavalier-Smith 2003). Because dinoflagellates and apicomplexa harbor remnant plastids of most likely red algal origin, it is self-evident to speculate about a loss of plastids in ciliates. The same might be true for the stramenopile-related oomycetes, which lack plastids (Stoebe and Maier 2002). However, for all of these examples, inferences from early branching lineages in phylogenetic trees are complicated and might lead to inaccurate predictions of plastid loss. Moreover, vestigal plastids may be overlooked, as indicated in the case of some stramenopiles (Sekiguchi *et al.* 2002).

Assuming that complete loss of plastids is possible, one can predict that other plastid-less lines had a phototrophic history. To identify such groups, the study of cyanobacterial genes known to have displaced their cytosolic counterparts is of major interest. One of these "marker" genes is the enzyme 6–phosphogluconate dehydrogenase (gnd), which was transferred to the nuclear genome in embryophytes and displaced the cytosolic copy after duplication (Krepinsky *et al.* 2001). Therefore, *gnd* is indeed a good indicator for the evaluation of a phototrophic history of heterotrophic organisms. Using this marker gene, phylogenetic analyses indicated that the human blood-stream parasite *Trypanosoma brucei*, which is closely related to the photosynthetically-active euglenophyte *Euglena gracilis*, possesses a *gnd* gene of cyanobacterial origin and might therefore be secondarily non-photosynthetic (Krepinsky *et al.* 2001; Martin and Borst 2003). In other studies, Andersson and Roger (2002) discovered that oomycetes and some heteroloboseid amoeboflagellates harbor cyanobacterial *gnd* genes, too; a lateral gene transfer event might explain these results. Alternatively, the *gnd* phylogeny could indicate that these organisms once harbored plastids of primary or secondary origin. If the latter is true, more genes of cyanobacterial origin should be found in ongoing genome projects.

Why are plastid genomes maintained?

The evolution of plastids from a cyanobacterial-like progenitor is accompanied by the transfer of plastid genes into the cell nucleus. Several suggestions have been made to explain these findings (e.g. Race *et al.* 1999; Timmis *et al.* 2004). Due to straightforward technologies in plastid transformation it was demonstrated that, in the case of tobacco, gene transfer from the plastid into the cell nucleus is a continuing process (Stegemann *et al.* 2003; Huang *et al.* 2003). Here it was shown that in 16 out of 250 000 seedlings tested, a gene inserted into the plastid genome by genetically engineering was afterwards transferred and expressed in the cell nucleus. However, similar experiments in *Chlamydomonas* failed, thereby indicating that high transfer rates are only possible in cells with more than one plastid (Lister *et al.* 2003). From mitochondria, a much higher transfer rate of genetic material into the cell nucleus is known (Thorsness and Fox 1990). Furthermore, genomics and analysis of the data indicated that the influx of plastid and mitochondrial material into the nuclear genome is common and a permanent process in *Arabidopsis* and rice (Richly and Leister 2004).

If no preference for preservation of plastid genes exists, plastid genomes should be eliminated quickly. That the actual situation is quite the converse was shown by a subset of approximately 40 genes maintained in all plastomes with the exception of dinoflagellates (Martin *et al.* 1998). One possible explanation is that some proteins are too hydrophobic to be reimported into the plastid. Nevertheless, because many nuclear-encoded proteins are both hydrophobic and reimported, this suggestion is unlikely. Beside genes for ribosomal subunits (Prechtl and Maier 2002), plastid genomes are enriched in genes encoding components of the photosystem core units. This fact led to a model put forward by Allen (see Allen *et al.* 2005). His central idea is that plastids have to maintain redox balance and avoid the generation of reactive oxygen species. Therefore, plastids have to control precisely the expression of core components of the electron transport chain. This may only be possible when these components are synthesized inside the plastid in a redox-dependent manner. Because redox-controlled transcription of these genes has indeed been shown (Allen and Pfannschmidt 2000; Pfannschmidt 2003), Allen's model may show a regulation hierarchy, which causes the manifestation of plastid-located genomes. However, a similar regulation network has to be postulated for ribosomal proteins, as a subset is also found in all plastid genomes (Prechtl and Maier 2002).

Conclusions

Genomics first produced the sequence of the organellar DNA of small mitochondrial genomes and now, due to more efficient methods, the determination of plastid genome sequences is no longer an enormous task. Until now approximately 45 plastid genomes have been sequenced, and others will soon be available. This extensive set of data is being used to reconstruct robust phylogenies of plastid-harboring organisms, with the result that the origin and evolution of primary, secondary, and tertiary plastids seems to be clarified in most cases. Today, we have – at least in land plants – a good picture of gene transfer from plastid to nucleus and the retargeting mechanisms of nuclear-encoded proteins into the plastid. Furthermore, nuclear-encoded genes for plastid functions emerged as important indicators for the understanding of the coevolution of organelle and the host cell. This was shown either by genes transferred into the cell nucleus and replacing cytosolic copies, or by proteins of host origin, which were relocated to the plastid. Hence, research on plastids, their genomes, and the host/ organelle interaction will remain a source for important insights into plant cell biology and the genomics of small genomes.

Acknowledgements

We are supported by the Deutsche Forschungs-gemeinschaft.

References

Allen, J. F. and Pfannschmidt, T. (2000). Balancing the two photosystems: photosynthetic electron transfer governs transcription of reaction centre genes in chloroplasts. *Phil. Trans. Roy. Soc. Lond. B Biol. Sci.*, **355**, 1351–1359.

Allen, J. F., Puthiyaveetil, S., Strom, J., and Allen, C. A. (2005). Energy transduction anchors genes in organelles. *Bioessays*, **27**, 426–435.

Andersson, J. O. and Roger, A. J. (2002). A cyanobacterial gene in nonphotosynthetic protists – an early chloroplast acquisition in eukaryotes? *Curr. Biol.*, **12**, 115–119.

Archibald, J. M., Longet, D., Pawowski, J., and Keeling, P. J. (2002). A novel polyubiquitin structure in Cercozoa and Foraminifera: evidence for a new eukaryotic supergroup. *Mol. Biol. Evol.*, **20**, 62–66.

Armbrust, E. V., Berges, J. A., Bowler, C. *et al.* (2004). The genome of the diatom *Thalassiosira pseudonana*: ecology, evolution, and metabolism. *Science*, **306**, 79–86.

Bachvaroff, T. R., Concepcion, G. T., Rogers, C. R., Herman, E. M., and Delwiche, C. F. (2004). Dinoflagellate expressed sequence tag data indicate massive transfer of chloroplast genes to the nuclear genome. *Protist*, **155**, 65–78.

Bachvaroff, T. R., Sanchez Puerta, M. V., and Delwiche, C. F. (2005). Chlorophyll c-containing plastid relationships based on analyses of a multigene data set with all four chromalveolate lineages. *Mol. Biol. Evol.*, **22**, 1772–1782.

Baldauf, S. L., Roger, A. J., Wenk-Siefert, I., and Doolittle, W. F. (2000). A kingdom-level phylogeny of eukaryotes based on combined protein data. *Science*, **290**, 972–977.

Barbrook, A. C. and Howe, C. J. (2000). Minicircular plastid DNA in the dinoflagellate *Amphidinium operculatum*. *Mol. Genet. Genomics*, **263**, 152–158.

Barbrook, A. C., Symington, H., Nisbet, R. E. R., Larkum, A., and Howe, C. J. (2001). Organisation and expression of the plastid genome of the dinoflagellate *Amphidinium operculatum*. *Mol. Genet. Genomics*, **266**, 632–638.

Bock, R. (2000). Sense from nonsense: how the genetic information of chloroplasts is altered by RNA editing. *Biochimie*, **82**, 549–557.

Boczar, B. A., Liston, J., and Catollico, R. A. (1991). Characterization of satellite DNA from three marine dinoflagellates (Dinophyceae): *Glenodinium* sp and two members of the toxic genus Protogonyaulax. *Plant Physiol.*, **97**, 613–618.

Brasier, M. D., Green, O. R., Jephcoat, A. P., *et al.* (2002). Questioning the evidence for Earth's oldest fossils. *Nature*, **416**, 76–81.

Butterfield, N. J. (2000). *Bangiomorpha pubescens* n gen, n sp: implications for the evolution of sex, multicellularity, and the Mesoproterozoic/Neoproterozoic radiation of eukaryotes. *Paleobiology*, **263**, 386–404.

Cavalier-Smith, T. (2000). Membrane heredity and early chloroplast evolution. *Trends Plant Sci.*, **5**, 174–182.

Cavalier-Smith, T. (2003). Genomic reduction and evolution of novel genetic membranes and protein-targeting machinery in eukaryote-eukaryote chimaeras (meta-algae). *Phil. Trans. Roy. Soc. Lond. B, Biol. Sci.*, **358**, 109–133.

Cavalier-Smith, T., Chao, E. E., and Allsopp, M. T. E. P. (1995). Ribosomal RNA evidence for chloroplast loss within Heterokonta: pedinellid relationships and a revised classification of ochristan algae. *Archiv Protistenkunde*, **145**, 209–220.

Cummings, M. P., Nugent, J. M., Olmstead, R. G., and Palmer, J. D. (2003). Phylogenetic analysis reveals five independent transfers of the chloroplast gene rbcL to the mitochondrial genome in angiosperms. *Curr. Genet.*, **43**, 131–138.

Deng, X. W., Wing, R. A., and Gruissem, W. (1989). The chloroplast genome exist in multimeric forms. *Proc. Natl. Acad. Sci. U S A*, **86**, 4156–4160.

Douglas, S. E. (1998). Plastid evolution: origins, diversity, trends. *Curr. Opin. Genet. Dev.*, **8**, 655–661.

Douglas, S. E. and Penny, S. L. (1999). The plastid genome of the cryptophyte alga, *Guillardia theta*: complete sequence and conserved synteny groups confirm its common ancestry with red algae. *J. Mol. Evol.*, **48**, 236–244.

Fast, N. M., Kissinger, J. C., Roos, D. S., and Keeling, P. J. (2001). Nuclear-encoded, plastid-targeted genes suggest a single common origin for apicomplexan and dinoflagellate plastids. *Mol. Biol. Evol.*, **18**, 418–426.

Foth, B. J. and McFadden, G. I. (2003). The apicoplast: a

plastid in *Plasmodium falciparum* and other Apicom-plexan parasites. *Int. Rev. Cytol.*, **224**, 57–110.

Gardner, M. J., Hall, N., Fung, E., *et al.* (2002). Genome sequence of the human malaria parasite *Plasmodium falciparum*. *Nature*, **419**, 498–511.

Gibbs, S. P. (1981). The chloroplast of some algal groups may have evolved from endosymbiotic eukaryotic algae. *Ann. New York Acad. Sci.*, **361**,193–207.

Gloeckner, G., Rosenthal, A., and Valentin, K. (2000). The structure and gene repertoire of an ancient red algal plastid genome. *J. Mol. Evol.*, **51**, 382–390.

Gockel, G. and Hachtel, W. (2000). Complete gene map of the plastid genome of the nonphotosynthetic euglenoid flagellate *Astasia longa*. *Protist*, **151**, 347–351.

Green, B. R. (2004). The chloroplast genome of dino-flagellates – a reduced instruction set? *Protist*, **155**, 23–31.

Grzebyk, D., Schofield, O., Vetriani, C., and Falkowski, P. G. (2003) The mesozoic radiation of eukaryotic algae: The portable plastid hypothesis. *J. Phycol.*, **39**, 259–267.

Hackett, J. D., Yoon, H. S., Soares, M. B., *et al.* (2004a). Migration of the plastid genome to the nucleus in a peridinin dinoflagellate. *Curr. Biol.*, **14**, 213–218.

Hackett, J. D., Anderson, D. M., Erdner, D. L., and Bhattacharya, D. (2004b). Dinoflagellates: a remarkable evolutionary experiment. *Am. J. Bot.*, **91**, 1523–1534.

Hallick, R. B., Hong, L., Drager, R. G., *et al.* (1993). Complete sequence of *Euglena gracilis* chloroplast DNA. *Nucleic Acids Res.*, **21**, 3537–3544.

Harper, J. T. and Keeling, P. J. (2003). Nucleus-encoded, plastid-targeted glyceraldehyde-3–phosphate dehy-drogenase (GAPDH) indicates a single origin for chromalveolate plastids. *Mol. Biol. Evol.*, **20**, 1730–1735.

Hedtke, B., Börner, T., and Weihe, A. (1997). Mitochon-drial and chloroplast phage-type RNA polymerases in *Arabidopsis*. *Science*, **277**, 809–811.

Herrmann, R. G., Bohnert, H. J., Kowallik, K. V., and Schmitt, J. M. (1975). Size, conformation and purity of chloroplast DNA of some higher plants. *Biochim. Bio-phys. Acta*, **378**, 305–307.

Hiller, R. G. (2001). "Empty" minicircles and petB/atpA and psbD/psbE (cytb559 alpha) genes in tandem in *Amphidinium carterae* plastid DNA. *FEBS Letters*, **505**, 449–452.

Hjorth, E., Hadfi, K., Gould, S. B., Kawach, O., Sommer, M. S., Zauner, S., and Maier, U. G. (2004). Zero, one, two, three, and perhaps four endosymbiosis and the gain and loss of plastids. *Endocytobiol. Cell Res.*, **15**, 459–468.

Huang, C. Y., Ayliffe, M. A., and Timmis, J. N. (2003). Organelle evolution meets biotechnology. *Nature Bio-tech.*, **21**, 489–490.

Ishida, K. and Green, B. R. (2002). Second- and third-hand chloroplasts in dinoflagellates: phylogeny of oxygen-evolving enhancer 1 (PsbO) protein reveals replacement of a nuclear-encoded plastid gene by that of a haptophyte tertiary endosymbiont. *Proc. Natl. Acad. Sci. U S A*, **99**, 9294–9299.

Jeffrey, S. W. and Vesk, M. (1976). Further evidence for a membrane-bound endosymbiont within the dino-flagellate *Peridinium foliaceum*. *J. Phycol.*, **12**, 450–455.

Keeling, P. J. (2004). Diversity and evolutionary history of plastids and their hosts. *Am. J. Bot.*, **91**, 1481–1493.

Köhler, S., Delwiche, C. F., Denny, P. W., *et al.* (1997). A plastid of probable green algal origin in Apicomplexan parasites. *Science*, **275**, 1485–1489.

Koumandou, V. L., Nisbet, R. E., Barbrook, A. C., and Howe, C. J. (2004). Dinoflagellate chloroplasts – where have all the genes gone? *Trends Genet.*, **20**, 261–267.

Kowallik, K. V., Stoebe, B., Schaffran, I., Kroth-Pancic, P. G., and Freier, U. (1995). The chloroplast genome of a chl a + c-containing alga, *Odontella sinensis*. *Plant Mol. Biol. Reporter*, **13**, 336–342.

Krepinsky, K., Plaumann, M., Martin, W., and Schnar-renberger, C. (2001). Purification and cloning of chloroplast 6–phosphogluconate dehydrogenase from spinach Cyanobacterial genes for chloroplast and cytosolic isoenzymes encoded in eukaryotic chromo-somes. *Eur. J. Biochem.*, **268**, 2678–2686.

Kugita, M., Yamamoto, Y., Fujikawa, T., Matsumoto, T., and Yoshinaga, K. (2003). RNA editing in hornwort chloroplasts makes more than half the genes func-tional. *Nucleic Acids Res.*, **31**, 2417–2423.

Laatsch, T., Zauner, S., Stoebe-Maier, B., Kowallik, K. V., and Maier, U. G. (2004). Plastid-derived single gene minicircles of the dinoflagellate *Ceratium horridum* are localized in the nucleus. *Mol. Biol. Evol.*, **21**, 1318–1322.

Lemieux, C., Otis, C., and Turmel, M. (2000). Ancestral chloroplast genome in *Mesostigma viride* reveals an early branch of green plant evolution. *Nature*, **403**, 649–652.

Lister, D. L., Bateman, J. M., Purton, S., and Howe, C. J. (2003). DNA transfer from chloroplast to nucleus is much rarer in *Chlamydomonas* than in tobacco. *Gene*, **316**, 33–38.

Löffelhardt, W., Bohnert, H. J., and Bryant, D. A. (1997). The complete sequence of the Cyanophora paradoxa cyanelle genome. In D. Bhattacharya, ed. *Origins of*

Algae and their Plastids. Springer-Verlag, Wien, pp. 149–162.

Maier, U. G., Douglas, S. E., and Cavalier-Smith, T. (2000). The nucleomorph genomes of cryptohytes and chlorarachniophytes. *Protist*, **151**, 103–109.

Manhart, J. R., Kelly, K., Dudock, B. S., and Palmer, J. D. (1998). Unusual characteristics of *Codium fragile* chloroplast DNA revealed by physical and gene mapping. *Mol. Genet. Genomics*, **216**, 417–421.

Margulis, L. (1970). *Origin of Eukaryotic Cells*. Yale University Press, New Haven.

Marin, B., Nowack, E. C. M., and Melkonian, M. (2005). A plastid in the making: Evidence for a second primary endosymbiosis. *Protist*, **156**, 425–432.

Martin, W. (2003). Gene transfer from organelles to the nucleus: frequent and in big chunks. *Proc. Natl. Acad. Sci. U S A*, **100**, 8612–8614.

Martin, W. and Borst, P. (2003). Secondary loss of chloroplasts in trypanosomes. *Proc. Natl. Acad. Sci. U S A*, **100**, 765–767.

Martin, W., Rujan, T., Richly, E., *et al.* (2002). Evolutionary analysis of Arabidopsis, cyanobacterial, and chloroplast genomes reveals plastid phylogeny and thousands of cyanobacterial genes in the nucleus. *Proc. Natl. Acad. Sci. U S A*, **99**, 12246–12251.

Martin, W., Stoebe, B., Goremykin, V., Hansmann, S., Hasegawa, M., and Kowallik, K. V. (1998). Gene transfer to the nucleus and the evolution of chloroplasts. *Nature*, **393**, 162–165.

Matsuzaki, M., Misumi, O., Shin-I. T., *et al.* (2004). Genome sequence of the ultrasmall unicellular red alga *Cyanidioschyzon merolae* 10D. *Nature*, **428**, 653–657.

McFadden, G. I. and Gilson, P. (1995). Something borrowed, something green: lateral transfer of chloroplasts by secondary endosymbiosis. *Trends Ecol. Evol.*, 12–17.

McFadden, G. I., Reith, M. E., Munholland, J., and Lang-Unnasch, N. (1996). Plastid in human parasites. *Nature*, **381**, 482.

Nozaki, H., Ohta, N., Matsuzaki, M., Misumi, O., and Kuroiwa, T. (2003). Phylogeny of plastids based on cladistic analysis of gene loss inferred from complete plastid genome sequences. *J. Mol. Evol.*, **57**, 377–382.

Obornik, M., Van de Peer, Y., Hypsa, V., Frickey, T., Slapeta, J. R., Meyer, A., and Lukes, J. (2002). Phylogenetic analyses suggest lateral gene transfer from the mitochondrion to the apicoplast. *Gene*, **285**, 109–118.

Oldenburg, D. J. and Bendich, A. J. (2004). Most chloroplast DNA of maize seedlings in linear molecules with defined ends and branched forms. *J. Mol. Evol.*, **335**, 953–970.

Palmer, J. D. (1983). Chloroplast DNA exists into two orientations. *Nature*, **301**, 92–93.

Palmer, J. D. and Delwiche, C. F. (1996). Second-hand chloroplasts and the case of the disappearing nucleus. *Proc. Natl. Acad. Sci. U S A*, **93**, 7432–7435.

Patron, N. J., Rogers, M. B., and Keeling, P. J. (2004). Gene replacement of fructose-1,6-bisphosphate aldolase supports the hypothesis of a single photosynthetic ancestor of Chromalveolates. *Eukaryot. Cell*, **3**, 1169–1175.

Pfannschmidt, T. (2003). Chloroplast redox signals: how photosynthesis controls its own genes. *Trends Plant Sci.*, **8**, 33–41.

Pfanzagl, B., Zenker, A., Pittenauer, E., *et al.* (1996). Primary structure of cyanelle peptidoglycan of *Cyanophora paradoxa*: a prokaryotic cell wall as part of an organelle envelope. *J. Bacteriol.*, **178**, 332–339.

Prechtl, J. and Maier, U. G. (2002). Zoology meets botany: establishing intracellular organelles by endosymbiosis. *Zoology*, **104**, 284–289.

Preisfeld, A., Busse, I., Klingberg, M., Talke, S., and Ruppel, H. G. (2001). Phylogenetic position and interrelationships of the osmotrophic euglenids based on SSU rDNA data, with emphasis on the Rhabdomonadales (Euglenozoa). *Int. J. System. Evol. Microbiol.*, **51**, 751–758.

Race, H. L., Herrmann, R. G., and Martin, W. (1999). Why have organelles retained genomes? *Trends Genet.*, **15**, 364–370.

Ralph, S. A., van Dooren, G. G., and Waller, R. F., *et al.* (2004). Tropical infectious diseases: metabolic maps and functions of the *Plasmodium falciparum* apicoplast. *Nat. Rev. Microbiol.*, **2**, 203–216.

Reith, M. and Munholland, J. (1995). Complete nucleotide sequence of the *Porphyra purpurea* chloroplast genome. *Plant Mol. Biol. Reporter*, **13**, 333–335.

Richly, E. and Leister, D. (2004). NUPTs in sequenced eukaryotes and their genomic organization in relation to NUMTs. *Mol. Biol. Evol.*, **21**, 1972–1980.

Saldarriaga, J. F., Taylor, F. J., Keeling, P. J., and Cavalier-Smith, T. (2001). Dinoflagellate nuclear SSU rRNA phylogeny suggests multiple plastid losses and replacements. *J. Mol. Evol.*, **53**, 204–213.

Sánchez Puerta, M. V., Bachvaroff, T. R., and Delwiche, C. F. (2005). The complete plastid genome seqeunce of the haptophyte *Emiliania huxleyi*: a comparison to other plastid genomes. *DNA Research* **12**, 151–156.

Schnepf, E. and Elbrächer, M. (1988). Cyrptophycean-like double membrane -bound chloroplasts in the dinoflagellate, *Dinophysis Ehrenb*: evolutionary, phylogenetic and toxicological implications. *Botanica Acta*, **101**, 196–203.

Schopf, J. W. and Packer, B. M. (1987). Early Archean (3.3–billion to 3.5–billion-year-old) microfossils from Warrawoona Group, Australia. *Science*, **237**, 70–73.

Schreiner, M., Geisert, M., Oed, M., *et al.* (1995). Phylogenetic relationship of the green alga *Nanochlorum eukaryotum* deduced from its chloroplast rRNA sequences. *J. Mol. Evol.*, **40**, 428–442.

Sekiguchi, H., Moriya, M., Nakayama, T., and Inouye, I. (2002). Vestigial chloroplasts in heterotrophic stramenopiles *Pteridomonas danica* and *Ciliophrys infusionum* (Dictyochophyceae). *Protist*, **153**, 157–167.

Shivji, M. S., Li, N., and Cattolico, R.A. (1992). Structure and organization of rhodophyte and chromophyte plastid genomes: implications for the ancestry of plastids. *Mol. Genet. Genomics*, **232**, 65–73.

Simpson, C. L. and Stern, D. B. (2002). The treasure trove of algal chloroplast genomes surprises in architecture and gene content, and their functional implications. *Plant Physiol.*, **129**, 957–966.

Soll, J. and Schleiff, E. (2004). Protein import into chloroplasts. *Nat. Rev. Mol. Cell Biol.*, **5**, 198–208.

Stegemann, S., Hartmann, S., Ruf, S., and Bock, R. (2003). High-frequency gene transfer from the chloroplast genome to the nucleus. *Proc. Natl. Acad. Sci. U S A*, **100**, 8828–8833.

Steiner, J. M. and Löffelhardt, W. (2002). Protein import into cyanelles. *Trends Plant Sci.*, **7**, 72–77.

Stoebe, B. and Kowallik, K. V. (1999). Gene-cluster analysis in chloroplast genomics. *Trends Genet.*, **15**, 344–347.

Stoebe, B. and Maier, U. G. (2002). One, two, three: nature's tool box for building plastids. *Protoplasma*, **219**, 123–130.

Takishita, K., Ishikura, M., Koike, K., and Maruyama, T. (2003). Comparison of phylogenies based on nuclear-encoded SSU rDNA and plastid-encoded psbA in the symbiotic dinoflagellate genus *Symbiodinium*. *Phycologia*, **42**, 285–291.

Takishita, K., Ishikura, M., and Maruyama, T. (2004). Phylogeny of nuclear-encoded plastid-targeted GAPDH gene supports separate origins for the peridinin- and the fucoxanthin derivative-containing plastids of dinoflagellates. *Protist*, **155**, 447–458.

Tengs, T., Dahlberg, O. J., Shalchian-Tabrizi, K., Klaveness, D., Rudi, K., Delwiche, C. F., and Jakobsen, K. S. (2000). Phylogenetic analyses indicate that the 19'hexanoyloxy-fucoxanthin-containing dinoflagellates have tertiary plastids of haptophyte origin. *Mol. Biol. Evol.*, **17**, 718–729.

Thorsness, P. E. and Fox, T. D. (1990). Escape of DNA from the mitochondrion to the nucleus in *Saccharomyces cerevisiae*. *Nature*, **346**, 376–379.

Tice, M. M. and Lowe, D. R. (2004). Photosynthetic microbial mats in the 3,416–Myr-old ocean. *Nature*, **431**, 549–552.

Timmis, J. N., Ayliffe, M. A., Huang, C. Y., and Martin, W. (2004). Endosymbiotic gene transfer: organelle genomes forge eukaryotic chromosomes. *Nat. Rev. Genet.*, **5**, 123–135.

Tsudzuki, T., Wakasugi, T., and Sugiura, M. (2001). Comparative analysis of RNA editing sites in higher plant chloroplasts. *J. Mol. Evol.*, **53**, 327–332.

Turmel, M., Otis, C., and Lemieux, C. (1999). The complete chloroplast DNA sequence of the green alga *Nephroselmis olivacea*: Insights into the architecture of ancestral chloroplast genomes. *Proc. Natl. Acad. Sci. U S A*, **96**, 10248–10253.

Turmel, M., Otis, C., and Lemieux, C. (2002). The chloroplast and mitochondrial genome sequences of the charophyte *Chaetosphaeridium globosum*: insights into the timing of the events that restructured organelle DNAs within the green algal lineage that led to land plants. *Proc. Natl. Acad. Sci. U S A*, **99**, 11275–11280.

Van de Peer, Y., Rensing, S. A., Maier, U. G., and De Wachter, R. (1996). Substitution rate calibration of small subunit ribosomal RNA identifies chlorarachniophyte endosymbionts as remnants of green algae. *Proc. Natl. Acad. Sci. USA*, **93**, 7732–7736.

Wakasugi, T., Nagai, T., Kapoor, M., *et al.* (1997). Complete nucleotide sequence of the chloroplast genome from the green alga *Chlorella vulgaris*: The existence of genes possibly involved in chloroplast division. *Proc. Natl. Acad. Sci. U S A*, **94**, 5967–5972.

Wang, Y., Jensen, L., Hojrup, P., and Morse, D. (2005). Synthesis and degradation of dinoflagellate plastid-encoded psbA proteins are light-regulated, not circadian-regulated. *Proc. Natl. Acad. Sci. USA*, **102**, 2844–28249.

Watanabe, M. M., Suda, S., Inouye, I., Sawaguchi, I., and Chihara, M. (1990). *Lepidodinium viride* gen et sp nov (Gymnodiniales, Dinophyta), a green dinoflagellate with chlorophyll a- and b- containing endosymbiont. *J. Phycol.*, **26**, 741–751.

Whatley, J. M., John, P., and Whatley, F. R. (1979). From extracellular to intracellular: the establishment of mitochondria and chloroplasts. *Proc. Roy. Soc. Lond. B Biol. Sci.*, **204**, 165–187.

Williams, B. A. and Keeling, P. J. (2003). Cryptic organelles in parasitic protists and fungi. *Adv. Parasitol.*, **54**, 9–68.

Wilson, R. J., Denny, P. W., Preiser, P. R., *et al.* (1997). Complete gene map of the plastid-like DNA of the malaria parasite *Plasmodium falciparum*. *J. Mol. Biol.*, **261**, 155–172.

Wolf, P. G., Rowe, C.A., and Hasebe, M. (2004). High levels of RNA editing in a vascular plant chloroplast genome: analysis of transcripts from the fern *Adiantum capillus-veneris*. *Gene*, **339**, 89–97.

Wolfe, K. H., Morden, C. W., and Palmer, J. D. (1992). Function and evolution of a minimal plastid genome from a nonphotosynthetic parasitic plant. *Proc. Natl. Acad. Sci. U S A*, **89**, 10648–10652.

Yoon, H. S., Hackett, J. D., Pinto, G., and Bhattacharya, D. (2002). The single, ancient origin of chromist plastids. *Proc. Natl. Acad. Sci. U S A*, **99**, 15507–15512.

Yoon, H. S., Hackett, J. D., van Dolah, F. M., Nosenko, T., Lidie, K. J., and Bhattacharya, D. (2005). Tertiary endosymbiosis driven genome evolution in dino-flagellate algae. *Mol. Biol. Evol.*, **22**, 1299–1308.

Zauner, S., Greilinger, D., Laatsch, T., Kowallik, K. V., and Maier, U. G. (2004). Substitutional editing of transcripts from genes of cyanobacterial origin in the dinoflagellate *Ceratium horridum*. *FEBS Letters*, **577**, 535–538.

Zhang, Z., Cavalier-Smith, T., and Green, B. R. (2001). A family of selfish minicircular chromosomes with jumbled chloroplast gene fragments from a dinoflagellate. *Mol. Biol. Evol.*, **18**, 1558–1565.

Zhang, Z., Green, B. R., and Cavalier-Smith, T. (1999). Single gene circles in dinoflagellates chloroplast genomes. *Nature*, **400**, 55–159.

CHAPTER 7

Genome evolution of anaerobic protists: metabolic adaptation via gene acquisition

Jan O. Andersson

Institute of Cell and Molecular Biology, Uppsala University, Biomedical Center, Uppsala, Sweden

Introduction

The textbook view of early eukaryote evolution is that the last common eukaryotic ancestor was a primitive anaerobic organism that lacked many key eukaryotic cellular functions (Prescott *et al.* 2005). The cell architecture has been thought to have become more complex in a step-wise fashion through the modification of existing genetic material and endosymbiotic events (introduction of mitochondria and plastids), finally resulting in multicellular eukaryotes such as animals and plants. Accordingly, some eukaryotes (i.e. anaerobic protists) are more "primitive" (i.e. more similar to prokaryotes) since they branched off the eukaryotic lineage early in evolution (Prescott *et al.* 2005). However, advances in cell biology and molecular phylogenetics during the last decade have proven this to be wrong (Baldauf 2003; Simpson and Roger 2004; Embley *et al.* 2003b) – all present-day eukaryotes seem to have most, if not all, features previously thought to be lacking in "primitive" eukaryotes; such as an organelle with mitochondrial ancestry (Roger 1999; Roger and Silberman 2002; Embley *et al.* 2003b; Tovar *et al.* 2003) (Fig. 7.1), Golgi apparatus (Dacks *et al.* 2003), and intron-containing genes (Russell *et al.* 2005). Furthermore, the current interpretation of the eukaryote phylogeny does not lend any phylogenetic support for the view that anaerobic protists are diverged early (Baldauf 2003; Simpson and

Roger 2004), suggesting that the absence of an aerobic mitochondrion is a derived feature in eukaryotes, not a primitive one (Roger 1999; Roger and Silberman 2002; Embley *et al.* 2003b; Tovar *et al.* 2003) (Fig. 7.1). This diverse group of organisms is commonly referred to as anaerobic protists, a widely used term that lacks a clear definition; many species in the group (such as *Giardia* and *Trichomonas*) live in fluctuating oxygen concentration (Lloyd 2004), and microsporidia even live inside cells of aerobic organisms as energy parasites (Katinka *et al.* 2001). However, the term anaerobic protists will be used in this chapter to refer to eukaryotes that lack aerobic mitochondria, only due to the lack of a better alternative.

The loss of aerobic mitochondria followed by adaptation to anaerobic or microaerophilic conditions has occurred multiple times, independently in eukaryotic evolution (Roger 1999; Roger and Silberman 2002; Embley *et al.* 2003b) (Fig. 7.1). For example: anaerobic ciliates have evolved several times, independently, from aerobic ancestors; microsporidia and chytrid fungi have independent origins from within the fungal lineage; and pelobionts and entamoebae share a common ancestor with aerobic slime moulds (Roger 1999; Embley *et al.* 2003b; Roger and Silberman 2002) (Fig. 7.1). This polyphyly of anaerobic protists indicates that eukaryotes are capable of remodeling their metabolism extensively in a relatively short

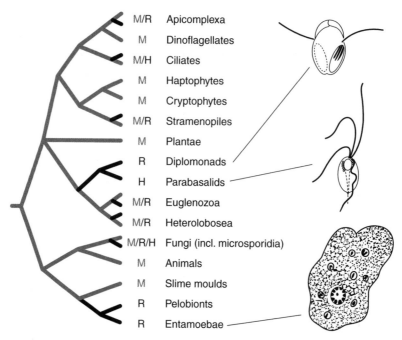

M/R	Apicomplexa
M	Dinoflagellates
M/H	Ciliates
M	Haptophytes
M	Cryptophytes
M/R	Stramenopiles
M	Plantae
R	Diplomonads
H	Parabasalids
M/R	Euglenozoa
M/R	Heterolobosea
M/R/H	Fungi (incl. microsporidia)
M	Animals
M	Slime moulds
R	Pelobionts
R	Entamoebae

Figure 7.1 Multiple origins of anaerobic protists. Schematic phylogeny of eukaryotes (Baldauf 2003; Simpson and Roger 2004). Lineages with typical aerobic mitochondria are shown in gray and indicated by M, and eukaryotic lineages that have lost aerobic mitochondrial functions are shown in black. A common feature for the latter group is that the organelle structure has been retained in a different form, either as hydrogenosomes (H) which make energy and excrete hydrogen, or as other double-membrane mitochondrial-related structures (R), often with poorly understood functions (sometimes called mitosomes) (Roger 1999; Roger and Silberman 2002; Embley *et al.* 2003b; Tovar *et al.* 2003; Riordan *et al.* 2003; Nasirudeen and Tan 2004). Line drawings of three anaerobic protists are shown; a diplomonad (*Trepomonas agilis*), and a parabasalid (*Pseudotrichomonas keilini*) reproduced from "Some free-living flagellates (Protista) from anoxic habitats" by Bernard, Simpson, and Patterson, from *Ophelia* 2000, **52**, 113–142 (Bernard *et al.*, 2000) by permission of Taylor and Francis AS, and *Entamoeba* reproduced by permission from the Microscope web page (http://www.microscope-microscope.org/).

evolutionary time and adapting to life in environments that were inaccessible to their ancestors. In prokaryotes, lateral gene transfer (LGT) has been identified as a very important mechanism for adaptation (Ochman *et al.* 2000; Boucher *et al.* 2003). In contrast, the impact of gene transfer on eukaryotic genome evolution has been thought to be very limited, with the exception of transfers from the ancestors of the mitochondria and chloroplasts (endosymbiotic gene transfer) (Timmis *et al.* 2004). However, analyses of, firstly, single gene phylogenies and, more recently, genome surveys and whole genome data sets, have revealed startling amounts of LGT in eukaryotes living in anaerobic and microaerophilic environments (Andersson 2005; Loftus *et al.* 2005). The acquisition of genes from organisms already adapted to these environments may indeed have facilitated metabolic adaptation in anaerobic protists. In this chapter these findings will be reviewed, mainly from the perspective of some of the most studied anaerobic protist lineages, such as *Entamoeba*, diplomonads, and parabasalids (Fig. 7.1), and in relation to the origin and evolution of the eukaryotic cell.

How is lateral gene transfer proven?

Most of the cases of gene transfer in anaerobic protists were initially identified as eukaryotic sequences found in anomalous positions in phylogenetic trees of individual proteins (Table 7.1). However, phylogenetic methods are imperfect and artefactual placements of sequences in trees, due to long branch attraction or other problems, may indeed lead to false interpretations of LGT events (Richards *et al.* 2003). For example gene sequences of anaerobic protists are very often found on long branches in phylogenetic trees, which have made it difficult to deduce their true positions in the tree of eukaryotes (see refs within Baldauf (2003) and Simpson and Roger (2004)). In addition, biological processes other than LGT may lead to phylogenetic relationships suggestive of transfer events (Richards *et al.* 2003). Therefore, caution is warranted before the conclusion that a protist gene has originated via LGT from a prokaryote is drawn solely based on an unexpected phylogenetic relationship between protist and prokaryotic sequences. Still, gene transfer scenarios should not be dismissed unless more parsimonious explanations exist.

To be able to distinguish between LGT and alternative scenarios questions such as the following need to be considered. Could the phylogeny be explained with a reasonable amount of gene duplication and loss events? What is the phylogenetic distribution of the gene; is it likely that the gene was present in the common ancestor of all eukaryotes? Is it possible that the putative donor and recipient lineages shared the same environment at the time of the putative transfer? Is there any plausible mechanism by which the foreign genetic material could have entered the eukaryotic cell? Is it likely that the gene provided a selective advantage in the recipient lineage? When this kind of information is taken into account, gene transfers from prokaryotes to anaerobic protists often appear much more parsimonious than alternative scenarios for unexpected phylogenetic positions for protist sequences.

Adaptation via gene acquisitions in anaerobic protists

Many anaerobic protists are phagotrophic, a mode of feeding that has been proposed to facilitate the introduction of prokaryotic genes into protist lineages (Doolittle 1998; Doolittle *et al.* 2003). Microbial eukaryotes feeding on prokaryotes indeed dominate among the reported cases of LGT, despite the bias towards non-phagotrophs such as plants, animals, and fungi among

completely sequenced eukaryotes (Andersson 2005). Thus, it seems likely that their phagotrophic lifestyle indeed exposes these protists to foreign genetic material, which may be incorporated into the genome and selected in the population. An expansion of the metabolic repertoire which makes the protist more fit to the environment often appears to be the basis for the selection of the incorporated DNA (Table 7.1).

One of the earliest claims of LGT in anaerobic protists was the acquisition of a prokaryotic *N*-acetylneuraminate lyase by *Trichomonas vaginalis* (de Koning *et al.* 2000). In the phylogenetic tree the parabasalid sequence branches with γ-proteobacterial epithelial parasites, but distantly from other eukaryotes. *N*-acetylneuraminate lyase is the final enzyme in the sialic acid degradation pathway, which is used to scavenge sialic acid from the host in parasitic bacteria. Therefore, it is likely that the transfer of this gene provided a metabolic adaptation to the epithelial parasitic lifestyle of *T. vaginalis* (de Koning *et al.* 2000). Similarly, the acquisition of group II large-subunit catalase by the microsporidian *Nosema locustae* from a proteobacterial donor lineage is likely to represent a metabolic adaptation (Fast *et al.* 2003). This version of the enzyme, which is more stable than the small subunit catalases, was shown to be expressed and the protein to be present in *N. locustae* spores. The absence of a gene for catalase in the *Encephalitozoon cuniculi* genome sequence (Katinka *et al.* 2001) and the proteobacterial affinity of the *Nosema* catalase suggest that the gene was lost after microsporidia diverged from fungi and regained in the lineage leading to *N. locustae*. The presence of the enzyme in this microsporidian lineage was speculated to confer a selective advantage by allowing the spore to survive long-term oxidative stress (Fast *et al.* 2003).

An example where phylogenetic analyses were combined with analyses of the phylogenetic distribution of the protein to show that LGT has occurred is the evolution of sulfide dehydrogenase. This protein is encoded by two adjacent genes in the genomes of some distantly unrelated prokaryotic lineages. The phylogenetic analysis recovered two diplomonad species – in which the two genes were fused – within a prokaryotic cluster

Table 7.1 LGT in anaerobic protists

Recipient lineage	Protein	Reference
Diplomonads	3-hydroxy-3-methylglutaryl coenzyme A reductase	Boucher and Doolittle 2000
Diplomonads	A dozen proteins	Andersson et al. 2003
Diplomonads	Fructose-1, 6-bisphosphate aldolase	Henze et al. 1998
Diplomonads	NADH oxidase	Nixon et al. 2002
Diplomonads	Phosphoenolpyruvate carboxykinase	Suguri et al. 2001
Diplomonads	Sulfide dehydrogenase	Andersson and Roger 2002
Entamoeba	96 proteins	Loftus et al. 2005
Entamoeba	Alanyl-tRNA synthetase	Andersson et al. 2005
Entamoeba	Eight proteins	Andersson et al. 2003
Entamoeba	Fe-hydrogenase	Nixon et al. 2003
Entamoeba	IscS and IscU	van der Giezen et al. 2004
Entamoeba	Malic enzyme, acetyl-CoA synthetase and alcohol dehydrogenase	Field et al. 2000
Entamoeba	NADH oxidase	Nixon et al. 2002
Entameoba and Mastigamoeba	A-type flavoprotein	Andersson et al. 2006
Mastigamoeba	Hybrid-cluster protein and alcohol dehydrogenase E	Andersson et al. 2006
Microsporidia	Catalase	Fast et al. 2003
Parabasalids	A-type flavoprotein glucosamine-6-phosphate isomerase hybrid-cluster protein	Andersson et al. 2006
Parabasalids	Alcohol dehydrogenase	Andersson et al. 2003
Parabasalids	Glyceraldehyde-3-phosphate dehydrogenase	Qian and Keeling 2001; Figge and Cerff 2001
Parabasalids	N-acetylneuraminate lyase	de Koning et al. 2000
Parabasalids	Potential surface protein	Hirt et al. 2002
Parabasalids and diplomonads	A-type flavoprotein	Andersson et al. 2006
Parabasalids and diplomonads	Alanyl-tRNA and prolyl-tRNA synthetase	Andersson et al. 2005
Parabasalids and diplomonads	Glucokinase and glucosephosphate isomerase	Henze et al. 2001
Parabasalids or animals (Hydra)	flp gene	Steele et al. 2004

(Andersson and Roger 2002). Two evolutionary scenarios could account for such a scattered phylogenetic distribution of genes encoding sulfide dehydrogenase. Either the genes were present in the last universal ancestor and subsequently lost from all sampled eukaryotic lineages, except diplomonads, and most prokaryotic lineages, or the gene was distributed via LGT between lineages in the three domains of life. The latter alternative seems more likely when additional information is considered – all lineages that do have the genes are found in oxygen-poor environments. Firstly, this observation indicates that the organisms that do have the gene could have been in physical contact since they share similar niches, which makes transfer events more likely. Secondly, the unique presence of a gene in a set of organisms that share a physical environment hints at a role for the gene in the adaptation to this niche. The prokaryotic homologs of this protein has been shown to catalyze the reduction of polysulfide to H_2S, although its biological role is unclear (Ma and Adams 1994). In the absence of functional experiments of the diplomonad enzyme, the role of sulfide dehydrogenase in diplomonads remain elusive (Andersson and Roger 2002). However, if it turns out that the protein is able to metabolize inorganic sulfur, it would be another example of such a function in eukaryotes; some anaerobic marine worms use sulfur compounds as an inorganic energy source by an unrelated mechanism (Grieshaber 1998). Anyhow, the observation that all organisms that encode sulfide dehydrogenase genes are anaerobes suggests a role related to a life in an oxygen-poor environment, and it is tempting to speculate that the acquisition of the gene by diplomonads provided a metabolic adaptation to such an environment (Andersson and Roger 2002).

In some cases the indications from a patchy phylogenetic distribution pattern have been strong enough to conclude that a gene has been acquired by LGT in anaerobic protists lineages, even in the absence of informative phylogenetic analyses. For example members of the TpLRR subfamily of leucine-rich repeats contain potential surface proteins that have only been found in the protist parasite *Trichomonas vaginalis* and some divergent eubacterial and archaeal lineages (Hirt *et al.* 2002). All species that encode the gene are either known to be parasites or commensals of mammalian mucosa, or have been isolated from such an environment. In evolutionary time this represent a relatively recent niche for microbes. The leucine-rich repeats are known to be involved in the binding to host proteins, which suggests that their presence may be important in pathogenicity (Hirt *et al.* 2002). These observations, in combination with the patchy phylogenetic distribution, are most easily explained by an acquisition of these leucine-rich repeats in the parabasalid lineage from a prokaryote found in the same niche.

In a test of the hypothesis that gene transfers have affected the genome evolution of diplomonads, we searched for sequences in the available genome sequences from *Giardia* and *Spironucleus* that showed higher similarity to prokaryotic sequences than to any eukaryotic sequence. Phylogenetic analyses revealed more than a dozen candidate genes that appeared to have been acquired from prokaryotes by diplomonads (Andersson *et al.* 2003). Most genes encoded metabolic enzymes, and about half of the genes appeared to encode functions connected to a microaerophilic lifestyle (i.e. genes involved in oxygen-stress response or anaerobic respiration). To our surprise, eight of the nine genes which were present in the genome of another anaerobic protist, *Entamoeba histolytica,* seemed to have been introduced by separate LGT events, suggesting independent adaptation by gene acquisitions in these distantly related anaerobic protists (Andersson *et al.* 2003) (Fig. 7.1). Four genes (encoding hybrid-cluster protein, A-type flavoprotein, glucosamine-6-phosphate isomerase, and alcohol dehydrogenase E) were selected to test whether these genes really were introduced multiple times into anaerobic protist lineages. Homologs were sampled from *Trichomonas* and *Mastigamoeba*, a parabasalid and a pelobiont which are related to diplomonads and *Entamoeba*, respectively (Fig. 7.1). Phylogenetic analyses using improved methods confirmed the earlier conclusion of multiple origins of the genes in eukaryotes; an increased number of gene transfers could indeed be inferred (Andersson *et al.* 2006). Interestingly, some of the genes have distinct origins in closely related anaerobic protists, such as diplomonads and parabasalids within Excavata, and *Entamoeba* and *Mastigamoeba* within Amoebozoa, while other genes appear to have been introduced in a common ancestor of the sampled lineages within the respective group. These observations are consistent with a model where genes are acquired continuously by anaerobic protists (Andersson *et al.* 2006), presumably as a consequence of the changing environments they experience.

Phylogenomic analyses of sequence data from genome surveys and complete genome sequences have indicated that LGT is not restricted to a few genes previously recognized in approaches designed to identify genes with prokaryotic origin. A phylogenetic screen of the genome sequence of *E. histolytica* indeed revealed 96 candidate genes that seemed to have been acquired relatively recently via LGT (Loftus *et al.* 2005). The vast majority of these encoded proteins with metabolic (58%) or unknown functions (41%). The acquisition of the metabolic genes most likely has expanded the repertoire of amino acids and sugars that this intestinal parasite is able to use as energy sources (Loftus *et al.* 2005). Members of the bacterial phylum lineage Bacteroidetes dominate among the putative donor lineages. As the authors point out, the impact of the 96 acquired genes on the metabolic capacity of *E. histolytica* is much larger than indicated by the relative small fraction of the total gene set (Loftus *et al.* 2005); the ability to acquire genes from prokaryotes in their environment most likely has been important in the evolution of this lineage of anaerobic protists.

A genome survey project of the diplomonad *Spironucleus barkhanus*, an Atlantic Salmon parasite, provided a similar picture. Phylogenomic analyses of the obtained Expressed Sequence Tag (EST) and Genome Survey Sequence (GSS) sequence data indicated about 80 cases of gene transfers affecting diplomonad sequences, which corresponded to ~12% of the genes suitable for phylogenetic analysis (Andersson *et al.*, manuscript in preparation). Similar to *E. histolytica*, genes with metabolic and unknown functions dominate among the candidate genes for LGT. Figure 7.2 shows the phylogenetic analyses of two conserved hypothetical proteins, indicating a common pattern where gene families are present only in a subset of prokaryotic and eukaryotic organisms with scrambled gene phylogenies with respect to organismal relationships. Such patterns are strongly suggestive that LGT has played an important role in the distribution of the gene family (Andersson 2005). The genes acquired by the *S. barkhanus* lineage most likely have expanded the metabolic repertoire of diplomonads. For example this protist group is known to be able to utilize arginine as an energy source via the arginine dehydrolase pathway (Adam 2001). Phylogenetic analyses of the three enzymes in the pathway strongly suggest that the genes were acquired from prokaryotic donors. About two-thirds of the candidate LGT genes in *S. barkhanus* were also present in the almost complete genome

sequence of *G. lamblia* (i.e. Fig. 7.2b), while the rest were absent (i.e. Fig. 7.2a) (Andersson *et al.*, manuscript in preparation). The latter genes probably have been acquired by *S. barkhanus* after the divergence from the *G. lamblia* lineage, in agreement, again, with a scenario where genes are introduced continuously into the genomes of diplomonads, on an evolutionary timescale.

Many anaerobic protists have hydrogenosomes, an organelle with mitochondrial ancestry that produces energy and hydrogen (Embley *et al.* 2003b; Hackstein 2005) (Fig. 7.1). Pyruvate:ferrodoxin oxidoreductase (PFO) and Fe-hydrogenase are two key enzymes in hydrogenosomes. Although these proteins, which are shared with eubacteria, have been studied intensively, their origins in eukaryotes remain unclear (Embley *et al.* 2003b; Hackstein 2005). Phylogenetic reconstructions of PFO have shown eukaryotes to be monophyletic, but without any affinity with α-proteobacteria, as expected if the enzyme originated from the ancestor of mitochondria. Similarly, monophyletic eukaryotes, with a single prokaryotic sequence (*Thermotoga maritima*) nested within, cannot be excluded in the hydrogenase phylogeny, although several unexpected relationships are observed (Embley *et al.* 2003b; Embley *et al.* 2003a) suggesting that LGT may indeed have been important in the evolution of this enzyme. Increased taxon sampling, a better understanding of the functional role of homologs in aerobic eukaryotes, and more sophisticated phylogenetic methods are needed to understand whether these key genes in anaerobic protists were present in the early evolution of eukaryotes, or if they are adaptive acquisitions in individual lineages (perhaps including transfer of genes between anaerobic eukaryotes). However, this uncertainty does not question the hypothesis that hydrogenosomes are modified mitochondria; the support for their shared origin is strong (Embley *et al.* 2003b; Martin 2005).

To conclude, when all putative transfers are considered collectively, there are indeed strong indications that gene transfer is an important genome evolution mechanism in anaerobic protists, mainly from comparative genomic analyses (Table 7.1). Firstly, metabolic genes are vastly

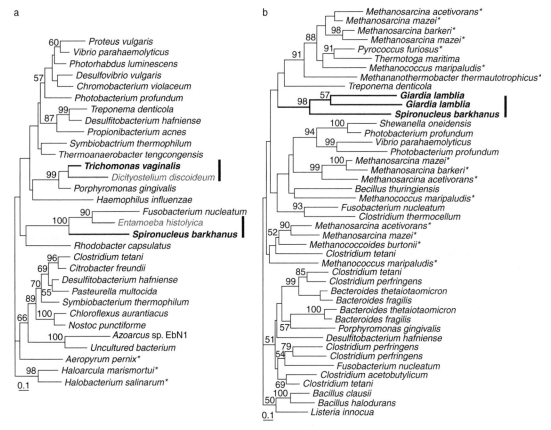

Figure 7.2 Maximum likelihood trees of two conserved hypothetical proteins. Phylogenomic analyses were performed on sequences obtained in a genome survey project on *Spironucleus barkhanus* (Andersson *et al.,* in preparation). The PHYLOGENIE software (Frickey and Lupas 2004) was used to automatically generate the data sets, which were analyzed using PHYML (Guindon and Gascuel 2003). Two of the ~80 sequences that showed phylogenetic relationships indicative of LGT events affecting the diplomonad lineage are shown. The WAG + I + G substitution models were applied with 100 bootstrap replicates. Support values > 50% are shown. The trees are arbitrarily rooted. Eukaryotes are indicated by bars and labeled gray and black, respectively, according to their classification into amoebozoa and excavates (Bapteste *et al.* 2002; Baldauf 2003; Simpson and Roger 2004). Archaea are indicated by asterisks. (a) Phylogenetic tree based on 160 aligned amino acid position of a conserved hypothetical protein. The gene is only present in a few eukaryotic lineages, and the eukaryotic sequences are nested within prokaryotic sequences indicative of prokaryote-to-eukaryote LGT events. However, the grouping of the eukaryotes does not follow the current eukaryote phylogeny, indicating that also gene transfers between eukaryotes may have occurred. (b) Phylogenetic tree based on 163 aligned amino acid position of a conserved hypothetical protein. The gene is only present in diplomonads within eukaryotes, strongly suggesting an introduction into this lineage from prokaryotes.

overrepresented, which make sense since such genes may provide a selective advantage in the recipient lineage. If the putative LGT events were the product of phylogenetic artifacts and/or gene duplication and loss events, a more random distribution among functionally categories would be expected. Secondly, anaerobic prokaryotes are often the donors for the gene transfer events affecting anaerobic protists, an observation which

is readily explained by a spread of genes by interdomain gene transfer events within oxygen-poor environment. This niche was most likely already inhabited by a wide diversity of prokaryotes at the time of the last common eukaryotic ancestor, and gene transfers from these prokaryotes have probably been an important evolutionary mechanism for the transition from aerobic to anaerobic protists.

Adaptation via gene acquisitions in other eukaryotes

Adaptation by acquisition of prokaryotic genes does not seem to be restricted to anaerobic protists within the eukaryotic domain. For example phylogenomic analyses of apicomplexan genome sequences – extensively covered in Chapter 8, Huang and Kissinger, of this book – have shown that LGT plays a role in the genome evolution of these parasites, including modification of the metabolic repertoire (Huang *et al.* 2004; Striepen *et al.* 2004). Like apicomplexans, kinetoplastids are parasitic protists that have evolved from free-living ancestors. Indeed, a recent phylogenomic analysis of the three available kinetoplastid genomes – *Trypanosoma brucei*, *Trypanosoma cruzi*, and *Leishmania major* – identified almost 50 genes showing strong support for prokaryote-to-eukaryote transfers (El-Sayed *et al.* 2005). The gene encoding dihydroorotate dehydrogenase of *Trypanosoma cruzi* is one kinetoplastid gene with a prokaryotic origin that has been studied in more detail (Annoura *et al.* 2005). This enzyme is localized in the cytosol and utilizes fumarate as electron acceptor, while the enzyme is linked to the mitochondrial membrane in most other eukaryotes. Thus, the *T. cruzi* enzyme, which was shown to be essential, is functional under both aerobic and anaerobic conditions since it is independent of the mitochondrial respiratory chain (Annoura *et al.* 2005). Phylogenetic analyses indicated that the kinetoplastids acquired this version of the gene after their divergence from *Euglena gracilis* (another euglenid). The authors suggest that this gene transfer may have enabled an ancestral kinetoplastid to adapt to anaerobiosis, which possibly represents the initial steps of parasitism (Annoura *et al.* 2005).

Interestingly, facultative anaerobic yeasts also have dihydroorotate dehydrogenase genes of prokaryotic origin, while yeasts unable to grow under anaerobic conditions have the typical eukaryotic mitochondrial version (Hall *et al.* 2005; Gojkovic *et al.* 2004). Intriguingly, experimental transfer of the *Saccharomyces cerevisiae* (a facultative anaerobic yeast) enzyme transformed the aerobic yeast *Pichia stipitis* into a facultative anaerobe (Shi and Jeffries 1998). Furthermore, the mitochondrial version of the enzyme from the yeast *Ashbya gossypii* complemented the deletion of the *S. cerevisiae* enzyme under aerobic, but not under anaerobic conditions (Hall *et al.* 2005). These observations suggest that the prokaryote-to-eukaryote LGT of the dihydroorotate dehydrogenase gene into the *S. cerevisiae* was a key event in the adaptation to anaerobic or microaerophilic conditions in these yeasts.

A different approach to test the hypothesis that LGT is important in the metabolic adaptation process in fungi was taken by Wenzl and co-workers (Wenzl *et al.* 2005). They reasoned that β-glucuronidase genes may be prone to LGT from bacteria since such a transfer would enable fungi living in soil to utilize glucuronides present in vertebrate urine as a carbon source. Using a functional screen they indeed identified two fungal species that expressed β-glucuronidase genes that enabled them to grow on glucuronides as the sole carbon source (Wenzl *et al.* 2005). Phylogenetic analyses and sequence analyses of 5′ regions showed that the fungal genes were most closely related to β-glucuronidase from *Arthrobacter* sp. which they had isolated using the same approach. They also identified homologous genes in two additional fungal genome sequences, although the gene is absent from most fungi (Wenzl *et al.* 2005). All these data are consistent with a relatively recent introduction of β-glucuronidase genes into a few fungal lineages from high G+C Gram-positive bacteria present in the soil, which provided these fungi with the advantage of being able to utilize glucuronides. These interdomain transfers possibly occurred soon after the colonization of land by vertebrates (Wenzl *et al.* 2005). Similarly, the colonization of the rumen of herbivorous animals by fungi was probably facilitated by gene transfers from prokaryotes already present in this niche; most enzymes involved in the degradation of cellulose present in rumen fungi–but absent in other fungi–were acquired from rumen bacteria (Garcia-Vallve *et al.* 2000).

To summarize, metabolic adaptation via gene acquisition appears to occur also in eukaryotes other than anaerobic protists, albeit, maybe, to a lesser extent. Since many of these lineages are non-phagotrophic other mechanisms must have facilitated the transfers. For example eubacterium-to-fungus conjugation and natural transformation have been invoked as plausible mechanisms by which the prokaryotic genes have entered fungal cells (Hall *et al.* 2005).

Gene replacements and eukaryote-to-eukaryote transfers

All gene transfers affecting eukaryotes do not have an apparent selective advantage–also replacement of genes does occur. For example two aminoacyl-tRNA synthetases were transferred from Nanoarchaeota to an ancestor of diplomonads and parabasalids (Andersson *et al.* 2005), members of the different subfamilies of glutamate dehydrogenase have been replaced several times in various protist lineages (Andersson and Roger 2003), and the complicated phylogeny of glyceraldehyde-3-phosphate dehydrogenase almost certainly involves homologous gene replacements of ancestral genes by bacterial genes in protists (Qian and Keeling 2001; Figge and Cerff 2001). Thus, essential genes that are widespread within eukaryotes may also be replaced with prokaryotic genes.

However, genes in the nucleus of eukaryotes may not necessary be replaced by genes with prokaryotic origin; foreign eukaryotic genes should indeed have a higher probability to be functional after integration into the genome. Such gene replacements are more difficult to identify than replacements with prokaryotic genes since eukaryote-to-eukaryote gene transfer events may be mediated by endosymbiotic gene transfer, such as transfers from the secondary plastid to the algal nucleus (Bhattacharya *et al.* 2004). Nevertheless, organisms that are capable of initiating secondary endosymbiosis should also be able to ingest eukaryotic cells for other reasons, which may lead to LGT events. An instructive example of this comes from a study of the algal group Chlorachniophytes, which are amoeboflagellates that have recruited their plastids secondarily by engulfing a green alga (Archibald *et al.* 2003). Phylogenetic analyses of 78 plastid-targeted proteins in the *Bigelowiella natans* (a chlorachniophyte) indicated that ~20% were not derived from the green algal endosymbiont, as expected, but

recruited via LGT, mostly from non-green algal eukaryotic lineages (Archibald *et al.* 2003).

Also eukaryotes that most likely never experience secondary endosymbiosis (i.e. many anaerobic protists) have been shown to be involved in intradomain gene transfer events. For example, the gene encoding alanyl-tRNA synthetase was transferred two times independently from the parabasalid lineage to the ciliates and *Entamoeba*, respectively (Andersson *et al.* 2003), and the unique presence of a homologous gene in the anaerobic protist *Trichomonas* and *Hydra*, a member of the metazoan phylum Cnidaria, has been explained by a intradomain transfer between the two lineages (Steele *et al.* 2004). Intriguingly, there are two phylogenetically distinct, but functional interchangeable, proteins involved in the eukaryotic translational machinery – the canonical EF-1α and an EF-like (EFL) protein present in eukaryotes (Keeling and Inagaki 2004). The EFL protein is found in diverse lineages of eukaryotes and its presence is almost always coupled with the absence of a canonical EF-1α. Although ancient duplication and differential losses of the two classes, in principle, could explain the distribution, a scenario with intradomain gene transfers of EFL coupled with losses of EF-1α appears more consistent with the observation (Keeling and Inagaki 2004).

The relative scarcity of cases of gene transfer between protists presently described in the literature could either be due to the fact that such events are very rare in evolution, or it could be due to a lack of analyses taking this possibility into account in combination with the difficulties to prove such events beyond reasonable doubt, since the eukaryote organismal phylogeny is far from established (Baldauf 2003; Simpson and Roger 2004). I favor the latter explanation and believe that many more gene transfers between protists will be described in the future when genome sequences from diverse protist lineages become available.

Fusion hypotheses in the light of gene transfers

Comparative genomic analyses have revealed that a large fraction of the protein-coding genes in eukaryotic genomes – probably the majority – show a closer relationship with eubacterial than archaeal homologs (Rivera *et al.* 1998; Esser *et al.* 2004). These observations have often been used as support for various fusion hypotheses for the origin of the eukaryotic cell (for example Gupta and Golding 1996; Moreira and Lopez-Garcia 1998; Martin and Müller 1998; Horiike *et al.* 2001; Rivera and Lake 2004). All these hypotheses assume that a eukaryotic gene with a eubacterial affinity originated via some kind of fusion or endosymbiotic event. However, the possibility that a substantial amount of the eukaryotic genes may have been introduced via multiple LGT events from diverse eubacteria (Doolittle 1998; Katz 2002; Doolittle *et al.* 2003) is usually not considered. This appears to be a problem in the light of the data presented in this chapter, that have shown that gene transfers probably have facilitated substantial metabolic adaptation in anaerobic protists in a relative short evolutionary time (Fig. 7.1). Therefore, it is not unreasonable that a considerable fraction of the eukaryotic genes with prokaryotic affinities may have originated via multiple LGT events (Doolittle 1998; Katz 2002; Doolittle *et al.* 2003), given that the evolutionary time since the divergence between prokaryotes and eukaryotes are much longer and interdomain gene transfer appears to be a continuous process. Only if this possibility is rigorously tested, and rejected, may the phylogenetic affiliations of eukaryotic genes be used to formulate fusion hypothesis for the origin of the eukaryotic cell.

The complete set of proteins from 60 different prokaryotic species was used to study the origin of the proteins encoded in the yeast genome in the most extensive analysis of the prokaryotic origin of eukaryotic genes presented to date (Esser *et al.* 2004). Using a method based on similarity searches, it was found that about 75% of yeast genes with detectable similarities with prokaryotic genes had greater amino acid identities to eubacterial than archaeal homologs. The authors argued that endosymbiotic gene transfer explained the vast majority of these similarities – multiple LGT events from eubacteria was dismissed since yeast is not phagotrophic, and only six genes (0.7% of the genes with prokaryotic homologs) were absent from all other eukaryotic

species. However, there are several problems with these arguments. Although phagotrophy may enhance the rate of gene transfers in eukaryotes (Doolittle 1998; Andersson 2005), LGT also occurs in yeast (Hall *et al.* 2005). Furthermore, the non-phagotrophic lifestyle of yeast is possibly a derived feature (Cavalier-Smith 2002) – their phagotrophic ancestors likely acquired genes from prokaryotes via gene transfers. The assumption that all yeast genes present in another eukaryote have been vertically inherited from the last common eukaryotic ancestor is also questionable; studies of gene transfer in eukaryotes indicate that prokaryotic genes very often are acquired several times independently by diverse eukaryotic lineages (Figge and Cerff 2001; Andersson *et al.* 2003; Andersson *et al.* 2006). Intriguingly, the analysis did not indicate any specific eubacterial group for the origin of the nuclear yeast genes – they indeed appear to have originated from different sources. In contrast, the majority of the proteins in the yeast mitochondrial genome grouped with α-proteobacterial proteins (Esser *et al.* 2004), as expected. This incongruity is noteworthy since it suggests that the mitochondrial and the majority of the nuclear eubacterial genes do not share the same history, although the observation may be explained by a number of scenarios which are very difficult to evaluate in the absence of taxon-rich gene phylogenies. Taken together, the conclusion that only a very tiny fraction of the yeast genes with eubacterial origin could be explained by LGT (Esser *et al.* 2004) is based on arguable assumptions; further analyses are needed to identify the origin of these eukaryotic genes.

A novel approach was used as the basis for a recent fusion hypotheses; the "ring of life" (Rivera and Lake 2004). The presence and absence of genes in complete genomes were analyzed, using the newly developed conditioned reconstruction method (Lake and Rivera 2004). Less than ten genomes from the three domains were included, with only two yeast genomes representing all eukaryotes (Rivera and Lake 2004). As expected from previous analyses, it was found that the yeast genomes consist of a mixture of eubacterial and archaeal proteins. The authors found that their data better explained the relationship between the three domains as a "ring of life" than a tree of life, indicating a fusion event between a eubacterial and an archaeal lineage (Rivera and Lake 2004). However, this conclusion is problematic since the origins of all eukaryotic members of a gene family may not necessarily be the same, given that eukaryotes acquire genes from prokaryotes via LGT. Therefore, fusion hypotheses cannot be based on the prokaryotic phylogenetic affinities of the genes in a single eukaryotic group, such as yeast (Rivera and Lake 2004). In addition, the results only imply that there have been contributions to the eukaryotic genomes both from eubacterial and archaeal sources; no strong signal indicating two distinct prokaryotic lineages was detected, as expected if the signal really were due to a single genome fusion event. The genes in the yeast genome may indeed have originated from a much larger number of sources – possibly via multiple LGT events (Doolittle 1998; Katz 2002; Doolittle *et al*. 2003) – making the relationships between the proteins encoded in individual genomes of prokaryotes and eukaryotes much more complex than a ring. For these reasons, further studies are needed, using whole genome data sets and including a much larger number of species covering a broader diversity. To be convincing, these analyses need to allow more complex relationships than trees and rings to explain the relationships of the gene content of individual genomes from the three domains of life.

As outlined above, the support for the various proposed fusion hypotheses (Gupta and Golding 1996; Moreira and Lopez-Garcia 1998; Martin and Müller 1998; Horiike *et al*. 2001; Rivera and Lake 2004) needs to be reanalyzed using broader data sets and more realistic assumptions. It is possible that their phylogenetic support still is present when the analyses accurately allow for multiple and continuous gene transfers between prokaryotes and eukaryotes. However, the studies of LGT in protists reviewed in this chapter question the existence of a strong phylogenomic signal indicating a fusion event between two cellular lineages for the origin of eukaryotes. The support for fusion hypotheses may indeed have to come from sources other than prokaryotic phylogenetic affinities of eukaryotic genes.

Concluding remarks

The identification of LGT as an evolutionary mechanism involved in the metabolic adaptation process in protists has provided some intriguing glimpses of how eukaryotes may adapt to their environment. However, further work is needed to understand how this process occurs in more detail: What are the barriers for transfer? What are the driving forces? Are genes in a metabolic pathway acquired one by one from diverse prokaryotes, or as a whole from a specific lineage with which the protists have a close relationship? How common are gene transfers between anaerobic protists? Are all protists equally prone to take up foreign genes, or do phagotrophs more readily incorporate foreign DNA into their genomes? What other mechanisms exist for exchange of genetic material in eukaryotes?

Obviously, the field of gene transfers in eukaryotes is in its infancy. Still, the accumulated data strongly argue that a deeper understanding of the process is needed not only to understand the genome evolution of anaerobic protists, but also to gain a deeper knowledge about the origin of the eukaryotic cell and the evolution of anaerobic metabolism within eukaryotes.

Acknowledgements

The collaboration with Andrew Roger on studies of gene transfers in anaerobic protists is greatly acknowledged. The *S. barkhanus* genome survey project is in collaboration with Andrew J. Roger (Dalhousie University, Halifax, Canada), Mark A. Ragan (University of Queensland, Brisbane, Australia), John M. Logsdon, Jr. (University of Iowa, USA), Robert P. Hirt (The University of Newcastle, UK), T. Martin Embley (The University of Newcastle, UK), and members of their laboratories. The author is supported by a Swedish Research Council (VR) Grant.

References

Adam, R. D. (2001). Biology of *Giardia lamblia*. *Clin. Microbiol. Rev.*, **14**, 447–475.

Andersson, J. O. (2005). Lateral gene transfer in eukaryotes. *Cell Mol. Life Sci.*, **62**, 1182–1187.

Andersson, J. O., Hirt, R. P., Foster, P. G., and Roger, A. J. (2006). Evolution of four gene families with patchy phylogenetic distribution: influx of genes into protist genomes. *BMC Evol. Biol.*, **6**, 27.

Andersson, J. O. and Roger, A. J. (2002). Evolutionary analyses of the small subunit of glutamate synthase: gene order conservation, gene fusions and prokaryote-to-eukaryote lateral gene transfers. *Eukaryot. Cell*, **1**, 304–310.

Andersson, J. O. and Roger, A. J. (2003). Evolution of glutamate dehydrogenase genes: evidence for lateral gene transfer within and between prokaryotes and eukaryotes. *BMC Evol. Biol.*, **3**, 14.

Andersson, J. O., Sarchfield, S. W., and Roger, A. J. (2005). Gene transfers from Nanoarchaeota to an ancestor of diplomonads and parabasalids. *Mol. Biol. Evol.*, **22**, 85–90.

Andersson, J. O., Sjögren, Å. M., Davis, L. A. M., Embley, T. M., and Roger, A. J. (2003). Phylogenetic analyses of diplomonad genes reveal frequent lateral gene transfers affecting eukaryotes. *Curr. Biol.*, **13**, 94–104.

Annoura, T., Nara, T., Makiuchi, T., Hashimoto, T., and Aoki, T. (2005). The origin of dihydroorotate dehydrogenase genes of kinetoplastids, with special reference to their biological significance and adaptation to anaerobic, parasitic conditions. *J. Mol. Evol.*, **60**, 113–127.

Archibald, J. M., Rogers, M. B., Toop, M., Ishida, K., and Keeling, P. J. (2003). Lateral gene transfer and the evolution of plastid-targeted proteins in the secondary plastid-containing alga *Bigelowiella natans*. *Proc. Natl. Acad. Sci. U S A*, **100**, 7678–7683.

Baldauf, S. L. (2003). The deep roots of eukaryotes. *Science*, **300**, 1703–1706.

Bapteste, E., Brinkmann, H., Lee, J. A., Moore, D. V., Sensen, C. W., Gordon, P., Durufle, L., Gaasterland, T., Lopez, P., Müller, M., and Philippe, H. (2002). The analysis of 100 genes supports the grouping of three highly divergent amoebae: *Dictyostelium*, *Entamoeba*, and *Mastigamoeba*. *Proc. Natl. Acad. Sci. U S A*, **99**, 1414–1419.

Bernard, C., Simpson, A. G. B., and Patterson, D. J. (2000). Some free-living flagellates (Protista) from anoxic habitats. *Ophelia*, **52**, 113–142.

Bhattacharya, D., Yoon, H. S., and Hackett, J. D. (2004). Photosynthetic eukaryotes unite: endosymbiosis connects the dots. *Bioessays*, **26**, 50–60.

Boucher, Y. and Doolittle, W. F. (2000). The role of lateral gene transfer in the evolution of isoprenoid biosynthesis pathways. *Mol. Microbiol.*, **37**, 703–716.

Boucher, Y., Douady, C. J., Papke, R. T., Walsh, D. A., Boudreau, M. E. R., Nesbø, C. L., Case, R. J., and Doolittle, W. F. (2003). Lateral gene transfer and the origins of prokaryotic groups. *Annu. Rev. Genet.*, **37**, 283–328.

Cavalier-Smith, T. (2002). The phagotrophic origin of eukaryotes and phylogenetic classification of Protozoa. *Int. J. Syst. Evol. Microbiol.*, **52**, 297–354.

Dacks, J. B., Davis, L. A. M., Sjögren, Å. M., Andersson, J. O., Roger, A. J., and Doolittle, W. F. (2003). Evidence for Golgi bodies in proposed "Golgi-lacking" lineages. *Proc. R. Soc. Lond. B Biol. Sci.*, **270**, S168–S171.

De Koning, A. P., Brinkman, F. S. L., Jones, S. J. M., and Keeling, P. J. (2000). Lateral gene transfer and metabolic adaptation in the human parasite *Trichomonas vaginalis*. *Mol. Biol. Evol.*, **17**, 1769–1773.

Doolittle, W. F. (1998). You are what you eat: a gene transfer ratchet could account for bacterial genes in eukaryotic nuclear genomes. *Trends Genet.*, **14**, 307–311.

Doolittle, W. F., Boucher, Y., Nesbø, C. L., Douady, C. J., Andersson, J. O., and Roger, A. J. (2003). How big is the iceberg of which organellar genes in nuclear genomes are but the tip? *Phil. Trans. R. Soc. Lond. B Biol. Sci.*, **358**, 39–58.

El-Sayed, N. M., Myler, P. J., Blandin, G., Berriman, M., Crabtree, J., Aggarwal, G., Caler, E., Renauld, H., Worthey, E. A., Hertz-Fowler, C., Ghedin, E., Peacock, C., Bartholomeu, D. C., Haas, B. J., Tran, A. N., Wortman, J. R., Alsmark, U. C., Angiuoli, S., Anupama, A., Badger, J., Bringaud, F., Cadag, E., Carlton, J. M., Cerqueira, G. C., Creasy, T., Delcher, A. L., Djikeng, A., Embley, T. M., Hauser, C., Ivens, A. C., Kummerfeld, S. K., Pereira-Leal, J. B., Nilsson, D., Peterson, J., Salzberg, S. L., Shallom, J., Silva, J. C., Sundaram, J., Westenberger, S., White, O., Melville, S. E., Donelson, J. E., Andersson, B., Stuart, K. D., and Hall, N. (2005). Comparative genomics of trypanosomatid parasitic protozoa. *Science*, **309**, 404–409.

Embley, T. M., van der Giezen, M., Horner, D. S., Dyal, P. L., Bell, S., and Foster, P. G. (2003a). Hydrogenosomes, mitochondria and early eukaryotic evolution. *IUBMB Life*, **55**, 387–395.

Embley, T. M., van der Giezen, M., Horner, D. S., Dyal, P. L., and Foster, P. (2003b). Mitochondria and hydrogenosomes are two forms of the same fundamental organelle. *Philos. Trans. R. Soc. Lond. B Biol. Sci.*, **358**, 191–201.

Esser, C., Ahmadinejad, N., Wiegand, C., Rotte, C., Sebastiani, F., Gelius-Dietrich, G., Henze, K., Kretschmann, E., Richly, E., Leister, D., Bryant, D., Steel, M. A., Lockhart, P. J., Penny, D., and Martin, W. (2004). A genome phylogeny for mitochondria among α-proteobacteria and a predominantly eubacterial ancestry of yeast nuclear genes. *Mol. Biol. Evol.*, **21**, 1643–1660.

Fast, N. M., Law, J. S., Williams, B. A., and Keeling, P. J. (2003). Bacterial catalase in the microsporidian *Nosema locustae*: implications for microsporidian metabolism and genome evolution. *Eukaryot. Cell*, **2**, 1069–1075.

Field, J., Rosenthal, B., and Samuelson, J. (2000). Early lateral transfer of genes encoding malic enzyme, acetyl-CoA synthetase and alcohol dehydrogenases from anaerobic prokaryotes to *Entamoeba histolytica*. *Mol. Microbiol.*, **38**, 446–455.

Figge, R. M. and Cerff, R. (2001). GAPDH gene diversity in spirochetes: a paradigm for genetic promiscuity. *Mol. Biol. Evol.*, **18**, 2240–2249.

Frickey, T. and Lupas, A. N. (2004). PhyloGenie: automated phylome generation and analysis. *Nucleic Acids Res.*, **32**, 5231–5238.

Garcia-Vallve, S., Romeu, A., and Palau, J. (2000). Horizontal gene transfer of glycosyl hydrolases of the rumen fungi. *Mol. Biol. Evol.*, **17**, 352–361.

Gojkovic, Z., Knecht, W., Zameitat, E., Warneboldt, J., Coutelis, J. B., Pynyaha, Y., Neuveglise, C., Moller, K., Loffler, M., and Piskur, J. (2004). Horizontal gene transfer promoted evolution of the ability to propagate under anaerobic conditions in yeasts. *Mol. Genet. Genomics*, **271**, 387–393.

Grieshaber, M. K. (1998). Animal adaptations for tolerance and exploitation of poisonous sulfide. *Annu. Rev. Physiol.*, **60**, 33–53.

Guindon, S. and Gascuel, O. (2003). A simple, fast, and accurate algorithm to estimate large phylogenies by maximum likelihood. *Syst. Biol.*, **52**, 696–704.

Gupta, R. S. and Golding, G. B. (1996). The origin of the eukaryotic cell. *Trends Biochem. Sci.*, **21**, 166–171.

Hackstein, J. H. (2005). Eukaryotic Fe-hydrogenases – old eukaryotic heritage or adaptive acquisitions? *Biochem. Soc. Trans.*, **33**, 47–50.

Hall, C., Brachat, S., and Dietrich, F. S. (2005). Contribution of horizontal gene transfer to the evolution of *Saccharomyces cerevisiae*. *Eukaryot. Cell*, **4**, 1102–1115.

Henze, K., Horner, D. S., Suguri, S., Moore, D. V., Sánchez, L. B., Müller, M., and Embley, T. M. (2001). Unique phylogenetic relationship of glucokinase and glucosephosphate isomerase of the amitochondriate eukaryotes *Giardia intestinalis*, *Spironucleus barkhanus* and *Trichomonas vaginalis*. *Gene*, **281**, 123–131.

Henze, K., Morrison, H. G., Sogin, M. L., and Müller, M. (1998). Sequence and phylogenetic position of a class II aldolase gene in the amitochondriate protist, *Giardia lamblia*. *Gene*, **222**, 163–168.

Hirt, R. P., Harriman, N., Kajava, A. V., and Embley, T. M. (2002). A novel potential surface protein in *Trichomonas vaginalis* contains a leucine-rich repeat shared by microorganisms from all three domains of life. *Mol. Biochem. Parasitol.*, **125**, 195–199.

Horiike, T., Hamada, K., Kanaya, S., and Shinozawa, T. (2001). Origin of eukaryotic cell nuclei by symbiosis of Archaea in Bacteria is revealed by homology-hit analysis. *Nat. Cell. Biol.*, **3**, 210–214.

Huang, J., Mullapudi, N., Lancto, C. A., Scott, M., Abrahamsen, M. S., and Kissinger, J. C. (2004). Phylogenomic evidence supports past endosymbiosis, intracellular and horizontal gene transfer in *Cryptosporidium parvum*. *Genome Biol.*, **5**, R88.

Katinka, M. D., Duprat, S., Cornillot, E., Metenier, G., Thomarat, F., Prensier, G., Barbe, V., Peyretaillade, E., Brottier, P., Wincker, P., Delbac, F., El Alaoui, H., Peyret, P., Saurin, W., Gouy, M., Weissenbach, J., and Vivares, C. P. (2001). Genome sequence and gene compaction of the eukaryote parasite *Encephalitozoon cuniculi*. *Nature*, **414**, 450–453.

Katz, L. A. (2002). Lateral gene transfers and the evolution of eukaryotes: theories and data. *Int. J. Syst. Evol. Microbiol.*, **52**, 1893–1900.

Keeling, P. J. and Inagaki, Y. (2004). A class of eukaryotic GTPase with a punctate distribution suggesting multiple functional replacements of translation elongation factor 1α. *Proc. Natl. Acad. Sci. U S A*, **101**, 15380–15385.

Lake, J. A. and Rivera, M. C. (2004). Deriving the genomic tree of life in the presence of horizontal gene transfer: conditioned reconstruction. *Mol. Biol. Evol.*, **21**, 681–690.

Lloyd, D. (2004). "Anaerobic protists": some misconceptions and confusions. *Microbiology*, **150**, 1115–1116.

Loftus, B., Anderson, I., Davies, R., Alsmark, U. C. M., Samuelson, J., Amedeo, P., Roncaglia, P., Berriman, M., Hirt, R. P., Mann, B. J., Nozaki, T., Suh, B., Pop, M., Duchene, M., Ackers, J., Tannich, E., Leippe, M., Hofer, M., Bruchhaus, I., Willhoeft, U., Bhattacharya, A., Chillingworth, T., Churcher, C., Hance, Z., Harris, B., Harris, D., Jagels, K., Moule, S., Mungall, K., Ormond, D., Squares, R., Whitehead, S., Quail, M. A., Rabbinowitsch, E., Norbertczak, H., Price, C., Wang, Z., Guillen, N., Gilchrist, C., Stroup, S. E., Bhattacharya, S., Lohia, A., Foster, P. G., Sicheritz-Ponten, T., Weber, C., Singh, U., Mukherjee, C., El-Sayed, N. M., Petri, W. A., Jr., Clark, C. G., Embley, T. M., Barrell, B., Fraser, C. M., and Hall, N. (2005). The genome of the protist parasite *Entamoeba histolytica*. *Nature*, **433**, 865–868.

Ma, K. and Adams, M. W. W. (1994). Sulfide dehydrogenase from the hyperthermophilic archaeon

Pyrococcus furiosus: a new multifunctional enzyme involved in the reduction of elemental sulfur. *J. Bacteriol.*, **176**, 6509–6517.

Martin, W. (2005). The missing link between hydrogenosomes and mitochondria. *Trends Microbiol.*, **13**, 457–459.

Martin, W. and Müller, M. (1998). The hydrogen hypothesis for the first eukaryote. *Nature*, **392**, 37–41.

Moreira, D. and Lopez-Garcia, P. (1998). Symbiosis between methanogenic archaea and delta-proteobacteria as the origin of eukaryotes: the syntrophic hypothesis. *J. Mol. Evol.*, **47**, 517–530.

Nasirudeen, A. M. and Tan, K. S. (2004). Isolation and characterization of the mitochondrion-like organelle from *Blastocystis hominis*. *J. Microbiol. Methods*, **58**, 101–109.

Nixon, J. E., Field, J., McArthur, A. G., Sogin, M. L., Yarlett, N., Loftus, B. J., and Samuelson, J. (2003). Iron-dependent hydrogenases of *Entamoeba histolytica* and *Giardia lamblia*: activity of the recombinant entamoebic enzyme and evidence for lateral gene transfer. *Biol. Bull.*, **204**, 1–9.

Nixon, J. E. J., Wang, A., Field, J., Morrison, H. G., McArthur, A. G., Sogin, M. L., Loftus, B. J., and Samuelson, J. (2002). Evidence for lateral transfer of genes encoding ferredoxins, nitroreductases, NADH oxidase, and alcohol dehydrogenase 3 from anaerobic prokaryotes to *Giardia lamblia* and *Entamoeba histolytica*. *Eukaryot. Cell* **1**, 181–190.

Ochman, H., Lawrence, J. G., and Groisman, E. A. (2000). Lateral gene transfer and the nature of bacterial innovation. *Nature*, **405**, 299–304.

Prescott, L. M., Harley, J. P., and Klein, D. A. (2005). *Microbiology, 6th edn*. McGraw-Hill, New York.

Qian, Q. and Keeling, P. J. (2001). Diplonemid glyceraldehyde-3-phosphate dehydrogenase (GAPDH) and prokaryote-to-eukaryote lateral gene transfer. *Protist*, **152**, 193–201.

Richards, T. A., Hirt, R. P., Williams, B. A., and Embley, T. M. (2003). Horizontal gene transfer and the evolution of parasitic protozoa. *Protist*, **154**, 17–32.

Riordan, C. E., Ault, J. G., Langreth, S. G., and Keithly, J. S. (2003). *Cryptosporidium parvum* Cpn60 targets a relict organelle. *Curr. Genet.*, **44**, 138–147.

Rivera, M. C., Jain, R., Moore, J. E., and Lake, J. A. (1998). Genomic evidence for two functionally distinct gene classes. *Proc. Natl. Acad. Sci. U S A*, **95**, 6239–6244.

Rivera, M. C. and Lake, J. A. (2004). The ring of life provides evidence for a genome fusion origin of eukaryotes. *Nature*, **431**, 152–155.

Roger, A. J. (1999). Reconstructing early events in eukaryotic evolution. *Am. Nat.*, **154**, S146–S163.

Roger, A. J. and Silberman, J. D. (2002). Cell evolution: Mitochondria in hiding. *Nature*, **418**, 827–829.

Russell, A. G., Shutt, T. E., Watkins, R. F., and Gray, M. W. (2005). An ancient spliceosomal intron in the ribosomal protein L7a gene (Rp17a) of *Giardia lamblia*. *BMC Evol. Biol.*, **5**, 45.

Shi, N. Q. and Jeffries, T. W. (1998). Anaerobic growth and improved fermentation of *Pichia stipitis* bearing a *URA1* gene from *Saccharomyces cerevisiae*. *Appl. Microbiol. Biotechnol.*, **50**, 339–345.

Simpson, A. G. and Roger, A. J. (2004). The real "kingdoms" of eukaryotes. *Curr. Biol.*, **14**, R693–696.

Steele, R. E., Hampson, S. E., Stover, N. A., Kibler, D. F., and Bode, H. R. (2004). Probable horizontal transfer of a gene between a protist and a cnidarian. *Curr. Biol.*, **14**, R298–299.

Striepen, B., Pruijssers, A. J., Huang, J., Li, C., Gubbels, M. J., Umejiego, N. N., Hedstrom, L., and Kissinger, J. C. (2004). Gene transfer in the evolution of parasite nucleotide biosynthesis. *Proc. Natl. Acad. Sci. U S A*, **101**, 3154–3159.

Suguri, S., Henze, K., Sánchez, L. B., Moore, D. V., and Müller, M. (2001). Archaebacterial relationships of the phosphoenolpyruvate carboxykinase gene reveal mosaicism of *Giardia intestinalis* core metabolism. *J. Eukaryot. Microbiol.*, **48**, 493–497.

Timmis, J. N., Ayliffe, M. A., Huang, C. Y., and Martin, W. (2004). Endosymbiotic gene transfer: organelle genomes forge eukaryotic chromosomes. *Nat. Rev. Genet.*, **5**, 123–135.

Tovar, J., Leön-Avila, G., Sánchez, L. B., Sutak, R., Tachezy, J., van der Giezen, M., Hernández, M., Müller, M., and Lucocq, J. M. (2003). Mitochondrial remnant organelles of *Giardia* function in iron-sulphur protein maturation. *Nature*, **426**, 172–176.

van der Giezen, M., Cox, S., and Tovar, J. (2004). The iron-sulfur cluster assembly genes *iscS* and *iscU* of *Entamoeba histolytica* were acquired by horizontal gene transfer. *BMC Evol. Biol.*, **4**, 7.

Wenzl, P., Wong, L., Kwang-Won, K., and Jefferson, R. A. (2005). A functional screen identifies lateral transfer of β-glucuronidase (*gus*) from bacteria to fungi. *Mol. Biol. Evol.*, **22**, 308–316.

CHAPTER 8

Horizontal and intracellular gene transfer in the Apicomplexa: The scope and fnctional consequences

Jinling Huang[1] and Jessica C. Kissinger[2]

[1] *Department of Biology, East Carolina University, Greenville, USA*
[2] *Department of Genetics and Center for Tropical and Emerging Global Diseases, University of Georgia, USA*

Introduction

Overview

The availability of completed genomes from diverse Apicomplexa, including the genera *Plasmodium, Theileria, Toxoplasma*, and *Cryptosporidium*, facilitates comparisons of the phylogenetic history of all genes encoded in these genomes. Surprisingly, genealogical analyses indicate that in numerous cases, metabolic genes in the Apicomplexa have been acquired by intracellular and lateral gene transfer. These acquired genes have provided the parasites with unexpected metabolic activities that have facilitated their evolution and pathogenesis. In some cases, these acquired genes may serve as therapeutic targets since they represent essential activities that are present in the parasite but not in the hosts they infect.

Gene transfer

Gene transfer, the exchange of genetic material between genomes, is an important force in the evolution of life. The introduction of novel genetic material from one organism into another can provide the recipient with unexpected capabilities. These new abilities, perhaps resistance to an antibiotic or an ability to metabolize a new food source, permit the organism to explore new ecological niches that were not previously accessible.

The importance of gene transfer in prokaryotic evolution has long been documented (Ochman *et al.* 2000; Gogarten *et al.* 2002; Gogarten and Townsend 2005). For example the spread of antibiotic resistance and nitrogen fixation abilities have been linked to gene transfer (Tauxe *et al.* 1989; Chen *et al.* 2003). It is estimated that gene transfer has contributed up to 24% of some bacterial genomes (Nelson *et al.* 1999).

Gene transfers are classified into two basic types, intracellular and lateral/horizontal (Table 8.1). Intracellular gene transfer (IGT) occurs within eukaryotic cells. IGT occurs between an endosymbiont (usually an organelle, such as a mitochondrion or chloroplast) and the host cell nucleus. With IGT, the donor of the genetic material literally resides within the same cell as the recipient of the genetic material (Fig. 8.1). In most cases, the donor organism came to reside inside another organism via phagocytosis, or engulfment, usually as a source of food, by a eukaryotic cell. In a few cases, phagocytosis has led to a long-term endosymbiotic relationship in which the donor organism becomes an organelle, such as a plastid or mitochondrion. In most instances of phagocytosis, the phagocytosed organism is destroyed by the host. In the case of lateral or horizontal gene transfer (LGT), genetic material is transferred between organisms that are not in a symbiotic relationship (this may also include organisms

Table 8.1 Summary of gene transfer types

Type of gene transfer	Directions of transfer	Known mechanisms
Intracellular gene transfer, IGT	Organelle to host eukaryote Nucleomorph to host eukaryote	Primary, secondary, tertiary . . . endosymbiosis
Lateral gene transfer, LGT	Prokaryote to prokaryote Prokaryote to eukaryote Eukaryote to eukaryote Eukaryote to prokaryote	Retroviral intermediates, DNA uptake, conjugation, transduction, other?

Organelle is defined here to be, or have been, genome containing and of prokaryotic origin, e.g. a mitochondrion or plastid. Nucleomorph is the term applied to the remnant eukaryotic nucleus contained within another eukaryotic cell created as the result of a secondary or higher degree endosymbiosis.

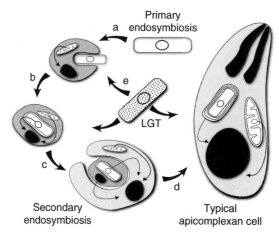

Figure 8.1 Apicomplexan evolution has been affected by endosymbiosis and gene transfer. Solid black rounded shapes are eukaryotic nuclei. Cristate and solid and spotted white rounded rectangles are mitochondria, cyanobacteria, and other prokaryotes respectively. (a) Primary endosymbiotic event in which a cyanobacterium is endosymbiosed by a eukaryotic cell to give rise to a photosynthetic algal cell. (b) Intracellular gene transfer occurs from the organelles (chloroplast and mitochondrion) to the nuclear genome and indicated by the fine black arrows. Organellar gene loss also occurs. (c) An algal cell is endosymbiosed by another eukaryotic cell. Gene transfer to the new host nuclear genome begins. The algal mitochondrion is lost. The vestigial algal nucleus becomes a nucleomorph. (d) The nucleomorph is lost giving rise to a modern apicomplexan cell in which the "apicoplast" is surrounded by four membranes. (e) Throughout the evolution of the Apicomplexa, genes have also been acquired horizontally from additional, diverse, prokaryotic (and potentially eukaryotic) sources via lateral gene transfer, LGT.

phagocytosed as food that exist very transiently within the host). The exact mechanisms of genetic exchange in IGT and LGT are not understood in all cases, but some organisms are known to take up foreign DNA (Mazodier and Davies 1991) and, in other cases, conjugation, transduction, and retroviral intermediates have been shown to facilitate the transfer of DNA from one organism to another (Benveniste 1985; Zambryski 1988).

LGT has been observed to occur between many different types of organisms (Table 8.1; Andersson,

Chapter 7). Thus far, the most widely observed and well-studied transfers are between prokaryotic organisms (Doolittle 1999; Ochman *et al.* 2000). However there are several examples of transfers from prokaryotes to eukaryotes (Striepen *et al.* 2002; Andersson and Roger 2003; Huang *et al.* 2004a; Van Der Giezen *et al.* 2004), eukaryotes to prokaryotes (Brown and Doolittle 1999), and, intriguingly, eukaryote to eukaryote (Keeling and Inagaki 2004). Gene transfers in eukaryotes are very prominent in phagotrophic organisms, and may represent, as

Ford Doolittle pointed out, "You are what you eat" (Doolittle 1998). An excellent example of this is provided by the phagotrophic amoeba *Entamoeba histolytica* (Van Der Giezen *et al.* 2004; Loftus *et al.* 2005).

The study of gene transfer in eukaryotes, particularly multicellular organisms, is largely an understudied area. As more eukaryotic genome sequences become available, however, there is increasing evidence that gene transfer may have played an important role in the evolution of eukaryotes, particularly unicellular eukaryotes (Martin *et al.* 2002; Andersson *et al.* 2003; Archibald *et al.* 2003; Huang *et al.* 2004a; Huang *et al.* 2004b; Loftus *et al.* 2005). See Andersson (2005) for a recent review.

Apicomplexans and their plastids

The clade Apicomplexa is comprised of unicellular eukaryotic protists characterized by an apical complex (Perkins *et al.* 2000). All of the estimated 5000 species of Apicomplexa are parasitic. The most notable members of the group include *Plasmodium falciparum*, the causative agent of human malaria, a disease that causes 2 to 3 million deaths annually; *Cryptosporidium parvum*, a significant AIDS-related pathogen that was responsible for more than 400 000 cases of gastrointestinal disease in Milwaukee 1993 (MacKenzie *et al.* 1995); *Toxoplasma gondii* a significant agent of morbidity and mortality, particularly among immunosuppressed individuals

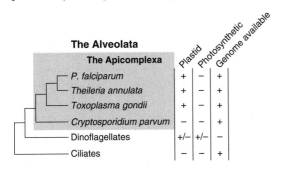

Figure 8.2 Relationships of the organisms within the Alveolata. "+", present; "−", absent; The four species shaded by the grey box belong to the clade Apicomplexa. Together, the Apicomplexa, Dinoflagellata, and the Ciliates comprise the supergroup, Alveolata. Comprehensive data are available for the multiple *Plasmodium* genome sequence projects, see Carlton, Chapter 2 this volume, for a review.

(Nash *et al.* 1994); *Eimeria tenella*, a prominent pathogen of chickens; and *Theileria parva* and *T. annulata*, two tick-borne parasites that infect millions of cattle, primarily in Africa (Brown 1990).

The closest relatives of apicomplexans are dinoflagellates (Hackett and Bhattacharya, Chapter 3) and ciliates (McGrath, Zufall and Katz, Chapter 4). Together these three phyla constitute the supergroup, Alveolata (Cavalier-Smith 1993) (Fig. 8.2). Many dinoflagellates are autotrophic protists that contain plastids (chloroplasts), which perform photosynthesis. Although none of the extant apicomplexans are photosynthetic, relict plastids (apicoplasts) have been found in this clade (Williamson *et al.* 1994; Wilson *et al.* 1994; Wilson *et al.* 1996). The apicoplast is the remnant of a plastid-containing algal cell that was engulfed by a phagocytic eukaryote. It is surrounded by four membranes that indicate its secondary endosymbiotic history (Fig. 8.1). This is in contrast to primary endosymbiosis, the event that created photosynthetic algae by the uptake and retention of a cyanobacterium by a heterotrophic eukaryotic cell. Primary plastids are surrounded by two membranes (Fig. 8.1). A single primary endosymbiotic event was responsible for the emergence of glaucophytes, red algae, and green plants, which are often collectively called primary photosynthetic eukaryotes (Nozaki 2005; Rodriguez-Ezpeleta *et al.* 2005).

The precise origin of the apicoplast is controversial. Several lines of evidence favor a red algal origin for the apicoplast, see Palmer (2003) and Keeling (2004) for reviews. Perhaps the strongest evidence comes from the nuclear-encoded apicoplast-targeted gene encoding glyceraldehyde-3-phosphate dehydrogenase (GAPDH). The plastid-targeted copies of the GAPDH gene in apicomplexans, dinoflagellates, cryptomonads, haptophytes, and heterokonts were all derived from a cytosolic homolog duplicate (Fast *et al.* 2001; Harper and Keeling 2003), supporting the chromalveolate hypothesis that posits a single origin for all plastids derived from a secondary red algal endosymbiont (Cavalier-Smith 1999). A recent study of chlorophyll-c containing plastids is also consistent with the chromalveolate hypothesis (Bachvaroff *et al.* 2005). Other evidence, including a shared *cox2* gene split between apicomplexans and green plants (Funes *et al.* 2002), and phylogenetic analysis of a plastid-encoded gene, *tuf*A, (Kohler *et al.* 1997) point to a possible green algal origin for the apicoplast. Regardless of the precise origin, the algal endosymbiont and its plastid provided a source of genetic material for IGT in this lineage. For a more thorough discussion of the origin of plastids and endosymbiosis, see Sommer *et al.*, Chapter 6, this volume.

Apicomplexan genome resources and methods for identifying gene transfers

Because of their medical and veterinary importance many apicomplexan genomes have been or are being sequenced. In fact, the Apicomplexa are among the most heavily sequenced eukaryotic groups with 19 genome projects either completed or in progress. See Carlton, Chapter 2, this volume, for a detailed discussion of the 13 *Plasmodium* genome projects. The large quantity of genome sequence data available for these organisms provides a rich source for data mining and comparative analyses of apicomplexan genome evolution.

Identification of transferred genes in a genome is not a straight-forward task. Phylogenetic analysis is considered the most reliable approach to infer the origin of a gene and assess whether it was likely to have been transferred (Genereux and Logsdon 2003). This approach is based on the incongruence between a gene tree (the evolutionary relationships of the genes themselves) and an accepted organismal tree (the relationships of the species from which the genes were isolated). For example a gene that originated in the mitochondrial genome but was transferred to a eukaryotic host nuclear genome would still show greater similarity to orthologous α-proteobacterial genes than it would to orthologous genes endogenous to the host nuclear genome. While this approach is powerful, analysis can be complicated by many factors such as incomplete taxonomic sampling, analytic methods, and differential gene losses. In particular, differential gene loss can almost always be invoked as an alternative explanation to gene transfer for observed incongruent tree topologies (Gogarten and Townsend 2005). Additionally, the degree of

relatedness between the donor and recipient can also complicate interpretation. If the donor and recipient are closely related, it is difficult to tell if the phylogenetic signal comes from vertical inheritance due to a shared ancestry or gene transfer. In the case of apicomplexans, given the lack of consensus agreement for the relationship between green plants and other eukaryotic groups (Baldauf 2003), the unambiguous identification of gene transfer events from the endosymbiont algal nucleus is difficult.

Intracellular gene transfer in apicomplexans

Endosymbionts and intracellular gene transfer

The acquisition and subsequent retention of cyanobacteria (plastids) and α-proteobacteria (mitochondria) within the cell set the stage for continuous gene transfer from organellar genomes to the eukaryotic nuclear genome. The scale and unidirectional nature of IGT is quite intriguing. There has been much speculation that perhaps gene transfer from organelles to the nucleus may lead to more effective gene regulation from the standpoint of the nucleus, but this remains to be proven (Martin and Herrmann 1998). Although data on the scale of intracellular gene transfer are still scarce, there is increasing evidence for frequent gene transfer from organelles to the nucleus. Analyses of *Arabidopsis thaliana* suggested that plastidic genes contribute about 18% of the nuclear genome (Martin *et al.* 2002). Another study indicated that up to 75% of yeast nuclear genome was likely derived from the mitochondria (Esser *et al.* 2004). Whether intracellular gene transfer is truly such a dominant force in genome evolution is still a matter of debate; however, the contribution of organelles and other sources to the gene content of the nuclear genome is beyond doubt. The continuous movement of genes from organelles/endosymbionts to their host nucleus leads to the creation of a mosaic nuclear genome consisting of genes from multiple evolutionary origins.

Since the Apicomplexa are the result of a secondary endosymbiosis, it is theoretically possible that the extant nuclear genome contains genes from as many a five distinct sources: the cyano-bactial/plastid genome, the algal nuclear genome, the algal mitochondrial genome, the apicomplexan mitochondrial genome, and the original apicomplexan nuclear genome source (Fig. 8.1). Given enough time and a marginal retention rate for transferred genes, several hundred organellar and algal symbiont genes could theoretically accumulate in the apicomplexan nuclear genome.

Fate of the endosymbiont's genome

Examinations of apicomplexan plastid genomes have shown them to be highly reduced relative to plant chloroplast and cyanobacterial genomes. The complete loss of photosynthetic capabilities contributes to some of this genome reduction; however, many other genes have also been lost. In the absence of a selective pressure to maintain genes relating to photosynthesis or other metabolic processes not utilized by an obligate intracellular parasite, it appears that the genes have simply been lost. Similar gene losses are observed in the genomes of non-parasitic symbionts as well (Ochman and Moran 2001). Additionally, as plastid genes were transferred to the nuclear genome, maintaining two copies of the same genes in different cellular compartments was not efficient for the regulation of their activities, and eventually the organellar copies were lost. However, this would not be possible until the nuclear copy could functionally substitute for the organellar product. The apicoplast has been reduced to 35 kb in the apicomplexan species, *Plasmodium falciparum* (Wilson *et al.* 1996). Many organellar genes that now reside in the nucleus have their protein products reimported into the organelles where they carry out their functions (Waller *et al.* 1998; Roos *et al.* 1999; Ishida 2005). This entails innovation or adaptation of importing machinery, including an organelle-targeting signal that facilitates the reimporting process. In apicomplexans, the apicoplast-targeting signal is a bipartite addition to the mature protein that consists of an N-terminal secretory signal peptide followed by a plastid transit peptide (Waller *et al.* 1998; Roos *et al.* 1999; Foth *et al.* 2003).

Analyses of genes transferred by intracellular gene transfer

The *Plasmodium falciparum* genome encodes 5268 proteins, 551 of which are predicted to contain an apicoplast-targeting signal (Gardner *et al.* 2002). However, not all apicoplast-targeting proteins are derived from the apicoplast genome. Phylogenetic analyses of 333 nuclear-encoded apicoplast targeted genes identified 26 genes of probable apicoplast ancestry and 35 additional genes of possible mitochondrial ancestry (Gardner *et al.* 2002). We performed automated phylogenomic analyses using PyPhy (Sicheritz-Ponten and Andersson 2001) to look for putative gene transfers. Unrooted gene trees (no origin specified, only relationships are shown) were successfully constructed for 1470 of the *P. falciparum* genes. More than 200 of these genes had cyanobacterial, red algal, and green plant homologs among their nearest neighbors (neighboring branches) on a phylogenetic tree (Huang *et al.* 2004b). While the number of plant-related genes in our automated analyses is certainly an overestimate because of the lack of directionality in an unrooted gene tree, the number of cyanobacterial genes identified in Gardner *et al.* (2002) is also likely a lower bound since they only considered genes with an apicoplast targeting signal. In an analysis of another apicomplexan species, *Cryptosporidium parvum*, which lacks an apicoplast, only a few genes of plastidic ancestry were identified in our phylogenomic analyses, suggesting that the method is fairly reliable and that artifacts are not a large problem in this particular analysis (Huang *et al.* 2004a).

As mentioned above, the number of nuclear genes of plastid origin is far fewer in *Cryptosporidium* than in *Plasmodium*. This may reflect the situation observed following organelle loss (if it had a plastid and subsequently lost it), or it may reflect eukaryote to eukaryote gene transfer of a few plastidic genes within this lineage (e.g. a plastid was never present). Currently, we favor the loss hypothesis over multiple intraeukaryote gene transfers of plastidic genes, but this interpretation awaits the analysis of other lineages more closely related to *Cryptosporidium*.

In *Plasmodium falciparum* the apicoplast genome is only about 35 kb and contains mostly housekeeping genes (Wilson *et al.* 1996) It has been shown to be essential in both *P. falciparum* (Jomaa *et al.* 1999) and *Toxoplasma gondii* (Fichera and Roos 1997). The apicoplast is the location for type II fatty acid, isoprenoid biosynthesis, and parts of the heme biosynthetic pathway (Waller *et al.* 1998; Jomaa *et al.* 1999; Gardner *et al.* 2002). These essential functions of the apicoplast organelle may have contributed to the retention of the genes encoding the proteins involved in these processes in the *Plasmodium* nuclear genome.

Old genes, new functions?

In contrast to the majority of examined apicomplexans, *Cryptosporidium* lacks a detectable plastid genome and apicoplast organelle (Zhu *et al.* 2000; Abrahamsen *et al.* 2004) and it only contains a relict mitochondrion that exists as an organelle, but does not function in oxidative respiration (Keithly *et al.* 2005). Yet, surprisingly, genes of mitochondrial and plastid origin were identified in the nuclear genome. While we favor an interpretation that these genes arrived by IGT, LGT cannot be ruled out at this juncture. However, regardless of the mechanism of acquisition, these genes appear to be functioning in altered or novel capacities. We argue that genes of organellar ancestry may contribute to functions unrelated to the original organelles. In the case of apicomplexans, these genes include a mitochondrial Hsp70 and a number of others related to the mitochondrial protein importing machinery (Abrahamsen *et al.* 2004; Slapeta and Keithly 2004; Xu *et al.* 2004) and a plastidic leucine aminopeptidase gene that does not contain a plastid targeting signal (Huang *et al.* 2004a). While the cellular activities and compartments they involve are still being determined (Striepen and Kissinger 2004), these genes likely function in different roles, on different substrates (in the case of transporters), or in a different environment. For example the gene encoding leucine aminopeptidase in *Cryptosporidium* shares the highest protein sequence similarity with those from other apicomplexans, plant chloroplast precursors, and cyanobacteria. The close relatedness of these sequences was also supported by phylogenetic analysis (Fig. 8.3) (Huang *et al.* 2004a). However, there is one significant difference, the leucine

aminopeptidase protein sequences from other apicomplexans possess a N-terminal extension that in *Plasmodium* is predicted to contain the apicoplast-targeting signal that targets this protein to the apicoplast organelle, where it presumably functions. This N-terminal targeting sequence is lacking in *Cryptosporidium*, as is the apicoplast, suggesting that the leucine aminopeptidase protein has been adapted to a different cellular environment.

Lateral gene transfer in eukaryotes

Ancient gene transfers are often difficult to identify because the phylogenetic signal (resolving power for evolutionary relationships) contained in the gene becomes obscured over evolutionary time (Huang *et al.* 2005). Therefore, most of the studies on lateral gene transfer in eukaryotes have focused on recent gene transfers that have occurred in specific, minor taxonomic groups. Genome sequence analysis of numerous eukaryotic genomes has reveal that lateral gene transfer is far more common than originally thought (Andersson 2005), especially among unicellular eukaryotes. One explanation for this observation is that most unicellular eukaryotes lack a separate germ line to protect their genome. The analyses of diplomonads and *Entamoeba*, both of which lack plastids and only contain a relict mitochondrion that does not contain a genome, offer a snapshot of the extent of lateral gene transfer in protists. The analyses of *Spironucleus barkhanus* and *Giardia lamblia* identified 15 genes that were possibly transferred from other species (Andersson *et al.* 2003; Andersson, Chapter 7). Nevertheless, since their data were prescreened by similarity searches, it is somewhat difficult to obtain an overall frequency of laterally transferred genes in these genomes. The *Entamoeba histolytica* nuclear genome contains 9938 genes, 96 of which were identified as transferred from prokaryotes (Loftus *et al.* 2005), suggesting lateral gene transfer contributes to at least 1% of the *Entamoeba* nuclear genome.

Lateral gene transfer in the Apicomplexa

By far, except for *Cryptosporidium*, the number of genes transferred from prokaryotes in other apicomplexan species is largely unclear. The

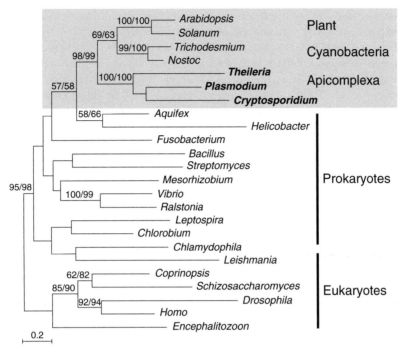

Figure 8.3 Relationships of leucine aminopeptidase sequences. Numbers above the branches indicate bootstrap support values for maximum likelihood and distance analyses respectively. The prokaryotic or eukaryotic origin of the genes analyzed is as indicated. The monophyletic group of sequences from taxa located inside the grey box were isolated from plants, cyanobacteria, or apicomplexan organisms, as indicated. Analyses were performed with PHYML and PHYLIP software. For maximum likelihood analyses, four gamma rates were chosen and the shape parameter was estimated from the data. Genbank accession numbers and parameters used for the distance with neighbor-joining analyses are as described in Huang *et al.* (2004a).

preliminary phylogenomic analyses of *Plasmodium falciparum* using PyPhy constructed gene trees for 1470 genes, 235 of which have non-cyanobacterial prokaryotes among their nearest neighbors (Huang *et al.* 2004b). Because of the lack of directionality for unrooted trees and the fact that each gene may have several nearest neighbors, this number of nearest neighbors does not translate into the number of genes transferred from prokaryotes. The true number of genes transferred from prokaryotes to each apicomplexan species should be significantly lower. In *Cryptosporidium parvum*, a detailed analysis suggests that 24 genes were acquired (Huang *et al.* 2004a). This would translate into a fraction of 0.5 to 2.5% of the nuclear genome originating from prokaryotic ancestry, depending on the analysis criteria. The laterally transferred genes in *Cryptosporidium parvum* are involved in a variety of biochemical activities including nucleotide metabolism, amino acid

biosynthesis, and energy production and conversion. Additionally, expression experiments on a subset of these transferred genes indicated that they are expressed and regulated throughout stages of the parasites life cycle (Huang *et al.* 2004a).

Lateral gene transfer and the evolution of new functionality

In principle, transferred genes may decay into pseudogenes and subsequently be deleted from the recipient genome unless they provide benefit to the recipients. In many cases, transferred genes are not found in closely related species, suggesting either independent acquisitions or secondary losses (Huang *et al.* 2004a; Huang *et al.* 2004b). In cases where the gene is present in ancestral lineages and then missing from more recent lineages, the loss scenario is favored. Interestingly however, the opposite may also occur. Laterally transferred genes may also further duplicate and differentiate into new functions. Lactate dehydrogenase (LDH) and malate dehydrogenase (MDH) are both oxidoreductases. LDH is typically tetrameric whereas MDH can be either dimeric or tetrameric (Madern *et al.* 2004). Phylogenetic analyses suggested that tetrameric MDH and LDH are closely related and likely derived from an ancient gene duplication event. The apicomplexan MDHs and LDHs appear to be cytosolic and lack an N-terminal extension that may contain an organelle-targeting signal. The apicomplexan MDHs group within α-proteobacterial tetrameric MDH and was very likely acquired from α-proteobacteria. Interestingly, apicomplexan LDHs were derived from two independent gene duplication events within the apicomplexan MDHs: the *Cryptosporidium* LDH was derived from a recent MDH duplication within the genus whereas all other LDHs resulted from a more ancient MDH duplication event within the clade (Madern *et al.* 2004). These independent gene duplication and functional divergence events suggest that the novel functions generated were highly desirable and beneficial to the apicomplexans containing them.

Mechanisms of lateral gene transfer

Very little is known about the mechanism of lateral gene transfer in the Apicomplexa. There are no identified retrotransposable elements in the *P. falciparum* genome (Gardner *et al.* 2002) nor several other apicomplexan genomes (Kissinger, unpublished observation); however, they may have been present in the past, involved in gene transfer, and subsequently lost from the genome. An analysis of transferred genes and their locations in apicomplexan genomes has shown that the acquired genes are distributed throughout the genome and suggested, in at least one case, that the transfer of a segment of prokaryotic genomic DNA has occurred (Huang *et al.* 2004a).

Two copies of the gene encoding 1,4-α-glucan-branching enzymes are found in *Cryptosporidium* and *Toxoplasma* (Fig. 8.4). These two copies are of bacterial ancestry and each corresponds to one of two copies found in multiple bacteria, suggesting

that they were transferred together following the bacterial duplication (Fig. 8.4) (Huang *et al.* 2004a). These two genes are located on the same chromosome in *Cryptosporidium parvum* and are approximately 110 kb apart. The fact that the gene encoding 1,4-α-glucan branching enzyme is not found in *Plasmodium* and *Theileria* suggests that these two copies were secondarily lost in these taxa since *Cryptosporidium* and *Toxoplasma* are ancestral to *Plasmodium* and *Theileria* (Fig. 8.2). More research is needed to help identify clues indicative of the mechanisms of both lateral and intracellular gene transfer in apicomplexans and eukaryotes in general.

Retention of transferred genes and implications for the evolution of apicomplexans

There is no question that gene transfer, either intracellular or lateral, can affect the recipient. Theoretically, if the foreign gene is a functional duplicate of an endogenous gene, gene displacement may occur without a significant impact to the existing biological process. If no functional homologs of the transferred gene exist, modification of the existing biochemical processes may be necessary to accommodate the new gene. Expression of the new gene will either be neutral, harmful, or beneficial to the recipient and selection will act upon the change accordingly.

In prokaryotes, genes of related biochemical activities often cluster into an operon, which could confer immediate benefits to recipient organisms upon lateral gene transfer if transferred as a complete operon unit (Lawrence and Roth 1996). Although operon-like gene structures are rarely found in eukaryotes and the extent of gene transfer in eukaryotes is debatable (Stanhope *et al.* 2001; Kurland *et al.* 2003), increasing evidence suggests that gene transfer can have significant impacts on the biochemistry, physiology, and thus the evolution of eukaryotes. For example *Entamoeba histolytica* acquired a large number of bacterial genes involved in amino acid and carbohydrate metabolism, which may significantly increase the range of substrates available for energy production and further enhance the biology and pathogenicity

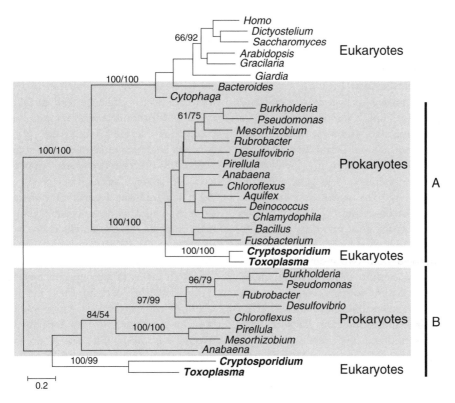

Figure 8.4 Relationships of 1,4-α-glucan branching enzymes. Numbers above the branches indicate bootstrap support values for maximum likelihood and distance analyses respectively. Genes isolated from prokaryotic organisms are shaded in grey. The duplicate gene clusters are indicated as "A" and "B". Note the apicomplexan lineage associated with each branch. Analyses were performed with PHYML and PHYLIP software. For maximum likelihood analyses, four gamma rates were chosen and the shape parameter was estimated from the data. Genbank accession numbers and parameters used for the distance with neighbor-joining analyses are as described in Huang *et al.* (2004a).

of the parasite (Loftus *et al.* 2005). As another example, analyses suggested that lateral gene transfer might have contributed to independent adaptation to an anaerobic lifestyle for *Entamoeba* and diplomonads (Andersson *et al.* 2003).

The apicoplast in *Plasmodium* and *Toxoplasma* is the location of isoprenoid and type II fatty acids biosyntheses; the latter may contribute to the establishment of parasitophorous vacuole membrane (Waller *et al.* 1998; Jomaa *et al.* 1999). The genome analyses of *Theileria parva* identified 345 apicoplast-targeting genes, but only 100 of them are shared with the congeneric *T. annulata*, suggesting that core functions of the apicoplast may vary even among closely related species (Gardner *et al.* 2005; Pain *et al.* 2005). Since the products of the intracellularly-transferred genes are often coupled with functions of the organelles,

the importance of intracellular gene transfer in apicomplexans appears obvious. Nevertheless, not all genes transferred from organelles are involved in organellar activities, some apparently contributing to other activities in the cell. For example, presuming the *Cryptosporidium* apicoplast was lost during evolution, the detection of genes of apicoplast and the secondary endosymbiont nuclear ancestry, including leucine aminopeptidase (Fig. 8.3), is significant. Questions of where and how these genes function in the cell remain intriguing areas to explore.

Many apicomplexan species, especially *Cryptosporidium* and *Theileria* species, have very compact genomes (<10 Mb) and reduced gene content ∼4000 genes. This situation is commonly observed in obligate intracellular pathogens and is believed to reflect gene loss, via natural selection, of genes that are no longer needed, given the intracellular environment (Ochman and Moran 2001). However, several cases suggest that lateral gene transfer has contributed to the adaptation of apicomplexans to an intracellular parasitic lifestyle.

Cryptosporidium parvum lacks the capacity for *de novo* nucleotide biosynthesis (all of the genes have been lost) and depends entirely on salvage from the host for both purines and pyrimidines. At least two genes of bacterial ancestry are involved in nucleotide salvage pathways. The gene encoding thymidine kinase (TK) is involved in pyrimidine salvage and is proteobacterial in origin (Striepen *et al.* 2004) whereas the gene encoding inosine 5′ monophosphate dehydrogenase (IMPDH), an enzyme involved in purine salvage, was acquired from an ε-proteobacterium (Striepen *et al.* 2002). Another gene in pyrimidine salvage, uridine kinase-uracil phosphoribosyltransferase (UK-UPRT), was likely transferred from the secondary algal endosymbiont (Fig. 8.5) (Striepen *et al.* 2004) or represents a lateral eukaryote to eukaryote gene

transfer. These three genes appear to be absent from *P. falciparum* and *T. annulata*, two species containing the *de novo* pyrimidine pathway.

It is interesting to note that apicomplexan parasites benefit from the retention of foreign genes that can meet or enhance their metabolic requirements. Mitochondria are important for energy production for most eukaryotes. *Cryptosporidium* has evolved an anaerobic life-style and has evolved to utilize other sources of energy, in part as a consequence of acquired genes. As a result, there is significant degeneration of the mitochondrion and loss of its associated activities. It has been suggested that amylopectin is an energy reserve for *Cryptosporidium* during the sporozoite stage of development (Petry and Harris 1999). At least two genes in amylopectin metabolism were transferred from bacteria; these including α-amylase for amylopectin degradation and 1,4-α-glucan branching enzymes for amylopectin biosynthesis (Huang *et al.* 2004a) (Fig. 8.4). As another example, the gene encoding fructose-1,6-bisphosphatase was

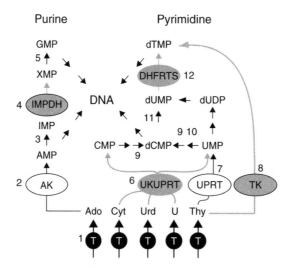

Figure 8.5 The *C. parvum* nucleotide biosynthetic pathway is a phylogenetic mosaic. Major enzymes within this pathway are indicated as ovals. Enzymes with a similar evolutionary history are indicated by the use of identical shading and outline, e.g. IMPDH and TK have similar histories that are distinct from DHFRTS and UKUPRT. The enzymes labeled 4 and 8 show strong a phylogenetic association with eubacteria, those labeled 6 and 12 show an association with plants. The remaining enzymes do not show indications of transfer. Analyses presented here are based on the *C. parvum* Type 2 IOWA genome. "T" = transporter. Two arrows indicate two or more enzymatic steps. Most nucleoside mono- and diphosphate kinase and phosphorylase steps have been omitted for simplicity. A complete set of these genes is present in the genome. 1. adenosine transporter, 2. adenosine kinase, 3. adenosinemonophosphate deaminase, 4. inosine monophosphate dedydrogenase, 5. guanosine monophosphate synthase, 6. uridine kinase/uracil phophoribosyltransferase, 7. uracil phosphoribosylltransferase, 8. thymidine kinase, 9 ribonucleotide diphosphate reductase, 10. cytosine triphosphate synthetase, 11. deoxycytosine monophosphate deaminase, 12. dihydrofolate reductase-thymidylate synthase.

transferred from a γ-proteobacterium to apicomplexan coccidians including *Toxoplasma*, *Eimeria*, and *Neospora* (Huang *et al.* 2004b). The recruitment of the fructose-1,6-bisphophatase gene may be related to the fact that *Eimeria* shunt fructose-1, 6-bisphosphate from the glycolytic pathway and convert it to mannitol via the mannitol cycle. Mannitol is the primary source of energy for *Eimeria* oocysts and makes up 25% of the dry weight of oocysts (Schmatz 1997).

Gene transfer and its implications for chemotherapy

For many researchers, an important goal of the pathogen genome sequencing projects was to discover new drug targets to tackle the diseases caused by these organisms. The Apicomplexa include many significant human and veterinary pathogens. Most of them have no effective treatments and many available antiapicomplexan drugs have serious side-effects (Smith and Corcoran 2004). The availability of genome sequence data for several apicomplexans has provided tremendous opportunities for drug and vaccine development against infections caused by apicomplexan parasites.

Traditionally, drug and vaccine design often targets parasite-specific pathways or structures that are essential for the survival and growth of the parasites (Fairlamb 2002; Agbo *et al.* 2003). The fact that apicoplasts provide essential functions (e.g. fatty acid and isoprenoid synthesis) and the fact that humans and other mammals to do not have apicoplasts, makes this organelle and its associated biochemical processes excellent antiapicomplexan drug targets (McFadden and Roos 1999; Ralph *et al.* 2004). Additionally, many of the laterally transferred genes also bear implications for chemotherapy. For example nucleotide biosynthesis is essential to parasite growth. Because *Cryptosporidium* lost both *de novo* purine and pyrimidine biosynthetic abilities, their dependence entirely on salvage pathways that assembled with several prokaryotic genes can be explored for anticryptosporidial drugs. Antagonists of IMPDH have already proven to affect *C. parvum* viability

and may provide an important weapon to tackle *Cryptosporidium* diseases (Striepen *et al.* 2004).

Acknowledgements

The authors would like to thank Boris Striepen for providing and permitting us to modify Fig. 8.1.

References

Abrahamsen, M. S., Templeton, T. J., Enomoto, S., Abrahante, J. E., Zhu, G., Lancto, C. A., Deng, M., Liu, C., Widmer, G., Tzipori, S., Buck, G. A., Xu, P., Bankier, A. T., Dear, P. H., Konfortov, B. A., Spriggs, H. F., Iyer, L., Anantharaman, V., Aravind, L., and Kapur, V. (2004). Complete genome sequence of the Apicomplexan, *Cryptosporidium parvum*. *Science*, **304**, 441–445.

Agbo, E. C., Majiwa, P. A., Buscher, P., Claassen, E., and Te Pas, M. F. (2003). *Trypanosoma brucei* genomics and the challenge of identifying drug and vaccine targets. *Trends Microbiol.*, **11**, 322–329.

Andersson, J. O. (2005). Lateral gene transfer in eukaryotes. *Cell Mol. Life Sci.*, **62**, 1182–1197.

Andersson, J. O. and Roger, A. J. (2003). Evolution of glutamate dehydrogenase genes: evidence for lateral gene transfer within and between prokaryotes and eukaryotes. *BMC Evol. Biol.*, **3**, 14.

Andersson, J. O., Sjogren, A. M., Davis, L. A., Embley, T. M., and Roger, A. J. (2003). Phylogenetic analyses of diplomonad genes reveal frequent lateral gene transfers affecting eukaryotes. *Curr. Biol.*, **13**, 94–104.

Archibald, J. M., Rogers, M. B., Toop, M., Ishida, K., and Keeling, P. J. (2003). Lateral gene transfer and the evolution of plastid-targeted proteins in the secondary plastid-containing alga *Bigelowiella* natans. *Proc. Natl. Acad. Sci. U S A*, **100**, 7678–7683.

Bachvaroff, T. R., Sanchez Puerta, M. V., and Delwiche, C. F. (2005). Chlorophyll c-containing plastid relationships based on analyses of a multigene data set with all four chromalveolate lineages. *Mol. Biol. Evol.*, **22**, 1772–1782.

Baldauf, S. L. (2003). The deep roots of eukaryotes. *Science*, **300**, 1703–1706.

Benveniste, R. E. (1985). The contribuitons of retroviruses to the study of mammalian evolution. In R. I. Macintyre, ed. *Molecular Evolutionary Genetics*. Plenum Press, New York.

Brown, C. G. (1990). Control of tropical theileriosis (*Theileria annulata* infection) of cattle. *Parassitologia*, **32**, 23–31.

Brown, J. R. and Doolittle, W. F. (1999). Gene descent, duplication, and horizontal transfer in the evolution of glutamyl- and glutaminyl-tRNA synthetases. *J. Mol. Evol.*, **49**, 485–495.

Cavalier-Smith, T. (1993). Kingdom protozoa and its 18 phyla. *Microbiol. Rev.*, **57**, 953–994.

Cavalier-Smith, T. (1999). Principles of protein and lipid targeting in secondary symbiogenesis: Euglenoid, Dinoflagellate, and Sporozoan plastid origins and the eukaryote family tree. *J. Eukaryot. Microbiol.*, **46**, 347–366.

Chen, W. M., Moulin, L., Bontemps, C., Vandamme, P., Bena, G., and Boivin-Masson, C. (2003). Legume symbiotic nitrogen fixation by beta-proteobacteria is widespread in nature. *J. Bacteriol.*, **185**, 7266–7272.

Doolittle, W. F. (1998). You are what you eat: a gene transfer ratchet could account for bacterial genes in eukaryotic nuclear genomes. *Trends Genet.*, **14**, 307–311.

Doolittle, W. F. (1999). Phylogenetic classification and the universal tree. *Science*, **284**, 2124–2129.

Esser, C., Ahmadinejad, N., Wiegand, C., Rotte, C., Sebastiani, F., Gelius-Dietrich, G., Henze, K., Kretschmann, E., Richly, E., Leister, D., Bryant, D., Steel, M. A., Lockhart, P. J., Penny, D., and Martin, W. (2004). A genome phylogeny for mitochondria among alpha-proteobacteria and a predominantly eubacterial ancestry of yeast nuclear genes. *Mol. Biol. Evol.*, **21**, 1643–1660.

Fairlamb, A. H. (2002). Metabolic pathway analysis in trypanosomes and malaria parasites. *Philos. Trans. R. Soc. Lond. B Biol. Sci.*, **357**, 101–107.

Fast, N. M., Kissinger, J. C., Roos, D. S., and Keeling, P. J. (2001). Nuclear-encoded, plastid-targeted genes suggest a single common origin for apicomplexan and dinoflagellate plastids. *Mol. Biol. Evol.*, **18**, 418–426.

Fichera, M. E. and Roos, D. S. (1997). A plastid organelle as a drug target in apicomplexan parasites. *Nature*, **390**, 407–409.

Foth, B. J., Ralph, S. A., Tonkin, C. J., Struck, N. S., Fraunholz, M., Roos, D. S., Cowman, A. F., and Mcfadden, G. I. (2003). Dissecting apicoplast targeting in the malaria parasite *Plasmodium falciparum*. *Science*, **299**, 705–708.

Funes, S., Davidson, E., Reyes-Prieto, A., Magallon, S., Herion, P., King, M. P., and Gonzalez-Halphen, D. (2002). A green algal apicoplast ancestor. *Science*, **298**, 2155.

Gardner, M. J., Bishop, R., Shah, T., De Villiers, E. P., Carlton, J. M., Hall, N., Ren, Q., Paulsen, I. T., Pain, A., Berriman, M., Wilson, R. J., Sato, S., Ralph, S. A., Mann, D. J., Xiong, Z., Shallom, S. J., Weidman, J., Jiang, L., Lynn, J., Weaver, B., Shoaibi, A., Domingo, A. R., Wasawo, D., Crabtree, J., Wortman, J. R., Haas, B.,

Angiuoli, S. V., Creasy, T. H., Lu, C., Suh, B., Silva, J. C., Utterback, T. R., Feldblyum, T. V., Pertea, M., Allen, J., Nierman, W. C., Taracha, E. L., Salzberg, S. L., White, O. R., Fitzhugh, H. A., Morzaria, S., Venter, J. C., Fraser, C. M., and Nene, V. (2005). Genome sequence of *Theileria parva*, a bovine pathogen that transforms lymphocytes. *Science*, **309**, 134–137.

Gardner, M. J., Hall, N., Fung, E., White, O., Berriman, M., Hyman, R. W., Carlton, J. M., Pain, A., Nelson, K. E., Bowman, S., Paulsen, I. T., James, K., Eisen, J. A., Rutherford, K., Salzberg, S. L., Craig, A., Kyes, S., Chan, M. S., Nene, V., Shallom, S. J., Suh, B., Peterson, J., Angiuoli, S., Pertea, M., Allen, J., Selengut, J., Haft, D., Mather, M. W., Vaidya, A. B., Martin, D. M., Fairlamb, A. H., Fraunholz, M. J., Roos, D. S., Ralph, S. A., Mcfadden, G. I., Cummings, L. M., Subramanian, G. M., Mungall, C., Venter, J. C., Carucci, D. J., Hoffman, S. L., Newbold, C., Davis, R. W., Fraser, C. M., and Barrell, B. (2002). Genome sequence of the human malaria parasite *Plasmodium falciparum*. *Nature*, **419**, 498–511.

Genereux, D. P. and Logsdon, J. M., Jr. (2003). Much ado about bacteria-to-vertebrate lateral gene transfer. *Trends Genet.*, **19**, 191–195.

Gogarten, J. P., Doolittle, W. F., and Lawrence, J. G. (2002). Prokaryotic evolution in light of gene transfer. *Mol. Biol. Evol.*, **19**, 2226–2238.

Gogarten, J. P. and Townsend, J. P. (2005). Horizontal gene transfer, genome innovation and evolution. *Nat. Rev. Microbiol.*, **3**, 679–687.

Harper, J. T. and Keeling, P. J. (2003). Nucleus-encoded, plastid-targeted glyceraldehyde-3-phosphate dehydrogenase (GAPDH) indicates a single origin for Chromalveolate plastids. *Mol. Biol. Evol.*, **20**, 1730–1735.

Huang, J., Mullapudi, N., Lancto, C. A., Scott, M., Abrahamsen, M. S., and Kissinger, J. C. (2004a). *Cryptosporidium parvum*: Phylogenomic evidence supports past endosymbiosis, intracellular and horizontal gene transfer. *Genome Biology*, **5**, (11), R88.

Huang, J., Mullapudi, N., Sicheritz-Ponten, T., and Kissinger, J. C. (2004b). A first glimpse into the pattern and scale of gene transfer in Apicomplexa. *Int. J. Parasitol.*, **34**, 265–274.

Huang, J., Xu, Y. and Gogarten, J. P. (2005). The presence of a haloarchaeal type tyrosyl-tRNA synthetase marks the opisthokonts as monophyletic. *Mol. Biol. Evol.*, **22**, 2142–2146.

Ishida, K. (2005). Protein targeting into plastids: a key to understanding the symbiogenetic acquisitions of plastids. *J. Plant Res.*, **118**, 237–245.

Jomaa, H., Wiesner, J., Sanderbrand, S., Altincicek, B., Weidemeyer, C., Hintz, M., Turbachova, I., Eberl, M.,

Zeidler, J., Lichtenthaler, H. K., Soldati, D., and Beck, E. (1999). Inhibitors of the nonmevalonate pathway of isoprenoid biosynthesis as antimalarial drugs. *Science*, **285**, 1573–1576.

Keeling, P. J. (2004). Diversity and evolutionary history of plastids and their hosts. *Am. J. Bot.*, **91**, 1481–1493.

Keeling, P. J. and Inagaki, Y. (2004). A class of eukaryotic GTPase with a punctate distribution suggesting multiple functional replacements of translation elongation factor 1alpha. *Proc. Natl. Acad. Sci. U S A*, **101**, 15380–15385.

Keithly, J. S., Langreth, S. G., Buttle, K. F., and Mannella, C. A. (2005). Electron tomographic and ultrastructural analysis of the *Cryptosporidium parvum* relict mitochondrion, its associated membranes, and organelles. *J. Eukaryot. Microbiol.*, **52**, 132–140.

Kohler, S., Delwiche, C. F., Denny, P. W., Tilney, L. G., Webster, P., Wilson, R. J., Palmer, J. D., and Roos, D. S. (1997). A plastid of probable green algal origin in Apicomplexan parasites. *Science*, **275**, 1485–1489.

Kurland, C. G., Canback, B., and Berg, O. G. (2003). Horizontal gene transfer: a critical view. *Proc. Natl. Acad. Sci. U S A*, **100**, 9658–9662.

Lawrence, J. G. and Roth, J. R. (1996). Selfish operons: horizontal transfer may drive the evolution of gene clusters. *Genetics*, **143**, 1843–60.

Loftus, B., Anderson, I., Davies, R., Alsmark, U. C., Samuelson, J., Amedeo, P., Roncaglia, P., Berriman, M., Hirt, R. P., Mann, B. J., Nozaki, T., Suh, B., Pop, M., Duchene, M., Ackers, J., Tannich, E., Leippe, M., Hofer, M., Bruchhaus, I., Willhoeft, U., Bhattacharya, A., Chillingworth, T., Churcher, C., Hance, Z., Harris, B., Harris, D., Jagels, K., Moule, S., Mungall, K., Ormond, D., Squares, R., Whitehead, S., Quail, M. A., Rabbinowitsch, E., Norbertczak, H., Price, C., Wang, Z., Guillen, N., Gilchrist, C., Stroup, S. E., Bhattacharya, S., Lohia, A., Foster, P. G., Sicheritz-Ponten, T., Weber, C., Singh, U., Mukherjee, C., El-Sayed, N. M., Petri, W. A., Jr., Clark, C. G., Embley, T. M., Barrell, B., Fraser, C. M., and Hall, N. (2005). The genome of the protist parasite *Entamoeba histolytica*. *Nature*, **433**, 865–868.

Mackenzie, W. R., Schell, W. L., Blair, K. A., Addiss, D. G., Peterson, D. E., Hoxie, N. J., Kazmierczak, J. J., and Davis, J. P. (1995). Massive outbreak of waterborne *cryptosporidium* infection in Milwaukee, Wisconsin: recurrence of illness and risk of secondary transmission. *Clin. Infect. Dis.*, **21**, 57–62.

Madern, D., Cai, X., Abrahamsen, M. S., and Zhu, G. (2004). Evolution of *Cryptosporidium parvum* lactate dehydrogenase from malate dehydrogenase by a very recent event of gene duplication. *Mol. Biol. Evol.*, **21**, 489–497. Epub 2003 Dec 23.

Martin, W. and Herrmann, R. G. (1998). Gene transfer from organelles to the nucleus: how much, what happens, and Why? *Plant Physiol.*, **118**, 9–17.

Martin, W., Rujan, T., Richly, E., Hansen, A., Cornelsen, S., Lins, T., Leister, D., Stoebe, B., Hasegawa, M., and Penny, D. (2002). Evolutionary analysis of *Arabidopsis*, cyanobacterial, and chloroplast genomes reveals plastid phylogeny and thousands of cyanobacterial genes in the nucleus. *Proc. Natl. Acad. Sci. U S A*, **99**, 12246–12251.

Mazodier, P. and Davies, J. (1991). Gene transfer between distantly related bacteria. *Annu. Rev. Genet.*, **25**, 147–171.

Mcfadden, G. I. and Roos, D. S. (1999). Apicomplexan plastids as drug targets. *Trends Microbiol.*, **7**, 328–333.

Nash, G., Kerschmann, R. L., Herndier, B., and Dubey, J. P. (1994). The pathological manifestations of pulmonary toxoplasmosis in the acquired immunodeficiency syndrome. *Hum. Pathol.*, **25**, 652–658.

Nelson, K. E., Clayton, R. A., Gill, S. R., Gwinn, M. L., Dodson, R. J., Haft, D. H., Hickey, E. K., Peterson, J. D., Nelson, W. C., Ketchum, K. A., Mcdonald, L., Utterback, T. R., Malek, J. A., Linher, K. D., Garrett, M. M., Stewart, A. M., Cotton, M. D., Pratt, M. S., Phillips, C. A., Richardson, D., Heidelberg, J., Sutton, G. G., Fleischmann, R. D., Eisen, J. A., Fraser, C. M., *et al.* (1999). Evidence for lateral gene transfer between Archaea and bacteria from genome sequence of *Thermotoga maritima*. *Nature*, **399**, 323–329.

Nozaki, H. (2005). A new scenario of plastid evolution: plastid primary endosymbiosis before the divergence of the "Plantae," emended. *J. Plant Res.*, **118**, 247–255.

Ochman, H., Lawrence, J. G., and Groisman, E. A. (2000). Lateral gene transfer and the nature of bacterial innovation. *Nature*, **405**, 299–304.

Ochman, H. and Moran, N. A. (2001). Genes lost and genes found: evolution of bacterial pathogenesis and symbiosis. *Science*, **292**, 1096–1099.

Pain, A., Renauld, H., Berriman, M., Murphy, L., Yeats, C. A., Weir, W., Kerhornou, A., Aslett, M., Bishop, R., Bouchier, C., Cochet, M., Coulson, R. M., Cronin, A., De Villiers, E. P., Fraser, A., Fosker, N., Gardner, M., Goble, A., Griffiths-Jones, S., Harris, D. E., Katzer, F., Larke, N., Lord, A., Maser, P., Mckellar, S., Mooney, P., Morton, F., Nene, V., O'neil, S., Price, C., Quail, M. A., Rabbinowitsch, E., Rawlings, N. D., Rutter, S., Saunders, D., Seeger, K., Shah, T., Squares, R., Squares, S., Tivey, A., Walker, A. R., Woodward, J., Dobbelaere, D. A., Langsley, G., Rajandream, M. A., Mckeever, D., Shiels, B., Tait, A., Barrell, B., and Hall, N. (2005). Genome of the host-cell transforming parasite *Theileria annulata* compared with *T. parva*. *Science*, **309**, 131–133.

Palmer, J. D. (2003). The symbiotic birth and spread of plastids: how many times and whodunit? *J. Phycol.*, **39**, 4–12.

Perkins, F. O., Barta, J. R., Clopton, R. E., Peirce, M. A., and Upton, S. J. (2000). Phylum Apicomplexa. In Society of Protozoologists, ed. *The Illustrated Guide to the Protozoa*, 2nd edn. Allen Press, Lawrence.

Petry, F. and Harris, J. R. (1999). Ultrastructure, fractionation and biochemical analysis of *Cryptosporidium parvum* sporozoites. *Int. J. Parasitol.*, **29**, 1249–1260.

Ralph, S. A., Van Dooren, G. G., Waller, R. F., Crawford, M. J., Fraunholz, M. J., Foth, B. J., Tonkin, C. J., Roos, D. S., and Mcfadden, G. I. (2004). Tropical infectious diseases: metabolic maps and functions of the *Plasmodium falciparum* apicoplast. *Nat. Rev. Microbiol.*, **2**, 203–216.

Rodriguez-Ezpeleta, N., Brinkmann, H., Burey, S. C., Roure, B., Burger, G., Loffelhardt, W., Bohnert, H. J., Philippe, H., and Lang, B. F. (2005). Monophyly of primary photosynthetic eukaryotes: green plants, red algae, and glaucophytes. *Curr. Biol.*, **15**, 1325–1330.

Roos, D. S., Crawford, M. J., Donald, R. G., Kissinger, J. C., Klimczak, L. J., and Striepen, B. (1999). Origin, targeting, and function of the apicomplexan plastid. *Curr. Opin. Microbiol.*, **2**, 426–432.

Schmatz, D. M. (1997). The mannitol cycle in Eimeria. *Parasitology*, **114** (Suppl.), S81–89.

Sicheritz-Ponten, T. and Andersson, S. G. (2001). A phylogenomic approach to microbial evolution. *Nucleic Acids Res.*, **29**, 545–552.

Slapeta, J. and Keithly, J. S. (2004). *Cryptosporidium parvum* mitochondrial-type HSP70 targets homologous and heterologous mitochondria. *Eukaryot. Cell*, **3**, 483–494.

Smith, H. V. and Corcoran, G. D. (2004). New drugs and treatment for cryptosporidiosis. *Curr. Opin. Infect. Dis.*, **17**, 557–564.

Stanhope, M. J., Lupas, A., Italia, M. J., Koretke, K. K., Volker, C., and Brown, J. R. (2001). Phylogenetic analyses do not support horizontal gene transfers from bacteria to vertebrates. *Nature*, **411**, 940–944.

Striepen, B. and Kissinger, J. C. (2004). Genomics meets transgenics in search of the elusive *Cryptosporidium* drug target. *Trends Parasitol.*, **20**, 355–358.

Striepen, B., Pruijssers, A. J. P., Huang, J., Li, C., Gubbels, M. J., Umejiego, N. N., Hedstrom, L., and Kissinger, J. C. (2004). Gene transfer in the evolution of parasite nucleotide biosynthesis. *Proc. Natl. Acad. Sci. U S A*, **101**, 3154–3159.

Striepen, B., White, M. W., Li, C., Guerini, M. N., Malik, S. B., Logsdon, J. M., Jr., Liu, C., and Abrahamsen, M. S. (2002). Genetic complementation in apicomplexan parasites. *Proc. Natl. Acad. Sci. U S A*, **99**, 6304–6309.

Tauxe, R. V., Cavanagh, T. R., and Cohen, M. L. (1989). Interspecies gene transfer in vivo producing an outbreak of multiply resistant shigellosis. *J. Infect. Dis.*, **160**, 1067–1070.

Van Der Giezen, M., Cox, S., and Tovar, J. (2004). The iron-sulfur cluster assembly genes iscS and iscU of *Entamoeba histolytica* were acquired by horizontal gene transfer. *BMC Evol. Biol.*, **4**, 7.

Waller, R. F., Keeling, P. J., Donald, R. G., Striepen, B., Handman, E., Lang-Unnasch, N., Cowman, A. F., Besra, G. S., Roos, D. S., and Mcfadden, G. I. (1998). Nuclear-encoded proteins target to the plastid in *Toxoplasma gondii* and *Plasmodium falciparum*. *Proc. Natl. Acad. Sci. U S A*, **95**, 12352–12357.

Williamson, D. H., Gardner, M. J., Preiser, P., Moore, D. J., Rangachari, K., and Wilson, R. J. (1994). The evolutionary origin of the 35 kb circular DNA of *Plasmodium falciparum*: new evidence supports a possible rhodophyte ancestry. *Mol. Gen. Genet.*, **243**, 249–252.

Wilson, R. J., Denny, P. W., Preiser, P. R., Rangachari, K., Roberts, K., Roy, A., Whyte, A., Strath, M., Moore, D. J., Moore, P. W., and Williamson, D. H. (1996). Complete gene map of the plastid-like DNA of the malaria parasite *Plasmodium falciparum*. *J. Mol. Biol.*, **261**, 155–172.

Wilson, R. J., Williamson, D. H., and Preiser, P. (1994). Malaria and other Apicomplexans: the "plant" connection. *Infect. Agents Dis.*, **3**, 29–37.

Xu, P., Widmer, G., Wang, Y., Ozaki, L. S., Alves, J. M., Serrano, M. G., Puiu, D., Manque, P., Akiyoshi, D., Mackey, A. J., Pearson, W. R., Dear, P. H., Bankier, A. T., Peterson, D. L., Abrahamsen, M. S., Kapur, V., Tzipori, S., and Buck, G. A. (2004). The genome of *Cryptosporidium hominis*. *Nature*, **431**, 1107–1112.

Zambryski, P. (1988). Basic processes underlying Agrobacterium-mediated DNA transfer to plant cells. *Annu. Rev. Genet.*, **22**, 1–30.

Zhu, G., Marchewka, M. J., and Keithly, J. S. (2000). *Cryptosporidium parvum* appears to lack a plastid genome. *Microbiol. Mol. Biol. Rev.*, **146**, 315–321.

PART III

Analyses of complete genomes

The nuts and bolts of sequencing protist genomes

Daniella C. Bartholomeu, Neil Hall, and Jane M. Carlton

The Institute for Genomic Research, Rockville, MD, USA

Introduction

The availability of the complete genome sequence of an organism is an essential step towards obtaining a comprehensive understanding of the organism's biology. In this regard, the species that make up the group Protista are important sequencing targets since: (i) many are human or animal pathogens of medical or veterinary importance, responsible for causing widespread morbidity and mortality; (ii) comparative genome analysis of free-living versus parasitic protists may provide a deeper understanding of the evolution of pathogenic processes; (iii) genome sequencing of eukaryotic taxa will yield insights into key events in eukaryote evolution, such as cellular compartmentalization and origin of cytoskeletal structures; and (iv) due to their paraphyletic nature, sequencing protists from diverse groups representing successive lineages is of tremendous benefit for inferences in comparative genomics.

Many protist genome sequencing projects have been launched in the last decade, primarily of parasitic protists but more recently of free-living protists too. Several have been completed and published, for example the genome of the human malaria parasite *Plasmodium falciparum* (Gardner *et al.* 2002), rodent malaria species *Plasmodium yoelii yoelii* (Carlton *et al.* 2002), *Plasmodium berghei* and *Plasmodium chabaudi chabaudi* (Hall *et al.* 2005), the enteric human pathogen *Entamoeba histolytica* (Loftus *et al.* 2005), two species of cattle parasite *Theileria parva* (Gardner *et al.* 2005) *Theileria annulata* (Pain *et al.* 2005), and two genotypes of *Cryptosporidium parvum* (Xu *et al.* 2004; Abrahamsen *et al.* 2004), and

three species of trypanosomatids, *Trypanosoma Cruzi* (El-Sayed *et al.*, 2005), *T. brucei.* (Berriman *et. al.,* 2005) and *Leishmania major* (Ivens *et al.*, 2005). Many others are close to completion, for example the sexually transmitted parasite *Trichomonas vaginalis* (Lyons and Carlton 2004), *Giardia lamblia*, *Tetrahymena thermophila*, and several more apicomplexans (for a comprehensive list of all completed and ongoing sequencing project, see http://www.genomesonline.org/).

Protist genomes are heterogeneous in many of their characteristics. Genome size can vary considerably, from the highly compact 2.9 Mb genome of *Encephalitozoon cuniculi* (Katinka *et al.* 2001) to hundred megabase genomes, as found in *Tetrahymena* and other free-living protists. Genome characteristics such as nucleotide bias, repeat content, and ploidy can also vary tremendously between species. Such characteristics affect the choice of genome sequencing strategy and are important to gauge at the outset. In addition, the choice of sequencing strategy involves an evaluation of the level of accuracy desired, as well as time to completion and availability of funds. The aim of this chapter is to introduce the reader to the nuts and bolts of the genome sequencing process, and to describe some of the pitfalls and circumventions used in current protist sequencing projects.

Choice of strain and genome characteristics

Before a genome sequencing project can commence, the strain or clone to be sequenced should be chosen and certain fundamental characteristics of the genome determined in order for the most suitable sequencing strategy to be identified. The chosen strain/clone is most frequently the standard or reference which has been developed by the community, for which a significant number of basic studies have been completed, and which can be propagated *in vivo* or *in vitro* in order to provide a stable source. For example the 3D7 clone of *P. falciparum* was chosen because the chromosomes of the parasite could be well separated by pulsed field gel electrophoresis (PFGE), enabling each chromosome to be assigned for sequencing to a particular sequencing center. In addition, good quality records for the isolation, cloning, and freezing of the parasite were available, and a significant body of research existed for various phenotypes of the clone. The merits of sequencing a lab. clone which may have been in continuous culture for many years and which may not be representative of field isolates, versus sequencing a clinical isolate for which little lab. data exists and which may not be clonal, need also to be discussed. The most important characteristic to then determine is the genome size. Methods of genome size determination include separation of chromosomes by PFGE (Leech *et al.* 2004), flow cytometry (Dolezel and Bartos 2005), Cot analysis (Wang and Wang 1985), Feulgen image analysis densitometry (Hardie *et al.* 2002), and chromosome spreading (Yuh *et al.* 1997). In addition, Cot analysis, which uses reassociation kinetics to determine the time for melted DNA to reanneal, gives an indication of the repeat content of a genome, which is a second important characteristic to be determined (see section Whole genome shotgun strategy, below). Finally, genome compositional bias needs to be determined. For example a genome that is highly [A + T]-rich will require special handling and specialized cloning procedures due to the instability of such DNA in regular plasmid vectors (Quail 2001; Gardner 2001).

Once a clone has been selected and genome characteristics have been determined as much as possible, the sequencing strategy can be chosen.

Whole genome sequencing basics

Since the current high-throughput sequencing technologies generate reads of ~600 to 800

nucleotides, the basic principle behind any sequencing strategy consists in fragmenting large DNA molecules into smaller pieces before the sequencing phase. Short fragments are used to construct distinct libraries (typically ~2- and ~10-kb insert size), which are end-sequenced. The sequences generated are reassembled into the original molecule by searching for overlaps. Three main strategies have been used to sequence complex genomes: (1) clone-by-clone strategy; (2) whole chromosome shotgun (WCS) strategy; and (3) whole genome shotgun (WGS) strategy (Fig. 9.1). Each one of these sequencing approaches relies on distinct assembly units: large insert libraries for clone-by-clone; chromosomes for WCS; and entire genomes for WGS. As discussed in more detail below, the clone-by-clone approach generates a more complete and accurate dataset, but requires an extensive physical mapping effort before the sequencing phase. The assembly, however, is much easier since it is performed locally at large insert clone level. WCS and WGS approaches do not require mapping data prior sequencing; therefore the data is generated much faster. On the other hand, the assembly is much more difficult since it is executed at chromosome and genome levels, respectively (Fig. 9.1).

Clone-by-clone strategy

The classical clone-by-clone strategy uses mapped, large insert clones as substrate to construct shotgun libraries. The initial step consists of digesting genomic DNA to pieces of 30 to 300 kb, which are cloned into vectors such as cosmid, fosmid, or

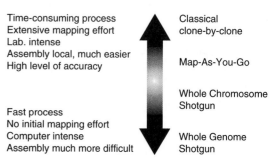

Figure 9.1 Comparison of sequencing strategies.

bacterial artificial chromosomes (BAC) systems. Physical mapping methods are used to identify overlapping clones and determine their relative order along the chromosome. Each clone is then submitted to shotgun sequencing, that is fragmenting the insert DNA into pieces of 1 to 3 kb and cloning them into plasmid vectors. Subclones are end-sequenced and algorithms are used to assemble the reads, thereby reconstituting the sequence of the large insert clone. The main advantage of the clone-by-clone method is that the local assembly of large insert clones minimizes complications due to conserved, genome-wide repeats and long, near identical segmental duplications. On the other hand, it requires an extensive mapping effort and the construction of a large number of libraries, in general two or three for each sequenced clone. A detailed explanation of each step is given below.

Physical mapping

The physical mapping step consists of identifying and ordering overlapping large insert clones that span a region of interest. The most common strategy used involves a combination of fingerprinting and marker hybridization. Briefly, the large insert size library is analyzed by restriction pattern and computer algorithms are used to compare the fingerprinting, order the clones into contigs, and infer the extent of clone overlap. The contigs can then be assigned to a chromosome by the hybridization of clone-derived markers to PFGE blots. Alternatively, sequence markers, such as EST (expressed sequence tag) and STS (sequence tagged site), previously assigned to a unique chromosome can be used to screen high density filters containing all clones from the library. The contig map formed by a set of ordered clones that confer around 10-fold coverage to a genomic region is known as a "sequence ready map". Once this map is constructed, the next step consists in selecting a minimally overlapping path of clones, called a "minimal tiling path", that together cover a region of interest. The selected clones are then submitted to shotgun sequencing (Fig. 9.2). Yeast Artificial Chromosome (YAC), BAC, and cosmids are the vector systems typically used to construct physical maps. Inserts up to 1 Mb in size can be cloned in YACs, up to 200 kb in BACs, and up to 40 kb in cosmid vectors. The genome size dictates *a priori* the choice of the cloning system. Bigger genomes require a larger average insert size to minimize the number of clones necessary to cover the entire genome. Other aspects are, however, also taken into account. BAC clones are more stable than YACs and cosmids, where the occurrence of deletions, chimeric inserts, and DNA rearrangements is common. Due to their low vector-to-insert ratio, BAC clones are excellent substrate for shotgun sequencing, while contamination with yeast DNA during the subcloning of short fragments of YAC inserts is a common occurrence. On the other hand, highly repetitive sequences seem to be under-represented into bacterial systems and yeast is able to maintain $[A + T]$ rich DNA more effectively than *Escherichia coli* (Gardner 2001).

Shotgun library

An essential step for the success of shotgun sequencing is the generation of random libraries. DNA fragmentation using physical shearing methods, such as nebulization, sonication, or hydrodynamic shearing are the most used approaches since enzymatic cleavage introduces bias inherent to a restriction pattern (Thorstenson *et al.* 1998). Since mechanical fragmentation generates DNA molecules containing $5'$ and $3'$ overhang ends, the fragments need to be end-repaired before cloning. Various enzymatic procedures can be used for this purpose. Single-strand exonucleases, such as mung bean nuclease, digest specifically single strand regions generating blunt ends. Alternatively, the DNA can be treated with T4 DNA polymerase, which removes $3'$ overhangs and has a polymerase activity that fills in $5'$ overhangs. It is also important to construct libraries with a narrow insert size range. To achieve this goal and to remove the enzymes of the end-repair reaction, the blunt-end fragments are size-selected by electrophoresis. Next, adaptor sequences are added to the end of the fragments, which are then cloned into the chosen plasmid vector. To assess library quality, a large number of clones are digested with appropriate restriction enzymes to check for the presence of the insert and to estimate

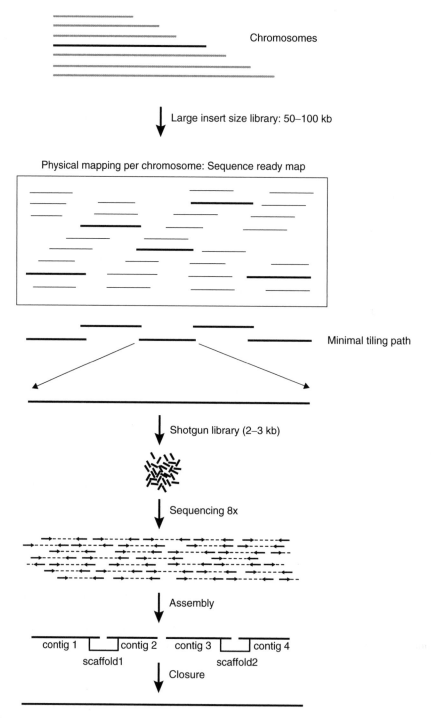

Figure 9.2 Schematic representation of clone-by-clone sequencing. This strategy follows a "map, then sequence" principle. Genomic DNA is fragmented into large pieces, which are cloned into suitable vectors. Physical maps are constructed for each chromosome producing a set of overlapping clones (sequence-ready maps), from which clones will be selected for sequencing (minimal tiling path). Each individual mapped clone is then submitted to shotgun sequencing, the reads are assembled, and gaps are closed.

the average and standard deviation of the insert size library. In addition, a few 384-well plates of clones are sequenced to evaluate the success rate and to confirm that the reads originate from the organism of interest, before high-throughput sequencing is implemented.

Sequencing and assembly

Clones from the shotgun library are randomly selected and sequenced at both ends. An initial assembly is then performed to achieve 6 to 8× sequence coverage. The Lander–Waterman model (Lander and Waterman 1988) is used to calculate the number of random clones that must be end-sequenced to obtain this level of coverage. According to this model, at ~8× sequence coverage, only about 1% of the target region remains unsequenced. In practice, however, the numbers of gaps are frequently greater than predicted. The assembly process has become highly accurate by the development of base-calling programs. Phred, a widely used base-calling program (Ewing and Green 1998; Ewing *et al.* 1998), reads the chromatograms generated by the automated sequencers, attributes a base for each identified peak and assigns a quality value for each base sequenced. The quality value is associated with the probability that a base is called incorrectly. One error per 1000 sequenced bases is acceptable for most sequencing projects. The sequence and the quality value files generated by base-calling programs are used by the assembler program to generate the contig sequence, which is not the consensus, but instead corresponds to the highest quality nucleotide at each position.

Genome finishing

The finishing phase is the most time-consuming of a sequencing project. During this step gaps between contigs are closed, misassemblies are corrected, and regions with low base quality ("low-coverage") are resequenced. These problems can be exacerbated by the presence of repetitive sequences, unclonable regions, high [A + T] content, or non-sequenced regions due to secondary structures. The use of alternative sequencing chemistries, such as dye primer, DMSO, and 7-deaza-dGTP analog nucleotide can reduce the effect of high [G + C] content, palindromic regions, and homopolymeric areas. Generally, misassemblies due to the presence of repeats and gaps are the most difficult problems to resolve during the finishing phase. "Sequence gaps", those for which a template is available, are usually easy to close by resequencing of clones pointing towards a gap using long-run sequencing conditions. On the other hand, "physical gaps" span regions that are not covered by an existing clone and result in contigs whose order with respect to each other is unknown. These gaps are more difficult to close and a variety of approaches are used to do so, including primer walking, transposon-mediated sequencing, combinatorial and multiplex-PCR. A wide variety of problems can appear and must be addressed on a case-by-case basis.

Map-as-you-go: an alternative clone-by-clone strategy

Since the physical mapping step is an extremely laborious and time consuming process, an alternative clone-by-clone strategy that eliminates the need for any prior mapping, called "map-as-you-go", has been proposed (Venter *et al.* 1996). This strategy uses BAC libraries as the main sequencing substrate and was used for sequencing *T. brucei* chromosomes 2 to 8 (Berriman *et al.* 2005; El-Sayed *et al.* 2000). The strategy has two main steps: generation of a BAC-ends database and chromosome walking. The BAC-ends database is the main source for markers used to select clones for sequencing. To construct the database, a large number of BAC clones are sequenced at both ends, generating paired end-sequence markers. Assuming a random BAC library, the end-sequences provide markers distributed uniformly throughout the genome. The marker density depends basically on the number of end-sequences generated, the library insert size, and the genome size. The first step to sequence an entire chromosome or genome using the map-as-you-go strategy is the identification of sequence markers, such as STSs or ESTs, that are specific for each chromosome. The chromosome-specific markers are identified by hybridizing ESTs and STSs sequences to PFGE blots. These markers are then used to search against the BAC-ends database or to screen a high density filter containing the gridded BAC library.

Positive BAC clones are then analyzed by finger-printing. The idea behind the fingerprinting process relies on the assumption that BACs showing similar restriction patterns are located in the same genomic region. Among those clones showing fingerprinting consistency, one is selected as a "seed" BAC and submitted to shotgun sequencing. Generally, two or more seed BACs anchored in distinct regions are selected per chromosome.

To generate BAC contigs, the sequence of a closed BAC is searched against the BAC-ends database to identify overlapping clones. Alternatively, overlapping clones can be identified through the screening of high density filters containing the gridded BAC library, using as probes DNA fragments amplified from the end of the BAC. The candidate clones for extension are fingerprinted and a new BAC clone showing internal consistency among the fingerprints and minimal overlap at the end of the closed BAC is then selected for sequencing. A problematic BAC extension is when the 5′ or 3′ end of a sequenced BAC is anchored in a repetitive region. In this case, the search against the BAC-ends database results in high depth coverage at the end of the closed BAC and the fingerprints reveal discrepancy among the candidates, indicating the BACs are located in distinct genomic regions. To overcome this, overlapping BAC clones anchored in the nearest non-repetitive region are selected for fingerprinting analysis.

Whole chromosome shotgun (WCS) strategy

The WCS strategy uses chromosome shotgun libraries as sequencing substrate. Chromosomal bands are gel-purified from PFGE gels, extracted by agarase digestion, sheared into 1 to 2-kb fragments, and cloned into plasmid vectors. Randomly-picked clones are then end-sequenced and the sequences assembled into contigs. Gaps between the contigs are closed using targeted closure procedures (see section Genome finishing, above). STSs, microsatellite markers, optical, and HAPPY (HAPloid DNA samples using the Pol Ymerase chain) mapping data can be used to order the contigs on the chromosome and to validate the genome assembly (see section Genome mapping and assembly validation, below). The WCS strategy is applicable to genomes whose molecular karyotype is well known. However, some adjustments to the PFGE protocol may be necessary, for example the use of different electrophoresis conditions to resolve all chromosomes (Melville *et al.* 2000; Ivens *et al.* 1998; Carlton *et al.* 1999), and there may be problems related to size variation between homologous chromosomes and comigration of non-homologous chromosomes. Contamination with DNA from other chromosomes may also hamper the assembly and closure processes, and requires a larger number of reads to produce the necessary sequence coverage. In addition, PFGE is only suitable for chromosomes less than ∼6 Mb in size, and some protist genomes, for example that of the parabasalid *T. vaginalis*, contain chromosomes of a much larger size (Yuh *et al.* 1997).

A modified WCS strategy, that exploits the merits of both clone-by-clone and WCS approaches, was adopted by the *P. falciparum* genome project (Gardner *et al.* 2002) and *T. brucei* chromosomes 1, 9, 10, and 11 (Hall *et al.* 2003). In addition to the sequencing of chromosome-specific libraries, low coverage shotgun sequences from a minimal tiling set of overlapping YAC (*P. falciparum*) or P1 and BAC (*T. brucei*) clones on the chromosome(s) were also obtained. Sequences from the large-insert clones are used to bin the corresponding WCS reads, providing a framework for local assembly. This reduced the complexity of the assembly and finishing processes, and enhanced the accuracy of the final assembly.

Whole genome shotgun (WGS) strategy

In WGS, the whole genome of an organism is sheared and libraries with different insert sizes are constructed. Clones from each library are picked at random and sequenced at both ends. All sequence reads are then compared against each other and overlapping information is used to build contigs. Paired end-sequences, an essential element in WGS, provide orientation and spacing information that are used to order the contigs and infer distance between them. A set of contigs that are ordered, oriented, and positioned with respect

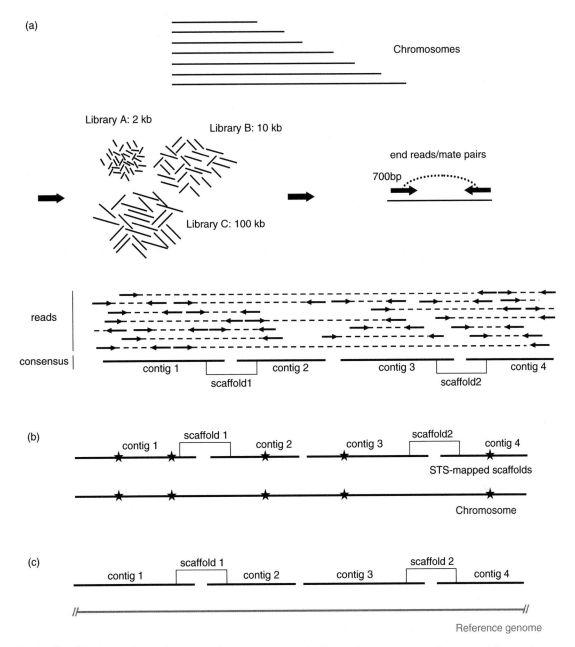

Figure 9.3 Overview of whole genome shotgun sequencing. This approach follows a "sequence, then map" principle. (a) The mapping phase is skipped and shotgun sequencing generates a large number of reads from different insert size libraries. Robust algorithm programs assemble the reads into contigs, which are ordered into scaffolds using mate pair information. (b) Scaffolds can be assigned to chromosomes using chromosome specific markers. (c) Assembled data can be used in whole genome comparisons with a closely related organism.

to each other by mate-pair reads is known as a scaffold. Scaffolds are the main product of WGS and can be assigned to chromosomes using chromosome-specific markers (Fig. 9.3). Linkage groups and optical maps, among other techniques, can be used to order the contigs and validate the genome assembly (see section Genome mapping and assembly validation, below). Gaps between the contigs are closed using targeted closure procedures (see section Genome finishing, above). WGS does not require any previous mapping effort before the sequencing phase and most of the protist genome projects launched recently have adopted this approach.

The use of distinct libraries is an important feature in WGS, since each library has a definite role during assembly. Typically, small (2–3 kbp), medium (10–15 kbp), and large (e.g. 50 and 100 kbp) insert size libraries are used as sequencing substrate. The end-sequences from small and medium insert size libraries provide the bulk of the sequence coverage, whereas end-sequences from large insert clones are essential for ordering of the contig.

The assembly of WGS data is extremely complex, since it involves all-against-all read comparison. Robust assembler algorithms have been developed to perform this task such as Celera Assembler (Myers et al. 2000), Arachne (Batzoglou et al. 2002), Jazz (Aparicio et al. 2002), RePS (Wang et al. 2002), and Phusion (Mullikin and Ning 2003). Due to the complexity of the assembly, the quality of the input sequences must be much higher than that used in other sequencing strategies. For instance, regions of low quality sequence need to be trimmed much more assertively to reduce the chance of false overlaps when comparing thousands or millions of reads in the initial phase of the assembly process. Also, most of the reads must be in mate pairs. Typically, more than 75% of reads in pairs is acceptable. In addition, the insert length variance in a library should be less than 10%. Libraries with narrow insert size distribution allow better assembly because of more accurate distance prediction between mate-pair sequences. This may represent a challenge especially when constructing large-insert size libraries, since large DNA fragments cannot be accurately resolved in ordinary agarose gels. The average and the variance of insert sizes previously estimated on agarose gel can be re-evaluated by mapping the reads on the assembled data. The corrected values are then used in the next round of the assembly.

A variation on the WGS strategy for the sequencing of highly repetitive DNA is Cot-based cloning and sequencing (CBCS), a synthesis of Cot analysis, DNA cloning, and high-throughput sequencing (Peterson et al. 2002). This method fractionates DNA into components that differ substantially in copy number, for example containing highly repetitive, moderately repetitive, and single/low copy sequences. Libraries of each component can then be generated and sequenced. CBCS reduces the number of clones that need to be sequenced to capture the sequencing complexity of eukaryotic genomes by approximately two-thirds (Whitelaw et al. 2003).

Genome mapping and assembly validation

For many projects, long-range genome maps have been generated to assist the assembly of a genome. For example the *T. brucei* genome project employed BAC maps, genetic maps, and optical maps, and the *P. falciparum* genome project utilized YAC maps, genetic maps, and HAPPY maps. Until recently a map of some form was considered a prerequisite for a genome sequencing project, but as sequencing and assembly technologies have improved and the number of genome projects has increased, genome mapping has become less common place. For the more complex genomes these maps are still beneficial in order to validate assemblies and to guide closure processes.

There are several strategies that can be employed to map a genome, depending upon the type of sequencing project and characteristics of the genome. The procedures for genetic mapping and clone mapping have been described in detail before (see the following and references therein McKusick 1991; McKusick and Ruddle 1987; Su et al. 1999; McPherson et al. 2001; Rubio et al. 1995; Tao et al. 2001) so here we will concentrate on two more recent technologies that have been used extensively for protist genomes, optical mapping and HAPPY mapping. Both are relatively cheap, rapid to deploy, and are independent of the particular physical properties of a genome; they are therefore well suited for almost any project.

HAPPY mapping

HAPPY mapping (Dear and Cook 1993; Dear and Cook 1989) is effectively an *in vitro* simulation of meiosis (Fig. 9.4). Haploid quantities of DNA are used as template for PCR reactions that detect the presence or absence of sequence markers, thereby allowing identification of cosegregating alleles. The method is independent from cloning or other biological steps and so makes it impervious to the peculiarities of different genomes and robust against artifacts and errors. It is the only technique to have made possible the accurate mapping – and subsequent finishing – of the genome of *Dictyostelium discoideum* (Eichinger *et al.* 2005). The method has also been applied successfully to a range of other genomes including the parasites *Plasmodium* (Hall *et al.* 2002), *Cryptosporidium* (Piper *et al.* 1998), and *Eimeria* (Shirley *et al.* 2004). The method can be broken down into three steps;

(1) generation of a mapping panel; (2) screening; and (3) analysis:

1. Generation of a mapping panel. High molecular weight DNA is sheared using physical methods. The size of DNA used depends on how fine the map will be: small fragments are used for fine maps, and large fragments for long-range maps. The DNA is then diluted and aliquoted into a microtiter plate so that each well contains roughly half a genome quantity of DNA. By adding linkers to the DNA fragments and amplifying the DNA this mapping panel can be replicated and used to place additional markers on the map.

2. Screening. The mapping panel is used as the template for screening each marker. Primers are designed to cDNA clones or shotgun sequences, and each well of the mapping panel is probed using these primers in a PCR reaction and then run on agrose gels to identify if PCR products were produced, thereby indicating that the well contained the target sequence.

3. Analysis. The objective of this process is to identify cosegregation of markers, that is markers that are

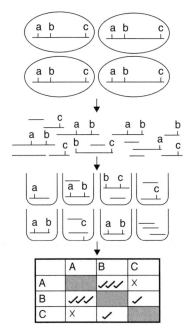

1. Cells containing high molecular weight DNA

2. DNA is sheared into fragments. The size of the fragments will depend on how fine the map is going to be.

3. DNA is diluted and aliquoted microtitre plate wells. So each well contains less than one haploid genome.

4. After PCR detection wells are scored for the presence and absence of markers.

Figure 9.4 HAPPY mapping technology. Haploid quantities of DNA are used as template for PCR reactions that detect the presence or absence of sequence markers, thereby allowing identification of co-segregating alleles.

amplified together more often than by chance. Hence markers which are often found in the same well are likely to be close together while markers that are randomly assorted are likely to be further apart. Because mapping points can be added throughout the closure process the map can be made as dense as is needed. Markers can be deliberately placed in positions that are difficult to assemble or on contigs that cannot be placed in the assembly.

Optical mapping

Optical mapping uses restriction digestion to create a long-range map of genomes and chromosomes (Samad *et al.* 1995). This protocol was originally applied to BAC and YAC clones (Cai *et al.* 1995; Cai *et al.* 1998), but was later extended to map the entire *P. falciparum* genome (Lai *et al.* 1999). The technique involves mounting high molecular weight DNA on glass surfaces, digesting the DNA with a restriction enzyme, and visualization of the restriction pattern using a light microscope (Fig. 9.5). Since the restriction pattern is measured *in situ*, the order of restriction fragments can be seen. By taking measurements using an automated image capture system, the restriction pattern across the whole genome can be assembled in a similar way to the assembly of sequencing reads. The result is an accurate restriction map of the genome which can be used to validate assemblies and order contigs. Although it does not generate dense mapping points like HAPPY mapping, the advantage of this system is that it requires no prior knowledge of the sequence and is completely independent of repeated sequences.

Genome sequence annotation

Once genome sequencing is complete, the most difficult component of the genomics process begins – that of gene prediction and annotation. A good biological description of a genome sequence greatly increases the utility of the sequence, and provides a framework for researchers for their analyses. However, there are different levels of annotation, most often dictated by the level of sequence coverage. Automated annotation uses a series of automated analyses with little manual

inspection of the results, with the consequence that annotation is error-prone and must be treated as such. Manual curation is most commonly used for gold-standard projects that have been sequenced to full coverage, and provides the best quality form of annotation. An appreciation of these differences is required by any researcher interpreting annotation data.

The subject of genome sequence annotation requires a chapter unto itself, so in the interests of space readers are recommended to read other more exhaustive reviews where indicated.

Gene prediction

Many factors influence the quality of gene predictions: partial genome sequence data, unusual genome composition, and the presence of introns can all influence the outcome of gene finding algorithms. The accuracy of gene predictions can only truly be determined once large-scale gene expression data is available. There are a number of *ab initio* algorithms, most of which use some form of statistical model such as Hidden Markov Models (HMMs). Other algorithms integrate multiple sources of evidence for gene prediction, such as JIGSAW (Allen and Salzberg 2005). For an in-depth discussion and further references, see Berriman and Harris (2004).

Functional annotation

Once genes have been predicted, assigning a function is the next step in the annotation pathway. Comparing sequences from different organisms is the center point of most predictions of gene function, and there are many different algorithms which have been developed to do this (for a review see Frazer *et al.* 2003). In addition, searches against curated protein domain databases and other sequence databases provides invaluable information, and can be used to infer homology relationships such as orthology and paralogy. Once again, the issues of assigning function based upon poor quality data must be realized (see Valencia 2005 for a review). For a more in-depth report of the common procedures used for functional annotation, see Berriman and Harris (2004).

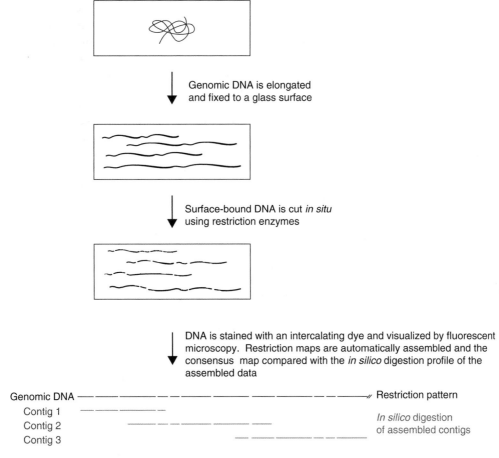

Figure 9.5 Optical mapping technology. High molecular weight DNA is elongated on a silanized glass surface and digested *in situ* with a restriction enzyme. DNA is stained with an intercalating dye to visualize the fragments, and inspected under fluorescence microscopy. The sizes of restriction fragments are estimated by measuring the relative fluorescence intensity and the distance between the cleavage sites. A consensus map for the whole genome is then obtained. The assembled contig sequences are digested *in silico* with the same enzyme used to generate the optical mapping data and the fragment sizes obtained from the two approaches are optimally aligned.

Protist sequencing projects and challenges to sequencing

This final section provides examples of several protist genome sequencing projects, which have presented particular problems or issues during their progression.

The *T. cruzi* sequencing project: a heterozygous genome

The draft sequence of the *T. cruzi* genome has been completed recently (El-Sayed *et al.* 2005, and presents a good example of the problems associated with a highly heterozygous genome. Initially, a map-as-you-go strategy was selected to sequence the CL Brener genome. However, the highly repetitive content of the genome imposed serious limitations on the generation of continuous BAC sequences and hampered the extension of closed BACs. An alternative sequencing strategy in the form of WGS was chosen in order to concentrate efforts on the sequencing of the non-repetitive part of the genome, and since WCS had to be excluded on the basis of the organism's complex karyotype. Moreover, the availability of complete genome sequences of two other trypanosomatids, *T. brucei* (Berriman *et al.*, 2005) and *L. Major* (Ivens *et al..*, 2005 and the possibility of performing three-way whole genome comparisons was another compelling rationale in favor of WGS. However, the genome was found to exhibit a high level of allelic variation, and this, in addition to the overall repeat content, meant that assembly of the WGS data was challenging. Standard assembly software is not able to handle highly polymorphic genomes, since ambiguities derived from allelic variations need to be discriminated from sequencing errors. Base-calling errors are frequently associated with low quality values and are not confirmed by other reads, so that in the case of highly polymorphic genomes, a much higher level of sequence coverage should be generated to ensure these inferences are reliable. In the case of *T. cruzi*, $1.4 \times$ coverage of the genome was generated in order to circumvent the problem of heterozygosity. Allelic variants were identified for approximately half of the *T. cruzi* genes.

The *Entamoeba histolytica* sequencing project: a highly repetitive genome

The *E. histolytica* genome project began simultaneously at two sequencing centers, TIGR and the Wellcome Trust Sanger Institute. Due to the high [A + T] content of the genome, large insert clones were difficult to construct and a WGS approach was used, with half of the sequence reads being generated by each center. The genome project had several technical hurdles to overcome; the genome is highly repetitive due to a high concentration of retrotransposons, and contains long arrays of tRNA genes. Additionally, the nucleus contains many circular DNA molecules encoding rDNA genes which were over-represented in shotgun libraries. No mapping information existed to guide the assembly and the genome currently exists as a raw assembly with many gaps and unresolved repeats (Loftus *et al.* 2005).

The *Dictyostelium discoidium* sequencing project: a compositionally biased genome

D. discoideum has a 33 Mb genome which is 77.6% [A + T] rich and highly repetitive, and was therefore one of the most difficult protist genomes to

sequence (Eichinger *et al.* 2005; Glockner *et al.* 2002). The genome was sequenced by a number of different genome sequencing centers using a WCS approach. As with the *P. falciparum* genome project, a number of different mapping strategies were employed to guide the assembly and closure project: a high density HAPPY map (Williams and Firtel 2000), YAC maps (Kuspa and Loomis 1996), and a genetic map (Welker and Williams 1982), as well as physical maps of genes to chromosomes (Kuspa *et al.* 1992). YACs were used to clone large fragments of DNA as the [G + C] content of the genome was so low that clones were unstable when cloned into *E. coli*. YAC clones were then ordered and orientated using restriction mapping. Mapped YAC clones were subject to shotgun sequencing and these YAC-derived reads added to the assembly of the chromosome-derived reads, so that regions of the assembly containing YAC-derived reads could be localized to specific mapped regions (Eichinger *et al.* 2005; Glockner *et al.* 2002). Primers were designed from the shotgun sequencing to produce the high density HAPPY map which, on certain chromosomes, generated a marker density of one marker per 6 kb. A major problem with the WCS approach is that chromosome preparations are never pure, and therefore sequence data will be contaminated with sequence derived from other chromosomes. This problem was overcome in *D. discoideum* using high resolution maps, and without this resource the problem could have been insurmountable. Although most of the repetitive sequence has now been assembled, there remains 23 kb in "floating contigs" that cannot been positioned in the genome since all of these contigs are bounded by repeats. In addition, there are roughly 300 gaps remaining in the genome, many caused by problems with repeats.

References

Abrahamsen, M. S., Templeton, T. J., Enomoto, S., Abrahante, J. E., Zhu, G., Lancto, C. A., Deng, M., Liu, C., Widmer, G., Tzipori, S., Buck, G. A., Xu, P., Bankier, A. T., Dear, P. H., Konfortov, B. A., Spriggs, H. F., Iyer, L., Anantharaman, V., Aravind, L., and Kapur, V. (2004). Complete genome sequence of the apicomplexan, *Cryptosporidium parvum*. *Science*, **304**, 441–445.

Allen, J. E. and Salzberg, S. L. (2005). JIGSAW: integration of multiple sources of evidence for gene prediction. *Bioinformatics*, **21**, 3596–3603.

Aparicio, S., Chapman, J., Stupka, E., Putnam, N., Chia, J. M., Dehal, P., Christoffels, A., Rash, S., Hoon, S., Smit, A., Gelpke, M. D., Roach, J., Oh, T., Ho, I. Y., Wong, M., Detter, C., Verhoef, F., Predki, P., Tay, A., Lucas, S., Richardson, P., Smith, S. F., Clark, M. S., Edwards, Y. J., Doggett, N., Zharkikh, A., Tavtigian, S. V., Pruss, D., Barnstead, M., Evans, C., Baden, H., Powell, J., Glusman, G., Rowen, L., Hood, L., Tan, Y. H., Elgar, G., Hawkins, T., Venkatesh, B., Rokhsar, D., and Brenner, S. (2002). Whole-genome shotgun assembly and analysis of the genome of *Fugu rubripes*. *Science*, **297**, 1301–1310.

Batzoglou, S., Jaffe, D. B., Stanley, K., Butler, J., Gnerre, S., Mauceli, E., Berger, B., Mesirov, J. P., and Lander, E. S. (2002). ARACHNE: a whole-genome shotgun assembler. *Genome Res.*, **12**, 177–189.

Berriman, M., Ghedin, E., Hertz-Fowler, C., Blandin, G., Renauld, H., Bartholomeu, D. C., Lennard, N. J., Caler, E., Hamlin, N. E., Haas, B., Bohme, U., Hannick, L., Aslett, M. A., Shallom, J., Marcello, L., Hou, L., Wickstead, B., Alsmark, U. C., Arrowsmith, C., Atkin, R. J., Barron, A. J., Bringaud, F., Brooks, K., Carrington, M., Cherevach, I., Chillingworth, T. J., Churcher, C., Clark, L. N., Corton, C. H., Cronin, A., Davies, R. M., Doggett, J., Djikeng, A., Feldblyum, T., Field, M. C., Fraser, A., Goodhead, I., Hance, Z., Harper, D., Harris, B. R., Hauser, H., Hostetler, J., Ivens, A., Jagels, K., Johnson, D., Johnson, J., Jones, K., Kerhornou, A. X., Koo, H., Larke, N., Landfear, S., Larkin, C., Leech, V., Line, A., Lord, A., Macleod, A., Mooney, P. J., Moule, S., Martin, D. M., Morgan, G. W., Mungall, K., Norbertczak, H., Ormond, D., Pai, G., Peacock, C. S., Peterson, J., Quail, M. A., Rabbinowitsch, E., Rajandream, M. A., Reitter, C., Salzberg, S. L., Sanders, M., Schobel, S., Sharp, S., Simmonds, M., Simpson, A. J., Tallon, L., Turner, C. M., Tait, A., Tivey, A. R., Van Aken, S., Walker, D., Wanless, D., Wang, S., White, B., White, O., Whitehead, S., Woodward, J., Wortman, J., Adams, M. D., Embley, T. M., Gull, K., Ullu, E., Barry, J. D., Fairlamb, A. H., Opperdoes, F., Barrell, B. G., Donelson, J. E., Hall, N., Fraser, C. M., Melville, S. E., El-Sayed, N. M. (2005). The genome of the African trypanosome *Trypanosoma brucei*. *Science*, **309**, 416–422.

Berriman, M. and Harris, M. (2004). Annotation of parasite genomes. In S. E. Melville, ed. *Parasite Genomics Protocols*. Humana Press, New Jersey.

Cai, W., Aburatani, H., Stanton, V. P., Jr., Housman, D. E., Wang, Y. K., and Schwartz, D. C. (1995). Ordered restriction endonuclease maps of yeast artificial chromosomes created by optical mapping on surfaces. *Proc. Natl. Acad. Sci. U S A*, **92**, 5164–5168.

Cai, W., Jing, J., Irvin, B., Ohler, L., Rose, E., Shizuya, H., Kim, U. J., Simon, M., Anantharaman, T., Mishra, B., and Schwartz, D. C. (1998). High-resolution restriction maps of bacterial artificial chromosomes constructed by optical mapping. *Proc. Natl. Acad. Sci. U S A*, **95**, 3390–3395.

Carlton, J. M., Angiuoli, S. V., Suh, B. B., Kooij, T. W., Pertea, M., Silva, J. C., Ermolaeva, M. D., Allen, J. E., Selengut, J. D., Koo, H. L., Peterson, J. D., Pop, M., Kosack, D. S., Shumway, M. F., Bidwell, S. L., Shallom, S. J., van Aken, S. E., Riedmuller, S. B., Feldblyum, T. V., Cho, J. K., Quackenbush, J., Sedegah, M., Shoaibi, A., Cummings, L. M., Florens, L., Yates, J. R., Raine, J. D., Sinden, R. E., Harris, M. A., Cunningham, D. A., Preiser, P. R., Bergman, L. W., Vaidya, A. B., van Lin, L. H., Janse, C. J., Waters, A. P., Smith, H. O., White, O. R., Salzberg, S. L., Venter, J. C., Fraser, C. M., Hoffman, S. L., Gardner, M. J., and Carucci, D. J. (2002). Genome sequence and comparative analysis of the model rodent malaria parasite *Plasmodium yoelii yoelii*. *Nature*, **419**, 512–519.

Carlton, J. M., Galinski, M. R., Barnwell, J. W., and Dame, J. B. (1999). Karyotype and synteny among the chromosomes of all four species of human malaria parasite. *Mol. Biochem. Parasitol.*, **101**, 23–32.

Dear, P. H. and Cook, P. R. (1989). Happy mapping: a proposal for linkage mapping the human genome. *Nucleic Acids Res.*, **17**, 6795–6807.

Dear, P. H. and Cook, P. R. (1993). Happy mapping: linkage mapping using a physical analogue of meiosis. *Nucleic Acids Res.*, **21**, 13–20.

Dolezel, J. and Bartos, J. (2005). Plant DNA flow cytometry and estimation of nuclear genome size. *Ann. Bot. (Lond.)*, **95**, 9–110.

Eichinger, L., Pachebat, J. A., Glockner, G., Rajandream, M. A., Sucgang, R., Berriman, M., Song, J., Olsen, R., Szafranski, K., Xu, Q., Tunggal, B., Kummerfeld, S., Madera, M., Konfortov, B. A., Rivero, F., Bankier, A. T., Lehmann, R., Hamlin, N., Davies, R., Gaudet, P., Fey, P., Pilcher, K., Chen, G., Saunders, D., Sodergren, E., Davis, P., Kerhornou, A., Nie, X., Hall, N., Anjard, C., Hemphill, L., Bason, N., Farbrother, P., Desany, B., Just, E., Morio, T., Rost, R., Churcher, C., Cooper, J., Haydock, S., van Driessche, N., Cronin, A., Goodhead, I., Muzny, D., Mourier, T., Pain, A., Lu, M., Harper, D., Lindsay, R., Hauser, H., James, K., Quiles, M., Madan Babu, M., Saito, T., Buchrieser, C., Wardroper, A., Felder, M., Thangavelu, M., Johnson, D., Knights, A., Loulseged, H., Mungall, K., Oliver, K., Price, C., Quail, M. A., Urushihara, H., Hernandez, J., Rabbinowitsch,

E., Steffen, D., Sanders, M., Ma, J., Kohara, Y., Sharp, S., Simmonds, M., Spiegler, S., Tivey, A., Sugano, S., White, B., Walker, D., Woodward, J., Winckler, T., Tanaka, Y., Shaulsky, G., Schleicher, M., Weinstock, G., Rosenthal, A., Cox, E. C., Chisholm, R. L., Gibbs, R., Loomis, W. F., Platzer, M., Kay, R. R., Williams, J., Dear, P. H., Noegel, A. A., Barrell, B., and Kuspa, A. (2005). The genome of the social amoeba *Dictyostelium discoideum*. *Nature*, **435**, 3–57.

El-Sayed, N. M., Hegde, P., Quackenbush, J., Melville, S. E., and Donelson, J. E. (2000). The African trypanosome genome. *Int. J. Parasitol.*, **30**, 29–345.

Ewing, B. and Green, P. (1998). Base-calling of automated sequencer traces using phred. II. Error probabilities. *Genome Res.*, **8**, 86–194.

Ewing, B., Hillier, L., Wendl, M. C., and Green, P. (1998). Base-calling of automated sequencer traces using phred. I. Accuracy assessment. *Genome Res.*, **8**, 75–185.

El-Sayed, N. M. A., Myler, P. J., Bartholomeu, D. C., Nilsson, D., Aggarwal, G., Tran, A., Ghedin, E., Worthey, E. A., Delcher, A. L., Blandin, G., Westenberger, S. J., Caler, E., Cerqueira, G., Branche, C., Haas, B., Anapuma, A., Arner, E., Éslund, L., Attipoe, P., Bontempi, E., Bringaud, F., Burton, P., Cadag, E., Campbell, D. A., Carrington, M., Crabtree, J., Darban, H., da Silveira, J. F., de Jong, P., Edwards, K., Englund, P. T., Fazelina, G., Feldblyum, T., Ferella, M., Frasch, A. C., Gull, K., Huang, Y., Kindlund, E., Klingbeil, M., Kluge, S., Koo, H., Lacerda, D., Levin, M., Lorenzi, H., Louie, T., Machado, C. R., McCulloch, R., McKenna, A., Mizuno, Y., Mottram, J.C., Nelson, S., Ochaya, S., Osoegawa, K., Pai, G., Parsons, M., Pentony, M., Pettersson, U., Pop, M., Ramirez, J. L., Rinta, J., Robertson, L., Salzberg, S. L., Seyler, A., Sharma, R., Sisk, E., Tammi, M. T., Tarleton, R., Teixeira, S., Aken, S. V., Vogt, C., Ward, P., Wickstead, B., Wortman, J., White, O., Fraser, C. M., Stuart K. D., Andersson, B. The genome sequence of Trypanosoma cruzi, etiological agent of Chagas disease. (2005). Science, **309**, 409-415.

Frazer, K. A., Elnitski, L., Church, D. M., Dubchak, I., and Hardison, R. C. (2003). Cross-species sequence comparisons: a review of methods and available resources. *Genome Res.*, **13**, –12.

Gardner, M. J. (2001). A status report on the sequencing and annotation of the P. falciparum genome. *Mol. Biochem. Parasitol.*, **118**, 33–138.

Gardner, M. J., Bishop, R., Shah, T., de Villiers, E. P., Carlton, J. M., Hall, N., Ren, Q., Paulsen, I. T., Pain, A., Berriman, M., Wilson, R. J., Sato, S., Ralph, S. A., Mann, D. J., Xiong, Z., Shallom, S. J., Weidman, J., Jiang, L., Lynn, J., Weaver, B., Shoaibi, A., Domingo, A. R., Wasawo, D., Crabtree, J., Wortman, J. R., Haas, B., Angiuoli, S. V., Creasy, T. H., Lu, C., Suh, B., Silva, J. C., Utterback, T. R., Feldblyum, T. V., Pertea, M., Allen, J.,

Nierman, W. C., Taracha, E. L., Salzberg, S. L., White, O. R., Fitzhugh, H. A., Morzaria, S., Venter, J. C., Fraser, C. M., and Nene, V. (2005). Genome sequence of *Theileria parva*, a bovine pathogen that transforms lymphocytes. *Science*, **309**, 34–137.

Gardner, M. J., Hall, N., Fung, E., White, O., Berriman, M., Hyman, R. W., Carlton, J. M., Pain, A., Nelson, K. E., Bowman, S., Paulsen, I. T., James, K., Eisen, J. A., Rutherford, K., Salzberg, S. L., Craig, A., Kyes, S., Chan, M. S., Nene, V., Shallom, S. J., Suh, B., Peterson, J., Angiuoli, S., Pertea, M., Allen, J., Selengut, J., Haft, D., Mather, M. W., Vaidya, A. B., Martin, D. M., Fairlamb, A. H., Fraunholz, M. J., Roos, D. S., Ralph, S. A., McFadden, G. I., Cummings, L. M., Subramanian, G. M., Mungall, C., Venter, J. C., Carucci, D. J., Hoffman, S. L., Newbold, C., Davis, R. W., Fraser, C. M., and Barrell, B. (2002). Genome sequence of the human malaria parasite *Plasmodium falciparum*. *Nature*, **419**, 98–511.

Glockner, G., Eichinger, L., Szafranski, K., Pachebat, J. A., Bankier, A. T., Dear, P. H., Lehmann, R., Baumgart, C., Parra, G., Abril, J. F., Guigo, R., Kumpf, K., Tunggal, B., Cox, E., Quail, M. A., Platzer, M., Rosenthal, A., and Noegel, A. A. (2002). Sequence and analysis of chromosome 2 of *Dictyostelium discoideum*. *Nature*, **418**, 9–85.

Hall, N., Berriman, M., Lennard, N. J., Harris, B. R., Hertz-Fowler, C., Bart-Delabesse, E. N., Gerrard, C. S., Atkin, R. J., Barron, A. J., Bowman, S., Bray-Allen, S. P., Bringaud, F., Clark, L. N., Corton, C. H., Cronin, A., Davies, R., Doggett, J., Fraser, A., Gruter, E., Hall, S., Harper, A. D., Kay, M. P., Leech, V., Mayes, R., Price, C., Quail, M. A., Rabbinowitsch, E., Reitter, C., Rutherford, K., Sasse, J., Sharp, S., Shownkeen, R., MacLeod, A., Taylor, S., Tweedie, A., Turner, C. M., Tait, A., Gull, K., Barrell, B., and Melville, S. E. (2003). The DNA sequence of chromosome I of an African trypanosome: gene content, chromosome organisation, recombination and polymorphism. *Nucleic Acids Res.*, **31**, 864–4873.

Hall, N., Karras, M., Raine, J. D., Carlton, J. M., Kooij, T. W., Berriman, M., Florens, L., Janssen, C. S., Pain, A., Christophides, G. K., James, K., Rutherford, K., Harris, B., Harris, D., Churcher, C., Quail, M. A., Ormond, D., Doggett, J., Trueman, H. E., Mendoza, J., Bidwell, S. L., Rajandream, M. A., Carucci, D. J., Yates, J. R., 3rd, Kafatos, F. C., Janse, C. J., Barrell, B., Turner, C. M., Waters, A. P., and Sinden, R. E. (2005). A comprehensive survey of the *Plasmodium* life cycle by genomic, transcriptomic, and proteomic analyses. *Science*, **307**, 2–86.

Hall, N., Pain, A., Berriman, M., Churcher, C., Harris, B., Harris, D., Mungall, K., Bowman, S., Atkin, R., Baker, S., Barron, A., Brooks, K., Buckee, C. O., Burrows, C., Cherevach, I., Chillingworth, C., Chillingworth, T.,

Christodoulou, Z., Clark, L., Clark, R., Corton, C., Cronin, A., Davies, R., Davis, P., Dear, P., Dearden, F., Doggett, J., Feltwell, T., Goble, A., Goodhead, I., Gwilliam, R., Hamlin, N., Hance, Z., Harper, D., Hauser, H., Hornsby, T., Holroyd, S., Horrocks, P., Humphray, S., Jagels, K., James, K. D., Johnson, D., Kerhornou, A., Knights, A., Konfortov, B., Kyes, S., Larke, N., Lawson, D., Lennard, N., Line, A., Maddison, M., McLean, J., Mooney, P., Moule, S., Murphy, L., Oliver, K., Ormond, D., Price, C., Quail, M. A., Rabbinowitsch, E., Rajandream, M. A., Rutter, S., Rutherford, K. M., Sanders, M., Simmonds, M., Seeger, K., Sharp, S., Smith, R., Squares, R., Squares, S., Stevens, K., Taylor, K., Tivey, A., Unwin, L., Whitehead, S., Woodward, J., Sulston, J. E., Craig, A., Newbold, C., and Barrell, B. G. (2002). Sequence of *Plasmodium falciparum* chromosomes, –9 and 13. *Nature*, **419**, 27–531.

Hardie, D. C., Gregory, T. R., and Hebert, P. D. (2002). From pixels to picograms: a beginners' guide to genome quantification by Feulgen image analysis densitometry. *J. Histochem. Cytochem.*, **50**, 35–749.

Ivens, A. C., Lewis, S. M., Bagherzadeh, A., Zhang, L., Chan, H. M., and Smith, D. F. (1998). A physical map of the *Leishmania major* Friedlin genome. *Genome Res.*, **8**, 35–145.

Ivens, A. C., Peacock, C. S., Worthey, E. A., Murphy, L., Aggarwal, G., Berriman, M., Sisk, E., Rajandream, M. A., Adlem, E., Aert, R., Anupama, A., Apostolou, Z., Attipoe, P., Bason, N., Bauser, C., Beck, A., Beverley, S. M., Bianchettin, G., Borzym, K., Bothe, G., Bruschi, C.V., Collins, M., Cadag, E., Ciarloni, L., Clayton, C., Coulson, R. M., Cronin, A., Cruz, A. K., Davies, R. M., De Gaudenzi, J., Dobson, D. E., Duesterhoeft, A., Fazelina, G., Fosker, N., Frasch, A. C., Fraser, A., Fuchs, M., Gabel, C., Goble, A., Goffeau, A., Harris, D., Hertz-Fowler, C., Hilbert, H., Horn, D., Huang, Y., Klages, S., Knights, A., Kube, M., Larke, N., Litvin, L., Lord, A., Louie, T., Marra, M., Masuy, D., Matthews, K., Michaeli, S., Mottram, J. C., Muller-Auer, S., Munden, H., Nelson, S., Norbertczak, H., Oliver, K., O'neil, S., Pentony, M., Pohl, T. M., Price, C., Purnelle, B., Quail, M.A., Rabbinowitsch, E., Reinhardt, R., Rieger, M., Rinta, J., Robben, J., Robertson, L., Ruiz, J. C., Rutter, S., Saunders, D., Schafer, M., Schein, J., Schwartz, D. C., Seeger, K., Seyler, A., Sharp, S., Shin, H., Sivam, D., Squares, R., Squares, S., Tosato, V., Vogt, C., Volckaert, G., Wambutt, R., Warren, T., Wedler, H., Woodward, J., Zhou, S., Zimmermann, W., Smith, D. F., Blackwell, J. M., Stuart, K. D., Barrell, B., Myler, P. J. The genome of the kinetoplastid parasite, Leishmania major. (2005). Science. 309: 436–442.

Katinka, M. D., Duprat, S., Cornillot, E., Metenier, G., Thomarat, F., Prensier, G., Barbe, V., Peyretaillade, E., Brottier, P., Wincker, P., Delbac, F., El Alaoui, H., Peyret, P., Saurin, W., Gouy, M., Weissenbach, J., and Vivares, C. P. (2001). Genome sequence and gene compaction of the eukaryote parasite *Encephalitozoon cuniculi*. *Nature*, **414**, 50–453.

Kuspa, A. and Loomis, W. F. (1996). Ordered yeast artificial chromosome clones representing the *Dictyostelium discoideum* genome. *Proc. Natl. Acad. Sci. U S A*, **93**, 562–5566.

Kuspa, A., Maghakian, D., Bergesch, P., and Loomis, W. F. (1992). Physical mapping of genes to specific chromosomes in *Dictyostelium discoideum*. *Genomics*, **13**, 9–61.

Lai, Z., Jing, J., Aston, C., Clarke, V., Apodaca, J., Dimalanta, E. T., Carucci, D. J., Gardner, M. J., Mishra, B., Anantharaman, T. S., Paxia, S., Hoffman, S. L., Craig Venter, J., Huff, E. J., and Schwartz, D. C. (1999). A shotgun optical map of the entire *Plasmodium falciparum* genome. *Nat. Genet.*, **23**, 09–313.

Lander, E. S. and Waterman, M. S. (1988). Genomic mapping by fingerprinting random clones: a mathematical analysis. *Genomics*, **2**, 31–239.

Leech, V., Quail, M. A., and Melville, S. E. (2004). Separation, digestion and cloning of intact parasite chromosomes embedded in Agarose. In S. E. Melville, ed. *Parasite Genomics Protocols*. Humana Press, New Jersey, pp. 335–351.

Loftus, B., Anderson, I., Davies, R., Alsmark, U. C., Samuelson, J., Amedeo, P., Roncaglia, P., Berriman, M., Hirt, R. P., Mann, B. J., Nozaki, T., Suh, B., Pop, M., Duchene, M., Ackers, J., Tannich, E., Leippe, M., Hofer, M., Bruchhaus, I., Willhoeft, U., Bhattacharya, A., Chillingworth, T., Churcher, C., Hance, Z., Harris, B., Harris, D., Jagels, K., Moule, S., Mungall, K., Ormond, D., Squares, R., Whitehead, S., Quail, M. A., Rabbinowitsch, E., Norbertczak, H., Price, C., Wang, Z., Guillen, N., Gilchrist, C., Stroup, S. E., Bhattacharya, S., Lohia, A., Foster, P. G., Sicheritz-Ponten, T., Weber, C., Singh, U., Mukherjee, C., El-Sayed, N. M., Petri, W. A., Jr., Clark, C. G., Embley, T. M., Barrell, B., Fraser, C. M., and Hall, N. (2005). The genome of the protist parasite *Entamoeba histolytica*. *Nature*, **433**, 65–868.

Lyons, E. J. and Carlton, J. M. (2004). Mind the gap: bridging the divide between clinical and molecular studies of the trichomonads. *Trends Parasitol.*, **20**, 04–207.

McKusick, V. A. (1991). Current trends in mapping human genes. *Faseb J.*, **5**, 2–20.

McKusick, V. A. and Ruddle, F. H. (1987). Toward a complete map of the human genome. *Genomics*, **1**, 03–106.

McPherson, J. D., Marra, M., Hillier, L., Waterston, R. H., Chinwalla, A., Wallis, J., Sekhon, M., Wylie, K., Mardis, E. R., Wilson, R. K., Fulton, R., Kucaba, T. A., Wagner-McPherson, C., Barbazuk, W. B., Gregory, S. G., Humphray, S. J., *et al.* (2001). A physical map of the human genome. *Nature*, **409**, 34–941.

Melville, S. E., Leech, V., Navarro, M., and Cross, G. A. (2000). The molecular karyotype of the megabase chromosomes of *Trypanosoma brucei* stock 427. *Mol. Biochem. Parasitol.*, **111**, 61–273.

Mullikin, J. C. and Ning, Z. (2003). The phusion assembler. *Genome Res.*, **13**, 1–90.

Myers, E. W., Sutton, G. G., Delcher, A. L., Dew, I. M., Fasulo, D. P., Flanigan, M. J., Kravitz, S. A., Mobarry, C. M., Reinert, K. H., Remington, K. A., Anson, E. L., Bolanos, R. A., Chou, H. H., Jordan, C. M., Halpern, A. L., Lonardi, S., Beasley, E. M., Brandon, R. C., Chen, L., Dunn, P. J., Lai, Z., Liang, Y., Nusskern, D. R., Zhan, M., Zhang, Q., Zheng, X., Rubin, G. M., Adams, M. D., and Venter, J. C. (2000). A whole-genome assembly of *Drosophila*. *Science*, **287**, 196–2204.

Pain, A., Renauld, H., Berriman, M., Murphy, L., Yeats, C. A., Weir, W., Kerhornou, A., Aslett, M., Bishop, R., Bouchier, C., Cochet, M., Coulson, R. M., Cronin, A., de Villiers, E. P., Fraser, A., Fosker, N., Gardner, M., Goble, A., Griffiths-Jones, S., Harris, D. E., Katzer, F., Larke, N., Lord, A., Maser, P., McKellar, S., Mooney, P., Morton, F., Nene, V., O'Neil, S., Price, C., Quail, M. A., Rabbinowitsch, E., Rawlings, N. D., Rutter, S., Saunders, D., Seeger, K., Shah, T., Squares, R., Squares, S., Tivey, A., Walker, A. R., Woodward, J., Dobbelaere, D. A., Langsley, G., Rajandream, M. A., McKeever, D., Shiels, B., Tait, A., Barrell, B., and Hall, N. (2005). Genome of the host-cell transforming parasite *Theileria annulata* compared with *T. parva*. *Science*, **309**, 31–133.

Peterson, D. G., Wessler, S. R., and Paterson, A. H. (2002). Efficient capture of unique sequences from eukaryotic genomes. *Trends Genet.*, **18**, 47–550.

Piper, M. B., Bankier, A. T., and Dear, P. H. (1998). A HAPPY map of *Cryptosporidium parvum*. *Genome Res.*, **8**, 299–1307.

Quail, M. A. (2001). M13 cloning of mung bean nuclease digested PCR fragments as a means of gap closure within A/T-rich, genome sequencing projects. *DNA Seq.*, **12**, 55–359.

Rubio, J. P., Triglia, T., Kemp, D. J., de Bruin, D., Ravetch, J. V., and Cowman, A. F. (1995). A YAC contig map of *Plasmodium falciparum* chromosome 4: characterization of a DNA amplification between two recently separated isolates. *Genomics*, **26**, 92–198.

Samad, A., Huff, E. F., Cai, W., and Schwartz, D. C. (1995). Optical mapping: a novel, single-molecule approach to genomic analysis. *Genome Res.*, **5**, 1–4.

Shirley, M. W., Ivens, A., Gruber, A., Madeira, A. M., Wan, K. L., Dear, P. H., and Tomley, F. M. (2004). The Eimeria genome projects: a sequence of events. *Trends Parasitol.*, **20**, 99–201.

Su, X., Ferdig, M. T., Huang, Y., Huynh, C. Q., Liu, A., You, J., Wootton, J. C., and Wellems, T. E. (1999). A

genetic map and recombination parameters of the human malaria parasite *Plasmodium falciparum*. *Science*, **286**, 351–1353.

Tao, Q., Chang, Y. L., Wang, J., Chen, H., Islam-Faridi, M. N., Scheuring, C., Wang, B., Stelly, D. M., and Zhang, H. B. (2001). Bacterial artificial chromosome-based physical map of the rice genome constructed by restriction fingerprint analysis. *Genetics*, **158**, 711–1724.

Thorstenson, Y. R., Hunicke-Smith, S. P., Oefner, P. J., and Davis, R. W. (1998). An automated hydrodynamic process for controlled, unbiased DNA shearing. *Genome Res*, **8**, 48–855.

Valencia, A. (2005). Automatic annotation of protein function. *Curr Opin Struct Biol*, **15**, 67–274.

Venter, J. C., Smith, H. O., and Hood, L. (1996). A new strategy for genome sequencing. *Nature*, **381**, 64–366.

Wang, A. L. and Wang, C. C. (1985). Isolation and cha acterization of DNA from *Tritrichomonas foetus* and *Trichomonas vaginalis*. *Mol. Biochem. Parasitol.*, **14**, 23–335.

Wang, J., Wong, G. K., Ni, P., Han, Y., Huang, X., Zhang, J., Ye, C., Zhang, Y., Hu, J., Zhang, K., Xu, X., Cong, L., Lu, H., Ren, X., He, J., Tao, L., Passey, D. A., Yang, H., Yu, J., and Li, S. (2002). RePS: a sequence assembler that masks exact repeats identified from the shotgun data. *Genome Res.*, **12**, 24–831.

Welker, D. L. and Williams, K. L. (1982). A genetic map of *Dictyostelium discoideum* based on mitotic recombination. *Genetics*, **102**, 91–710.

Whitelaw, C. A., Barbazuk, W. B., Pertea, G., Chan, A. P., Cheung, F., Lee, Y., Zheng, L., van Heeringen, S., Karamycheva, S., Bennetzen, J. L., SanMiguel, P., Lakey, N., Bedell, J., Yuan, Y., Budiman, M. A., Resnick, A., Van Aken, S., Utterback, T., Riedmuller, S., Williams, M., Feldblyum, T., Schubert, K., Beachy, R., Fraser, C. M., and Quackenbush, J. (2003). Enrichment of gene-coding sequences in maize by genome filtration. *Science*, **302**, 118–2120.

Williams, J. G. and Firtel, R. A. (2000). HAPPY days for the *Dictyostelium* genome project. *Genome Res.*, **10**, 658–1659.

Xu, P., Widmer, G., Wang, Y., Ozaki, L. S., Alves, J. M., Serrano, M. G., Puiu, D., Manque, P., Akiyoshi, D., Mackey, A. J., Pearson, W. R., Dear, P. H., Bankier, A. T., Peterson, D. L., Abrahamsen, M. S., Kapur, V., Tzipori, S., and Buck, G. A. (2004). The genome of *Cryptosporidium hominis*. *Nature*, **431**, 107–1112.

Yuh, Y. S., Liu, J. Y., and Shaio, M. F. (1997). Chromosome number of *Trichomonas vaginalis*. *J. Parasitol.*, **83**, 551–553.

CHAPTER 10

Comparative genomics of the trypanosomatids

Kenneth D. Stuart[1,2] and Peter J. Myler[1,2,3]

[1] *Seattle Biomedical Research Institute, Seattle, WA, USA*
[2] *Department of Pathobiology, University of Washington, Seattle, WA, USA*
[3] *Division of Biomedical and Health Informatics, University of Washington, Seattle, WA, USA*

Introduction

The protozoan clade Trypanosomatidae contains a number of parasitic genera and species with various similarities and differences in morphology, life cycle, insect vector, and disease pathology in humans and animals. The draft genome sequences and annotation of three "Tritryp" species, *Trypanosoma brucei*, *T. cruzi*, and *Leishmania major*, have been recently completed (Berriman *et al.* 2005; El-Sayed *et al.* 2005a; Ivens *et al.* 2005), and the variation among the life cycles and disease pathologies of each species is generally reflected in the similarities and differences between their genomes. The different complexities of the individual genomes have resulted in various degrees of completion for each sequencing project. *L. major* Friedlin has the simplest genome with few repetitive sequences and negligible differences between the homologous chromosomes of each haplotype, and consequently its sequence is essentially complete (Ivens *et al.* 2005). The situation is similar for the large chromosomes of *T. brucei* TREU927/4, which can be ordered into a haploid mosaic genome, although large subtelomeric repeat arrays of variant surface glycoprotein (VSG) genes and VSG expression site associated genes (ESAGs) involved in antigenic variation within the mammalian host are not yet completely represented in the published sequence (Berriman *et al.* 2005). In addition, the numerous small and intermediate sized (50–700 kb) chromosomes, which contain mostly repetitive sequence, have not been sequenced. The *T. cruzi* CL Brener genome is characterized by being at least 50% tandem and dispersed gene repeats, and having substantial divergence between haplotypes (see section *T. cruzi* is a hybrid, below). This resulted in assembly of the draft genome into two separate haplotypes, each containing many sequence gaps (i.e. lack of contiguity), and misassembled repetitive regions (El-Sayed *et al.* 2005a). Nevertheless, the sequences of all three genomes are essentially complete in terms of critical genes, allowing a comprehensive comparison of their content and organization (El-Sayed *et al.* 2005b).

Genome structure

The trypanosomatid genomes are generally diploid, albeit with varying degrees of aneuploidy (see below), and the nuclear DNA contents of *T. brucei*, *T. cruzi*, and *L. donovani* have been estimated to be 83, 255, and 91 megabase pairs (Mb), respectively, based on cytophotometric studies (Swindle and Tait 1996). These values are somewhat larger than the diploid genome sizes determined by sequencing (Table 10.1), due to factors such as aneuploidy, the presence of mini-chromosomes in *T. brucei*, and limited accuracy of cytophotometry.

Table 10.1 Tritryp genome statistics

	T. brucei	*T. cruzi*	*L. major*
DNA content (Mb)	83	255	91
Haploid genome size(Mb)	25[a]	~55	33
Chromosomes	11[a]	~28	36
Genome sequence (Mb)	26.1	60.4	32.8
G + C content	46.4%	53.4%	59.7%
RNA genes	556	2035[b]	911
Protein-coding genes	8164	~12 000[b]	8272
Pseudogenes	904	2590[b]	39
Average CDS length (bp)	1592	1513	1901
Average inter-CDS size	1279	1024	2045
Gene density (gene/Mb)	317	385	252
Gene families	582[c]	1052	662
Genes in families	27%[c]	36%	37%
Function known	5%	n.d.[d]	4%
Function inferred	38%	43%	28%
Hypothetical, conserved	51%	48%	56%
Hypothetical, species-specific	6%	9%	8%
In all Tritryps (3-ways COGs)	73%	54%	80%
Tb + Tc only	5%	4%	–
Tb + Lm only (2-way COGs)	1%	–	1%
Tc + Lm only	–	4%	6%
Species-specific	21%	38%	13%

[a] Not including intermediate- and mini-chromosomes
[b] Total number in both haplotypes is 19 980
[c] Not including VSG (pseudo)genes
[d] Not determined

The Old Word *Leishmania* (including *L. major*) genomes are comprised of 36 chromosome pairs (numbered 1–36 from smallest to largest) ranging in size from 0.28 to 2.8 Mb (Bastien *et al.* 1998). Chromosome 1 is triploid in the Friedlin strain and one homolog apparently underwent a size change during laboratory cultivation, presumably due to unequal recombination in the repetitive sub-telomeric region at one end (Sunkin *et al.* 2000). Indeed, this chromosome appears to be particularly prone to copy number variation, since tetraploidy has also been observed (Martinez-Calvillo *et al.* 2005). In addition, at least two other chromosomes (20 and 31) also appear to be tetraploid, based on fluorescence staining intensity of chromosomal DNA on pulse-field gel electrophoresis (PFGE) and over-representation of BAC clones in whole genome shotgun libraries (unpublished

data). Interestingly, the New World *Leishmania* species have 35 (*L. braziliensis* complex) or 34 (*L. mexicana* complex) chromosomes, due to fusion of chromosomes 20 + 34 or 20 + 36 and 8 + 29, respectively (Britto *et al.* 1998).

T. cruzi is estimated to contain about 28 chromosome pairs, ranging in size from 0.3 to 2 Mb, but the exact number and ploidy is still unknown (Cano *et al.* 1999). While the sizes of *Leishmania* chromosome pairs are relatively constant, in *T. cruzi* the homologous chromosomes vary widely both within and between strains (Vargas *et al.* 2004). This is likely a reflection of frequent recombination between the large number of repetitive sequences and the hybrid nature of many isolates (see below).

T. brucei has 11 chromosomes pairs (numbered I to XI, smallest to largest) with sizes ranging from

0.9 to 6 Mb, several intermediate (150–900 kb) chromosomes, and a larger number of mini-chromosomes (50–150 kb) (Melville *et al.* 1998; Turner *et al.* 1997). The intermediate and mini-chromosomes are aneuploid, their numbers and size vary widely among strains, and they appear to contain only 177-bp repeats, VSG genes, and ESAGs (see below). The megabase-sized chromosomes are diploid, although their sizes also vary widely, both between homologs and among strains. They contain all the housekeeping genes and appear to have arisen as a result of fusion of smaller ancestral chromosomes (as seen in *L. major* and *T. cruzi*) during evolution (El-Sayed *et al.* 2005b).

Genome organization

Gene clusters

Perhaps the most characteristic feature of the Tritryp genomes is their organization into large polycistronic gene clusters (PGCs) consisting of tens-to-hundreds of protein-coding genes on the same DNA strand (Andersson *et al.* 1998; Berriman *et al.* 2005; El-Sayed *et al.* 2003; El-Sayed *et al.* 2005b; El-Sayed *et al.* 2005a; Hall *et al.* 2003; Ivens *et al.* 2005; Myler *et al.* 1999; Worthey *et al.* 2003a) (Fig. 10.1). Based on nuclear run-on analyses of chromosomes 1 and 3 in *L. major*, it appears that RNA polymerase II-mediated transcription initiates bidirectionally in the strand-switch regions between divergent gene clusters and extends polycistronically through each gene cluster before terminating in the strand-switch region separating convergent clusters (Martinez-Calvillo *et al.* 2003; Martinez-Calvillo *et al.* 2004). Thus, there appear to be only a few initiation sites (promoters) on each chromosome. This organization is possible because the polycistronic RNA is subsequently processed by concomitant *trans*-splicing and polyadenylation, which results in a 39-nt capped spliced leader (SL) sequence being added to the 5′ end of each mRNA and a poly(A) tail to the 3′ end of the upstream mRNA (Perry and Agabian 1991). Unlike polycistronic operons in prokaryotes, trypanosomatid PGCs generally do not contain only genes with related functions, and there is currently no evidence for regulation of their expression at the level of transcription initiation. Consequently, trypanosomatid gene expression appears to be regulated post-transcriptionally, at the level of RNA turnover, translation, and/or protein stability (Clayton 2002).

The directional gene clusters observed in all three Tritryp genomes are associated with distinctive strand asymmetry in nucleotide bias, particularly purine excess, GC bias, and AT skew (McDonagh *et al.* 2000; Nilsson and Andersson 2005). Curiously, while the AT skew is the same in all three organisms, the purine excess and GC bias have the opposite orientation relative to the coding strand in *L. major* compared to *T. brucei* and *T. cruzi*. Since this skew is thought to reflect transcription and/or DNA repair processes, this difference may reflect a fundamental difference between these two genera, possibly indicating different roles of recombination and repair. An intriguing possibility is that recombination may serve to maintain near identity between chromosome homologs in *Leishmania*, but function to ensure heterogeneity of VSG genes in *T. brucei* and repeated genes for surface antigens in *T. cruzi*, thereby facilitating antigenic variation and diversity in the trypanosomes.

The Tritryp genomes show a remarkable degree of synteny, with 68 and 65% of the protein-coding genes in *T. brucei* and *L. major*, respectively, maintaining the same genomic context (El-Sayed *et al.* 2005b; Ghedin *et al.* 2004). Interestingly, many synteny breaks are associated with expansion of repeated gene families, retroelements, and/or RNA genes, and usually occur at, or near, the strand-switch regions separating gene clusters (Fig.10.1). These synteny blocks can be large (>1 Mb), often encompassing several clusters. As a consequence the chromosomes of *T. brucei* appear as mosaics composed of several smaller *L. major* chromosomes. While the current lack of contiguity in the *T. cruzi* genome sequence due to the numerous repeated sequences does not yet permit a precise comparison of its genome organization, it appears to be more similar to that of *L. major* than *T. brucei*, which is unexpected given its closer phylogenetic relationship to the latter.

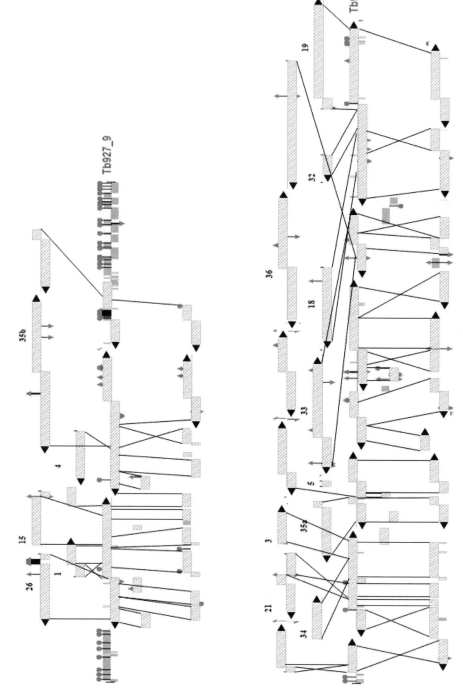

Figure 10.1 Trityp synteny. *T. brucei* chromosomes 9 (upper panel) and 10 (lower panel) are aligned with the corresponding syntenic regions of *L. major* (LmjF) and *T. cruzi* (TcBr). The *L. major* chromosomes are indicated by the numbers above each line, with the arrow indicating their orientation relative to the published sequence. The *T. cruzi* scaffolds represent only one of the two haplotypes, and arrows indicate those which are reverse relative to the published sequences. Regions of synteny are shown by lines connecting each species map. Individual genes are not shown, but polycistronic gene clusters (PGCs) are demarcated by hatched boxes with arrowheads indicating the direction of transcription. RNA genes are indicated by black triangles and retrotransposons are shown by grey circles.

Table 10.2 RNA genes

RNA type[a]	T. brucei		L. major		T. cruzi	
	RNAs[b]	genes[c]	RNAs[d]	genes[b]	RNAs[a]	genes[c]
rRNA	9	103	9	63	9	110/114
tRNA	47	65	47	83	47	58/59
SL RNA	1	28	1	63	1	96/96
snRNA	6	6	6	6	6	11/18
snoRNA	91	353	93	695	108	770/713
SRP RNA	1	1	1	1	1	1/1
TOTAL	115	556	157	911	172	1046/991

[a] RNA types: rRNA = ribosomal RNA; +RNA = transfer RNA; +RNA; SLRNA = spliced leader RNA; snRNA = small nucleolar RNA; SRP RNA = small nuclear RNA; SRP RNA = signal recognition particle RNA

[b] Number of different RNA molecules

[c] Number of different genes in haploid genome

[d] Number of genes in Es haplotype/Number of genes in non-Es haplotype

RNA genes

Hundreds of genes encoding structural RNAs have been identified in each of the Tritryp genome (Table 10.2), and they have characteristic genomic distributions in each organism (Ivens *et al.* 2005). There is a single tandem array of the nine ribosomal RNA (rRNA) genes on *L. major* chr27 (Fig. 10.1), but several loci on different chromosomes in *T. brucei* and *T. cruzi*. In contrast, the 5S RNA genes occur at multiple loci in *L. major*, but in a single tandem array in the trypanosomes. All three genomes encode 27 transfer RNAs (tRNAs) with 45 of the 61 possible anticodons, but *L. major* has somewhat more gene copies. In all three genomes, the tRNA genes usually occur in small groups of two to five genes, oriented in a head-to-head fashion. These often occur in the strand-switch regions between convergent clusters of protein-coding genes, where they may act as transcription terminators (see above). However, many are also located internally within a cluster of genes, and thus may represent transcriptional boundaries of gene clusters that are transcribed in the same direction. The spliced leader RNA (SLRNA) genes occur in a single tandem array in all three genomes, but *T. cruzi* and *L. major* appear to have more copies than *T. brucei* (Table 10.2). Genes encoding the U1–U6 small nuclear RNAs (snRNAs) are found in some of the tRNA clusters, and *T. cruzi* has multiple copies of the U2 genes, while the other species do not. Like the

tRNAs, the snRNAs are transcribed by RNA polymerase III (Fantoni *et al.* 1994; Nakaar *et al.* 1995; Nakaar *et al.* 1994; Nakaar *et al.* 1997). There are several hundred copies per haploid genome of ~100 different small nucleolar RNA (snoRNA) genes, which guide numerous rRNA modifications. These generally occur in tandem arrays within polycistronic clusters of protein-coding genes, where they are transcribed by RNA polymerase II (Xu *et al.* 2001). Genes for other RNAs, such as the 7SLRNA, the RNA in the signal recognition protein (SRP) complex, and telomerase have also been identified.

Telomeres

While there are substantial similarities among the Tritryp genomes, there are characteristic differences in addition to those described above. One key difference can be found in the sub-telomeric regions (i.e. between terminal hexameric repeats and the first syntenic protein-coding genes), which are dramatically different among these three species. The subtelomeric regions of *L. major* chromosomes are relatively uncomplicated, with most containing 1 to 6 kb of tandem repeats of two different telomere associated sequences (TAS), interspersed with the hexameric repeats, except for six that contain 0.7 to 25 kb of distinctive combinations of subtelomeric repeats (LST-R). In contrast, the subtelomeric regions of *T. brucei* and

T. cruzi contain a variety of protein-coding genes and related pseudogenes, as well as retroelements or derivatives thereof. Many of these genes encode families of surface proteins that are involved in antigenic variation and immune evasion.

About half of the 44 telomeres on the 11 large diploid chromosomes of *T. brucei* appear to contain bloodstream form VSG expression sites (BESs), which are characterized by a telomere-proximal *VSG* gene, followed by tandem (70-bp) repeats, and unique combinations of several different expression site associated genes (ESAGs) under the control of an RNA polymerase I promoter (Pays *et al.* 2004). Because these regions proved difficult to clone and/ or assemble correctly during the *T. brucei* 927 genome project, characterization of *T. brucei* telomeres is currently incomplete. However, it is clear that many contain subtelomeric arrays of tens-to-hundreds of VSG and ESAG pseudogenes, along with numerous *Ingi*/RIME/DIRE retroelements (Berriman *et al.* 2005) (Fig. 10.1). This organization reflects the extensive antigenic variation that *T. brucei* can undergo. Only one BES is expressed at a time with the resultant VSG covering the cell surface and determining the antigenic type. However, transcription can periodically switch from one BES to another and recombination occurs between BESs, as well as between *VSG* genes in BESs and the subtelomeric loci. Since most of the latter are pseudogenes – this suggests that partial and mosaic recombination contributes to the extensive variant antigen repertoire. The ESAGs encode a number of different proteins that are expressed in the bloodstream stage, including transferrin receptors and adenylate kinases. It is thought that sequence variation in least some of these genes is involved in selective adaptation to different animal hosts (Pays *et al.* 2001).

The subtelomeric regions of *T. cruzi* contain a variety of different surface antigen genes and pseudogenes, as well as retroelements. These include members of the dispersed gene family 1 (DGF-1), trans-sialidase (TS) superfamily, mucins, mucin-associated surface proteins (MASPs), and recombination hotspot (RHS) pseudogenes; along with vestigial interposed retroelements (VIPER), short interspersed repetitive elements (SIRE), non-autonomous non-LTR retrotransposons (NARTc), and degenerate *ingi*/L1Tc-related retroelements (DIRE).

Many of these genes also occur in large repeat arrays at chromosome-internal loci. Analysis of expression of these subtelomeric genes is less documented than in *T. brucei*, but it appears that they do not possess the same type of system for generation of antigenic variation. Instead, many of the genes are expressed simultaneously, and recombination between gene copies generates antigenic diversity over time.

Retroelements

The genomes of all three Tritryps contain retroelements (El-Sayed *et al.* 2005b). However, while *L. major* contains only a few inactive DIRE retroelements, the trypanosomes contain hundreds of retroelements of several different types, including some apparently capable of active retrotransposition. Both *T. brucei* and *T. cruzi* contain VIPER retrotransposons and its non-autonomous SIRE derivative, which have long terminal repeats (LTRs). In addition, both contain SLRNA site-specific non-LTR retroelements, although they differ somewhat in size and sequence, with *T. brucei* containing SLACS and *T. cruzi* containing CZAR. *T. brucei* also contains numerous copies of the *ingi*/ RIME retroelement, while *T. cruzi* contains even more copies of L1Tc/NARTc elements, all of which lack LTRs. Sequence similarity in the reverse transcriptase domains of *ingi*, L1Tc, and DIRE, which is present in all three Tritryp genomes, implies that they have a common ancestor that presumably was active in retrotransposition. Interestingly, the retroelements are often found in regions that appear to undergo frequent recombination. Indeed, the interspersion of these retroelements in the large subtelomeric surface antigen gene arrays in *T. brucei* and *T. cruzi*, and the internal arrays of *T. cruzi*, implies that they may function as recombination hotspots and have a role in generation of antigenic diversity. Retroelements are also often present at synteny break points in the Tritryp genomes, suggesting that they may be involved in chromosome evolution. In particular, their frequent presence at chromosome-internal sites in *T. brucei* that correspond to the telomeres of *L. major*/*T. cruzi* chromosomes, suggests that they may have played a role in chromosome fusion.

T. cruzi is a hybrid

While the homologous chromosomes of *L. major* Friedlin show very low levels (< 0.1%) of allelic variation (Ivens *et al.* 2005; Myler *et al.* 1999), and *T. brucei* strain 927 has ~1% sequence difference between the two haplotypes (Berriman *et al.* 2005; Hall *et al.* 2003), the two haplotypes of *T. cruzi* CL-Brener differ by 5.4% (El-Sayed *et al.* 2005a). Most polymorphisms are due to insertions and deletions in intergenic or subtelomeric regions (resulting in only 2.2 % sequence variation in protein-coding regions), and amplification/deletion of repetitive sequences. One haplotype of CL-Brener (a member of *T. cruzi* subgroup IIe) is very similar to the Esmeraldo strain (from subgroup IIb), providing evidence that CL-Brener is a hybrid resulting from fusion of subgroups IIb and IIc (Sturm *et al.* 2003; Westenberger *et al.* 2005). While analysis of isoenzyme patterns provides evidence for frequent sexual recombination in *T. brucei* (Mihok *et al.* 1990; Mihok and Otieno 1990; Tait 1980), this is not the case for *T. cruzi* or *Leishmania*, which appear to have clonal population structures (Tibayrenc *et al.* 1986; Tibayrenc *et al.* 1990; Tibayrenc 1992). Thus, how and when the *T. cruzi* genome became hybrid is unresolved.

Gene content

Core proteome

The haploid genomes of *T. brucei*, *T. cruzi*, and *L. major* encode about 9100, 12 000, and 8300 protein-coding genes, respectively (Table 10.1). The gene densities of the two trypanosome genomes are quite similar and somewhat higher than that of *Leishmania*. The average coding sequence (CDS) is slightly larger in *Leishmania*, as a result of small sequence insertions relative to trypanosomes, but the lower gene density in *Leishmania* is mostly explained by its larger inter-CDS regions. Each species contains a number of gene families of varying size (Table 10.1). Only ~5% of the predicted proteins have had their functions experimentally confirmed, ~28 to 43% have functions predicted on the basis of sequence homology, ~51 to 56% encode hypothetical proteins conserved in other genomes, and ~6 to 9% encode species-specific hypothetical proteins (Table 10.1).

Approximately 50% of the predicted proteins in *T. brucei* are found only in Trypanosomatids, but this figure is somewhat lower in *L. major* (Fig. 10.2). This is largely due to the greater number of species-specific proteins in *T. brucei*, most of which are VSG (pseudo)genes (see below). Interestingly, while ~22% of the Tritryp proteins are restricted to eukaryotes, ~2 to 3% are not found in other eukaryotes, but are related to prokaryotic proteins (Fig. 10.2). Three-way BLAST comparison of the predicted protein sequences from each genome (El-Sayed *et al.* 2005b) revealed 6158 "clusters of orthologous genes" (COGs), which were present in all three genomes, and another 1014 COGs shared by two of the three species. These predicted proteins of this Tritryp "core" proteome showed an average 57% identity between *T. brucei* and *T. cruzi*, and 44% identity between *L. major* and the two other trypanosomes, reflecting the expected phylogenetic relationships (Haag *et al.* 1998; Stevens *et al.* 2001). Interestingly, while *T. brucei* and *L. major* share substantially fewer two-way COGs than *T. brucei* and *T. cruzi*, as expected from the greater evolutionary distance between the former pair, *L. major* and *T. cruzi* share slightly more, perhaps reflecting the common intracellular environment of their mammalian stage.

Species-specific differences

The remainder of each proteome consists of genes which are found in only one of the three genomes (Table 10.1). The proportion of such species-specific genes is highest (38%) in *T. cruzi* and lowest (13%) in *L. major* (Table 10.1). Although some of these encode proteins which are present in other organisms and carry out distinct metabolic and physiological functions, and many encode hypothetical proteins of unknown function, most (especially in *T. brucei* and *T. cruzi*) are members of large surface antigen gene families. These are exemplified by the VSGs, ESAGs, and procyclic acidic repetitive proteins (EP/PARP/procyclin) of *T. brucei*; the trans-sialidases, DGF-1, mucins, and MASPs of *T. cruzi*; as well as the amastins and promastigote surface antigens (PSA-2) of *L. major*. In addition to the species-restricted genes described above, there are a number of interesting examples of gene family expansion or contraction.

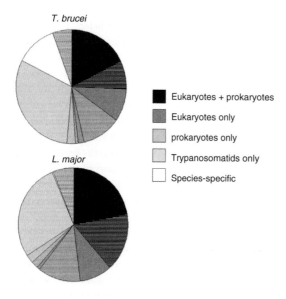

Figure 10.2 Conservation of Tritryp protein sequences in other organisms. The pie charts show the proportions of *T. brucei* and *L. major* proteins with Blastp sequence similarity (with e-values $<10^{-8}$) to proteins from other eukaryotes and prokaryotes (black), eukaryotes only (dark grey), prokaryotes only (medium grey), and other trypanosomatids only (light grey), and those that are species-specific (white). The proportion of each category represented by proteins with no ascribed function is shown by the striped regions.

These include expansion of the ESAG4 adenylate/guanylate cyclases and leucine-rich repeat proteins in *T. brucei*; GP63 proteases and RHS proteins in *T. cruzi*; and mitochondrial carrier protein, ATP-binding cassette (ABC) transporters, and heat shock protein (HSP)90 gene families in *L. major*. Interestingly, many of these genes occur in telomeric and subtelomeric gene clusters (see above) and their species-specificity and organization is presumably related to the different strategies used in each organism for immune evasion.

Transcription and RNA processing

It has long been known that transcription and RNA processing in the trypanosomatids is quite different from that in other eukaryotes (Campbell *et al.* 2003), with such oddities as large polycistronic gene clusters (Martinez-Calvillo *et al.* 2003; Martinez-Calvillo *et al.* 2004; Myler *et al.* 1999), RNA polymerase I-mediated transcription of some protein-coding genes (Lodes *et al.* 1995; Vanhamme and Pays 1995), *trans*-splicing (Perry and Agabian 1991), and little evidence for gene regulation at the level of transcription (Clayton 2002). Eukaryotic RNA polymerase (RNAP) core complexes I, II, and III typically contain about 14, 12, and 17 protein subunits, respectively. Five proteins are shared between all three polymerases, and the Tritryp orthologs of all but one are obvious in the Tritryp

genomes, although they appear to contain additional copies of two of the smaller three subunits, suggesting that they may not be shared in trypanosomatids (Ivens *et al.* 2005). Another seven proteins are similar, but not identical, in all three polymerases, and once again orthologs of all but one can be found in the Tritryps. Interestingly, the largest RNAP II subunit (RBP1) in Tritryps lacks the heptad repeats found in the higher eukaryotic C-terminal domain (CTD). The RNAP I and RNAP III complexes are less conserved, with neither of the two RNAP I-specific proteins, and only one (RPC34) of five RNAP III-specific proteins identified by sequence homology, although this may merely reflect sequence divergence, since putative orthologs of three additional subunits have been identified by mass spectrometry of purified RNAP III complexes (Martinez-Calvillo, Leland, Saxena, and Myler, manuscript in preparation).

Annotation of the Tritryp genomes revealed a dearth of typical eukaryotic transcription factors that are normally involved in regulation of transcription initiation, elongation, termination, and transcriptional (Ivens *et al.* 2005). The only RNAP II basal transcription factor definitively identified was a divergent ortholog of the TATA binding protein (TBP – a component of TFIIH and TFIIIB), which has been called TBP-related factor 4 (TRF4) since it is most related to metazoan TRF1 and TRF2 (Ruan *et al.* 2004). Several subunits of TFIIH were also identified, but these also function in DNA repair. An ortholog of BTF3b, which binds the RNAP II complex in metazoans, was also found by sequence similarity, as were the BRF and B$''$ subunits of TFIIIB, La antigen, and two subunits of the snRNA-activating protein complex SNAP$_c$, which are all involved in RNAP III transcription. Recent experiments using tandem affinity purification and mass spectrometry have identified several more likely transcription factors in *T. brucei*, including two additional SNAP$_c$ subunits and two subunits of TFIIA (Das *et al.* 2005; Schimanski *et al.* 2005).

Several proteins related to transcription elongation were found in the Tritryp genomes, including putative orthologs of yeast ELP2, ELP3, SPT5, SPT16, and CHD1, as well as three or four (depending on species) proteins containing a TFIIS zinc ribbon domain, which occurs in RNAP II transcription elongation factor TFIIS/DST1 and the RPA12 and RPC11 subunits of the RNAP I and RNAP III core complexes, respectively. Proteins associated with transcriptional regulation were also conspicuously absent, with only three candidates identified: two related to human NF-X1 and *Drosophila* shuttle craft proteins, and a SNF2-like DNA helicase which may function via chromatin remodeling. The absence (or substantial divergence) of Tritryp transcription factors involved in promoter recognition, transcription initiation and transcriptional regulation, but apparent presence of those involved in transcriptional elongation, is probably a reflection of the unusual RNAP II transcription mechanisms in these organisms, as described above. The paucity of genes for transcriptional regulators is offset by an abundance of proteins with RNA binding motifs (Anantharaman *et al.* 2003), implying a reliance on post-transcriptional control of gene expression (Clayton 2002).

Trypanosomatid mRNA processing is also distinctive, with a 39-nt capped RNA added post-transcriptionally to the 5$'$ end of all mRNAs by *trans*-splicing (Clayton 2002; Perry and Agabian 1991), and the site of polyadenylation determined by *trans*-splicing of the downstream mRNA, rather than by an AAUAAA and downstream G/U rich tract (LeBowitz *et al.* 1993; Matthews *et al.* 1994). The Tritryp spliceosome catalyzes both *cis*- and *trans*-splicing, which are mechanistically related,

and most of the snRNAs and spliceosomal proteins have been identified on the basis of similarity to their eukaryotic orthologs (Liang *et al.* 2003). The Tritryp genomes encode two, quite different, poly(A) polymerases, suggesting they may have distinct functional roles. The Tritryp cleavage and polyadenylation specificity factor (CPF/CPSF) and mRNA degradation (exosome) complexes appear relatively similar to those in yeast and mammals (Ivens *et al.* 2005). Trypanosomatid mitochondrial (kinetoplast) mRNAs undergo RNA editing that inserts and deletes uridylate residues (Stuart *et al.* 1997), and the genes for many proteins that function in editing have been identified and are conserved among the Tritryps (Worthey *et al.* 2003b). This unusual process is reflected in the unusual kinetoplast DNA organization, in which the fairly conventional mitochondrial DNA (called the maxicircle) is intercatenated with thousands of minicircles that encode guide RNAs (gRNAs).

Metabolism

Analysis of the Tritryp genomes has provided a comprehensive view of the parasites' metabolic potential by identifying common metabolic and transport processes and species-specific differences that likely reflect the different life-styles of the three organisms (Berriman *et al.* 2005). In general, the *T. brucei* genome has the lowest metabolic potential, with *T. cruzi* possessing somewhat more genes encoding enzymes and transporters and the *L. major* genome encoding the most. Tritryps lack biosynthetic pathways for 10 amino acids and possess a larger number of amino acid permeases (Berriman *et al.* 2005). As expected the Tritryps possess a full complement of genes for uptake and degradation of glucose *via* glycolysis or the pentose phosphate shunt and tricarboxylic acid (TCA) cycle, but have a limited number of hexose transporters. Only *T. cruzi* possesses hexose phosphate transporters, presumably to allow uptake from the host cell cytosol. In contrast, only *Leishmania* appears capable of hydrolyzing disaccharides. *T. brucei* lacks several sugar kinases present in *T. cruzi* and *L. major*. This is consistent with the biology of the insect vectors: tsetse flies and triatomines are obligate blood feeders, whereas sand flies also feed on nectar and aphid honeydew.

The Tritryps lack lactate dehydrogenase and consequently degrade glucose only to ethanol, alanine, glycerol, and several TCA intermediates. In many insects, including tsetse and sand flies, proline is a highly abundant energy source used in flight muscles, and the Tritryp insect stages degrade proline to alanine and succinate. A fully functional mitochondrial electron transport system and ATP synthase is present in the Tritryps, with several ubiquinone-linked dehydrogenases, although its activity is regulated during the life cycle, especially in *T. brucei*, which also contains a mitochondrial alternative oxidase that is not found in *L. major* or *T. cruzi* (Clarkson *et al.* 1990).

The Tritryp genomes lack almost all the genes necessary for *de novo* purine synthesis, but encode a number of nucleotide and nucleoside transporters and numerous enzymes for the interconversion of purine bases and nucleosides (Berriman *et al.* 2005). Conversely, trypanosomatids can synthesize pyrimidines *de novo* from aspartate, glutamine, and bicarbonate, and interestingly the six enzymatic activities in this pathway are located in a single cluster of five genes in all three genomes. Reduction of ribonucleotides to corresponding deoxyribonucleotides is carried out by ribonucleotide reductase using trypanothione or tryparedoxin (Dormeyer *et al.* 2001). Synthesis of thymidylate from dUMP involves the bifunctional dihydrofolate reductase–thymidylate synthase, which is essential in *Leishmania* (Titus *et al.* 1995).

Trypanothione, the parasite's equivalent of glutathione, is central to defense against chemical and oxidant stress, with the trivalent arsenical and antimonial drugs being conjugated with trypanothione and eliminated from the cell via ATP-binding cassette transporters (Fairlamb 2003). The Tritryps rely on trypanothione-dependent peroxidases for removal of peroxides and these enzymes use tryparedoxin instead of thioredoxins. Trypanosomatids are also unusual in that trypanothione *S*-transferase activity is associated with the eukaryotic elongation factor complex, eEF1B (Vickers *et al.* 2004b; Vickers *et al.* 2004a).

Lipid metabolism in the trypanosomatids has aspects reminiscent of both mammals and fungi. Like mammals, they are capable of fatty acid biosynthesis and β-oxidation of fatty acids in two separate cellular compartments, glycosomes (equivalent to mammalian peroxisomes) and the mitochondrion. However, like the fungi, they synthesize ergosterol instead of cholesterol and are susceptible to antifungal agents that inhibit this pathway. The preferred source of 3-hydroxy-3-methylglutaryl coenzyme-A appears to be leucine rather than acetate or acetyl CoA (Ginger *et al.* 2001). Intermediates of this pathway, namely farnesyl- and geranyl pyrophoshosphate, are also precursors for dolichols and the side chain of ubiquinone, and are promising targets for drug development against these organisms.

Trypanosomatid surface membranes contain a variety of lipophosphoglycan (LPG), glycoinositolphospholipids (GIPLs), proteophosphoglycans (PPG), and glycosylphosphatidylinositol (GPI)-anchored glycoproteins that are involved in immune evasion, attachment, or invasion. The genes for the ether phospholipid synthetic pathway and GPI synthesis have been identified (Berriman *et al.* 2005), but identification of the genes for the entire LPG synthetic pathway remains incomplete (Dobson *et al.* 2003). Many, but not all, of these genes that have been identified appear to be *L. major*-specific, although *T. brucei* and *T. cruzi* both contain a number of genes encoding glycosyltransferases with different specificities. These often appear in telomeric or subtelomeric locations, suggesting a possible linkage with recombinational processes underlying antigenic variation and diversity. For example, *T. brucei* contains a large family of UDP-galactose- or UDP-*N*-acetylglucosamine-dependent glycosyltransferase genes, nine of which appear to be pseudogenes associated with arrays of *VSG* pseudogenes; while seven PG-galactosyltransferase genes and three potential GIPL β-galactofuranosyltransferase genes have telomeric locations in *L. major* (Dobson *et al.* 2003; Dobson *et al.* 2006; Zhang *et al.* 2004).

The GPI linkage to *T. cruzi* mucins occurs via aminoethylphosphonate. While only *T. cruzi* contains genes for all the enzymes needed for its synthesis from phosphoenolpyruvate, *L. major* also contains the third enzyme (2-aminoethylphosphonate: pyruvate transaminase) of the pathway and the second step (phosphonopyruvate decarboxylase) is found in all three parasites. There is evidence for their acquisition from bacteria by horizontal gene transfer (Berriman *et al*. 2005). There are several other examples of apparent gene transfer from bacteria, most notably the last three enzymes (coproporphyrinogen III oxidase, protoporphyrinogen IX oxidase, ferrochelatase) of heme biosynthesis, which are present only in *L. major*, despite all three Tritryps being heme auxotrophs.

Other processes

Based on analysis of the Tritryp genomes, DNA replication in trypanosomatids appears to differ significantly from that of higher eukaryotes. This is most marked for the replication origin complex, where only one of the six eukaryotic subunits can be found (El-Sayed *et al*. 2005a), suggesting that initiation may resemble that in the Archaea. There are also substantial differences in the replication machinery for kDNA from its mitochondrial counterpart in higher eukaryotes. The complexity of the kDNA structure (see above) dictates an unusual replication mechanism, with six nuclear-encoded DNA polymerases finding their way to the mitochondrion (Klingbeil *et al*. 2002), instead of the single one (DNA polymerase γ) in yeast and mammalian mitochondria. These polymerases differ in that four are related to bacterial DNA polymerase I, and two are related to DNA polymerase β, an enzyme located in the nucleus of other eukaryotes and involved in base excision repair. There also appear to be multiple DNA ligases and helicases. DNA repair pathways in the Tritryps appear similar to those in other eukaryotes, with a few notable exceptions. The enzymatic machinery for non-homologous end-joining has not been found and there are multiple genes encoding DNA polymerase κ. The latter is interesting because of its potential role in spontaneous mutagenesis during replication of tandem gene arrays (Hubscher *et al*. 2002).

Trypanosomatids have complicated life cycles in different hosts, and thus need to co-ordinate and regulate their response to changes in these different environments. However, analysis of Tritryp genomes indicate that they lack several classes of signaling molecules found in other eukaryotes, including serpentine receptors, heterotrimeric G proteins, most classes of catalytic receptors, SH2 and SH3 interaction domains, and regulatory transcription factors. They do, however, possess a large and complex set of protein kinases and protein phosphatases (El-Sayed *et al*. 2005a; Parsons *et al*. 2005). Indeed, the trypanosomatid kinome is larger than that of other single-celled organisms. The distribution of protein kinase classes differs from that in other organisms, with no tyrosine kinases (other than dual specificity kinases); few, if any, receptor kinases; and no TKL or RGC group kinases. They do, however, contain numerous CMGC, STE, and NEK kinases, as well as many that could not be classified into any established group. In addition, the Tritryp kinases generally lack identifiable accessory domains (such as SH2, SH3, FN-III, and immunoglobulin-like domains) found in other organisms, although a number contain PH domains, suggesting a role of phosphoinositide metabolism in regulatory networks in the parasite. Despite the paucity of recognizable domains, the vast majority of Tritryp kinases are much larger than a simple catalytic domain, suggesting novel interaction or regulatory elements exist in the parasites.

Implications for drugs/diagnostics and vaccine development

The availability of the genome sequence of these three parasitic trypanosomatids should accelerate progress in developing effective clinical interventions for these important diseases that affect large, vulnerable populations. In particular, differences in transcription, RNA processing, DNA replication, nucleotide excision/repair, and signaling pathways all hold promise for identifying novel drug targets. Likewise, the finding that a number of metabolic enzymes appear to have bacterial origin adds considerably to the likelihood that effective, new chemotherapeutic agents can be developed. On the down side, however, the genome sequences have clearly indicated the complexity of parasite mechanisms for immune evasion, making it unlikely that effective vaccines will be developed, at least for *T. brucei* and *T. cruzi*.

References

Anantharaman, V., Aravind, L., and Koonin, E. V. (2003). Emergence of diverse biochemical activities in evolutionarily conserved structural scaffolds of proteins. *Curr. Opin. Chem. Biol.*, **7**, 12–20.

Andersson, B., Åslund, L., Tammi, M., Tran, A.-N., Hoheisel, J. D., and Pettersson, U. (1998). Complete sequence of an 93.4 kb contig from chromosome 3 of *Trypanosoma cruzi* containing a strand switch region. *Genome Res.*, **8**, 809–816.

Bastien, P., Blaineau, C., Britto, C. *et al.* (1998). The complete chromosomal organization of the reference strain of the *Leishmania* genome project, *L. major* "Friedlin". *Parasitol. Today*, **14**, 301–303.

Berriman, M., Ghedin, E., Hertz-Fowler, C. *et al.* (2005). The genome of the African trypanosome, *Trypanosoma brucei*. *Science*, **309**, 416–422.

Britto, C., Ravel, C., Bastien, P. *et al.* (1998). Conserved linkage groups associated with large-scale chromosomal rearrangements between Old World and New World *Leishmania* genomes. *Gene*, **222**, 107–117.

Campbell, D. A., Thomas, S., and Sturm, N. (2003). Transcription in kinetoplastid protozoa: why be normal? *Microbes Infect.*, **5**, 1231–1240.

Cano, M. I., Gruber, A., Vazquez, M., *et al.* (1999). Molecular karyotype of clone CL Brener chosen for the *Trypanosoma cruzi* genome project. *Mol. Biochem. Parasitol.*, **71**, 273–278.

Clarkson, A. B., Bienen, E. J., Pollakis, G., Saric, M., McIntosh, L., and Grady, R. W. (1990). Trypanosome alternative oxidase. In A. S. Peregrine, ed. *Chemotherapy for Trypanosomiasis: Proceedings of a workshop held at ILRAD, Nairobi, Kenya*. Man Graphics, Nairobi, Kenya, pp. 43–48.

Clayton, C. E. (2002). Life without transcriptional control? From fly to man and back again. *EMBO J.*, **21**, 1881–1888.

Das, A., Zhang, Q., Palenchar, J. B., Chatterjee, B., Cross, G. A., and Bellofatto, V. (2005). Trypanosomal TBP functions with the multisubunit transcription factor tSNAP to direct spliced-leader RNA gene expression. *Mol. Cell. Biol.*, **25**, 7314–7322.

Dobson, D. E., Scholtes, L. D., Myler, P. J., Turco, S. J., and Beverley, S. M. (2006). Genomic organization and expression of the expanded SCG/L/R gene family of *Leishmania major*: internal clusters and telomeric localization of SCGS mediating species-specific LPG modifications. *Mol. Biochem. Parasitol*, 231–241.

Dobson, D. E., Scholtes, L. D., Valdez, K. E., *et al.* (2003). Functional identification of galactosyltransferases (SCGs) required for species-specific modifications of the lipophosphoglycan adhesin controlling *Leishmania major*-sand fly interactions. *J. Biol. Chem.*, **278**, 15523–15531.

Dormeyer, M., Reckenfelderbaumer, N., Ludemann, H., and Krauth-Siegel, R. L. (2001). Trypanothione-dependent synthesis of deoxyribonucleotides by *Trypanosoma brucei* ribonucleotide reductase. *J. Biol. Chem.*, **276**, 10602–10606.

El-Sayed, N. M. A., Ghedin, E., Song, J., *et al.* (2003). The sequence and analysis of *Trypanosoma brucei* chromosome II. *Nucleic Acids Res.*, **31**, 4856–4863.

El-Sayed, N. M. A., Myler, P. J., Bartholomeu, D., *et al.* (2005a). The genome sequence of *Trypanosoma cruzi*, etiological agent of Chagas' disease. *Science*, **309**, 409–415.

El-Sayed, N. M. A., Myler, P. J., Blandin, G., *et al.* (2005b). Comparative genomics of trypanosomatid parasitic protozoa. *Science*, **309**, 404–409.

Fairlamb, A. H. (2003). Chemotherapy of human African trypanosomiasis: current and future prospects. *Trends Parasitol.*, **19**, 488–494.

Fantoni, A., Dare, A. O., and Tschudi, C. (1994). RNA polymerase III-mediated transcription of the trypanosome U2 small nuclear RNA gene is controlled by both intragenic and extragenic regulatory elements. *Mol. Cell. Biol.*, **14**, 2021–2028.

Ghedin, E., Bringaud, F., Peterson, J., *et al.* (2004). Gene synteny and evolution of genome architecture in trypanosomatids. *Mol. Biochem. Parasitol.*, **134**, 183–191.

Ginger, M. L., Chance, M. L., Sadler, I. H., and Goad, L. J. (2001). The biosynthetic incorporation of the intact leucine skeleton into sterol by the trypanosomatid *Leishmania mexicana*. *J. Biol. Chem.*, **276**, 11674–11682.

Haag, J., O'hUigin, C., and Overath, P. (1998). The molecular phylogeny of trypanosomes: evidence for an early divergence of the Salivaria. *Mol. Biochem. Parasitol.*, **91**, 37–49.

Hall, N., Berriman, M., Lennard, N. J., *et al.* (2003). The DNA sequence of chromosome I of an African trypanosome: gene content, chromosome organisation, recombination and polymorphism. *Nucleic Acids Res.*, **31**, 4864–4873.

Hubscher, U., Maga, G., and Spadari, S. (2002). Eukaryotic DNA polymerases. *Annu. Rev. Biochem.*, **71**, 133–163.

Ivens, A. C., Peacock, C. S., Worthey, E. A., *et al.* (2005). The genome of the kinetoplastid parasite, *Leishmania major*. *Science*, **309**, 436–442.

Klingbeil, M. M., Motyka, S. A., and Englund, P. T. (2002). Multiple mitochondrial DNA polymerases in *Trypanosoma brucei*. *Mol. Cell*, **10**, 175–186.

LeBowitz, J. H., Smith, H. Q., Rusche, L., and Beverley, S. M. (1993). Coupling of poly(A) site selection and *trans*-splicing in *Leishmania*. *Genes Devel.*, **7**, 996–1007.

Liang, X. H., Haritan, A., Uliel, S., and Michaeli, S. (2003). *trans* and *cis* splicing in trypanosomatids: mechanism, factors, and regulation. *Eukaryot. Cell*, **2**, 830–840.

Lodes, M. J., Merlin, G., deVos, T., *et al.* (1995). Increased expression of *LD1* genes transcribed by RNA polymerase I in *Leishmania donovani* as a result of duplication into the *rRNA* gene locus. *Mol. Cell. Biol.*, **15**, 6845–6853.

Martinez-Calvillo, S., Nguyen, D., Stuart, K., and Myler, P. J. (2004). Transcription initiation and termination on *Leishmania major* chromosome 3. *Eukaryot. Cell*, **3**, 506–517.

Martinez-Calvillo, S., Stuart, K., and Myler, P. J. (2005). Ploidy changes associated with disruption of two adjacent genes on *Leishmania major* chromosome 1. *Int. J. Parasitol.*, **35**, 419–429.

Martinez-Calvillo, S., Yan, S., Nguyen, D., Fox, M., Stuart, K. D., and Myler, P. J. (2003). Transcription of *Leishmania major* Friedlin chromosome 1 initiates in both directions within a single region. *Mol. Cell*, **11**, 1291–1299.

Matthews, K. R., Tschudi, C., and Ullu, E. (1994). A common pyrimidine-rich motif governs trans-splicing and polyadenylation of tubulin polycistronic pre-mRNA in trypanosomes. *Genes Devel.*, **8**, 491–501.

McDonagh, P. D., Myler, P. J., and Stuart, K. D. (2000). The unusual gene organization of *Leishmania major* chromosome 1 may reflect novel transcription processes. *Nucleic Acids Res.*, **28**, 2800–2803.

Melville, S. E., Leech, V., Gerrard, C. S., Tait, A., and Blackwell, J. M. (1998). The molecular karyotype of the megabase chromosomes of *Trypanosoma brucei* and the assignment of chromosome markers. *Mol. Biochem. Parasitol.*, **94**, 155–173.

Mihok, S. and Otieno, L. H. (1990). Trypanosome diversity in Lambwe Valley, Kenya-Sex or selection? Reply. *Parasitol. Today*, **6**, 353.

Mihok, S., Otieno, L. H., and Darji, N. (1990). Population genetics of *Trypanosoma brucei* and the epidemiology of human sleeping sickness in the Lambwe Valley, Kenya. *Parasitol.*, **100**, 219–233.

Myler, P. J., Audleman, L., deVos, T., *et al.* (1999). *Leishmania major* Friedlin chromosome 1 has an unusual distribution of protein-coding genes. *Proc. Natl. Acad. Sci. U S A*, **96**, 2902–2906.

Nakaar, V., Dare, A. O., Hong, D., Ullu, E., and Tschudi, C. (1994). Upstream tRNA genes are essential for expression of small nuclear and cytoplasmic RNA genes in trypanosomes. *Mol. Cell. Biol.*, **14**, 6736–6742.

Nakaar, V., Günzl, A., Ullu, E., and Tschudi, C. (1997). Structure of the *Trypanosoma brucei* U6 snRNA gene promoter. *Mol. Biochem. Parasitol.*, **88**, 13–23.

Nakaar, V., Tschudi, C., and Ullu, E. (1995). An unusual liaison: Small nuclear and cytoplasmic RNA genes team up with tRNA genes in trypanosomatid protozoa. *Parasitol. Today*, **11**, 225–228.

Nilsson, D. and Andersson, B. (2005). Strand asymmetry patterns in trypanosomatid parasites. *Exp. Parasitol.*, **109**, 143–149.

Parsons, M., Worthey, E. A., Ward, P. N., and Mottram, J. C. (2005). Comparative analysis of the kinomes of three pathogenic trypanosomatids: *Leishmania major*, *Trypanosoma brucei*, and *Trypanosoma cruzi*. *BMC Genomics*, **6**, 127.

Pays, E., Lips, S., Nolan, D., Vanhamme, L., and Perez-Morga, D. (2001). The VSG expression sites of *Trypanosoma brucei*: multipurpose tools for the adaptation of the parasite to mammalian hosts. *Mol. Biochem. Parasitol.*, **114**, 1–16.

Pays, E., Vanhamme, L., and Perez-Morga, D. (2004). Antigenic variation in *Trypanosoma brucei*: facts, challenges and mysteries. *Curr. Opin. Microbiol.*, **7**, 369–374.

Perry, K. and Agabian, N. (1991). messenger RNA processing in the Trypanosomatidae. *Experientia*, **47**, 118–128.

Ruan, J. P., Arhin, G. K., Ullu, E., and Tschudi, C. (2004). Functional characterization of a *Trypanosoma brucei* TATA-binding protein-related factor points to a universal regulator of transcription in trypanosomes. *Mol. Cell. Biol.*, **24**, 9610–9618.

Schimanski, B., Nguyen, T. N., and Günzl, A. (2005). Characterization of a multisubunit transcription factor complex essential for spliced-leader RNA gene transcription in *Trypanosoma brucei*. *Mol. Cell. Biol.*, **25**, 7303–7313.

Stevens, J. R., Noyes, H. A., Schofield, C. J., and Gibson, W. (2001). The molecular evolution of Trypanosomatidae. *Adv. Parasitol.*, **48**, 1–56.

Stuart, K., Allen, T. E., Heidmann, S., and Seiwert, S. D. (1997). RNA Editing in kinetoplastid protozoa. *Microbiol. Mol. Biol. Rev.*, **61**, 105–120.

Sturm, N. R., Vargas, N. S., Westenberger, S. J., Zingales, B., and Campbell, D. A. (2003). Evidence for multiple hybrid groups in *Trypanosoma cruzi*. *Int. J. Parasitol.*, **33**, 269–279.

Sunkin, S. M., Kiser, P., Myler, P. J., and Stuart, K. D. (2000). The size difference between *Leishmania major* Friedlin chromosome one homologues is localized to sub-telomeric repeats at one chromosomal end. *Mol. Biochem. Parasitol.*, **109**, 1–15.

Swindle, J. and Tait, A. (1996). Trypanosomatid genetics. In D. F. Smith and M. Parsons, eds. *Molecular Biology of Parasitic Protozoa*. Oxford University Press, Oxford, UK, pp. 6–34.

Tait, A. (1980). Evidence for diploidy and mating in trypanosomes. *Nature*, **287**, 536–538.

Tibayrenc, M. (1992). *Leishmania*: Sex, karyotypes and population genetics. *Parasitol. Today*, **8**, 305–306.

Tibayrenc, M., Kjellberg, F., and Ayala, F. J. (1990). A clonal theory of parasitic protozoa: The population structures of *Entamoeba, Giardia, Leishmania, Naegleria, Plasmodium, Trichomonas*, and *Trypanosoma* and their medical and taxonomical consequences. *Proc. Natl. Acad. Sci. U S A*, **87**, 2414–2418.

Tibayrenc, M., Ward, P., Moya, A., and Ayala, F. J. (1986). Natural populations of *Trypanosoma cruzi*, the agent of Chagas' disease, have a complex multiclonal structure. *Proc. Natl. Acad. Sci. U S A*, **83**, 115–119.

Titus, R. G., Gueiros-Filho, F. J., De Freitas, L. A. R., and Beverley, S. M. (1995). Development of a safe live *Leishmania* vaccine line by gene replacement. *Proc. Natl. Acad. Sci. U S A*, **92**, 10267–10271.

Turner, C. M., Melville, S. E., and Tait, A. (1997). A proposal for karyotype nomenclature in Trypanosoma brucei. *Parasitol. Today*, **13**, 5–6.

Vanhamme, L. and Pays, E. (1995). Control of gene expression in trypanosomes. *Microbiol. Rev.*, **59**, 223–240.

Vargas, N., Pedroso, A., and Zingales, B. (2004). Chromosomal polymorphism, gene synteny and genome size in T. cruzi I and T. cruzi II groups. *Mol. Biochem. Parasitol.*, **138**, 131–141.

Vickers, T. J. and Fairlamb, A. H. (2004a). Trypanothione S-transferase activity in a trypanosomatid ribosomal elongation factor 1B. *J. Biol. Chem.*, **279**, 27246–27256.

Vickers, T. J., Wyllie, S. H., and Fairlamb, A. H. (2004b). *Leishmania major* elongation factor 1B complex has trypanothione S-transferase and peroxidase activity. *J. Biol. Chem.*, **279**, 49003–49009.

Westenberger, S. J., Barnabe, C., Campbell, D. A., and Sturm, N. R. (2005). Two hybridization events define the population structure of *Trypanosoma cruzi*. *Genetics*, **171**, 527–543.

Worthey, E., Martinez-Calvillo, S., Schnaufer, A., et al. (2003a). *Leishmania major* chromosome 3 contains two long "convergent" polycistronic gene clusters separated by a tRNA gene. *Nucleic Acids Res.*, **31**, 4201–4210.

Worthey, E. A., Schnaufer, A., Mian, I. S., Stuart, K., and Salavati, R. (2003b). Comparative analysis of editosome proteins in trypanosomatids. *Nucleic Acids Res.*, **31**, 6392–6408.

Xu, Y., Liu, L., Lopez-Estrano, C., and Michaeli, S. (2001). Expression studies on clustered trypanosomatid box C/D small nucleolar RNAs. *J. Biol. Chem.*, **276**, 14289–14298.

Zhang, K., Barron, T., Turco, S. J., and Beverley, S. M. (2004). The LPG1 gene family of *Leishmania major*. *Mol. Biochem. Parasitol.*, **136**, 11–23.

CHAPTER 11

The genome of *Entamoeba histolytica*

C. Graham Clark

Department of Infectious and Tropical Diseases, London School of Hygiene and Tropical Medicine, London, UK

Introduction

Entamoeba histolytica

Among the parasitic microbial eukaryote, *Entamoeba histolytica* ranks second behind the malaria parasites as a cause of human mortality. The WHO (1998) estimates that as many as 70 000 people die each year from the diseases it causes – primarily amoebic colitis and amoebic liver abscess. The number of people infected is much higher, perhaps as high as 50 million. The reason why so few infections become symptomatic is unknown and all infected individuals are considered to be at risk of developing disease.

Entamoeba histolytica, like almost all other species of *Entamoeba*, has only two life-cycle stages, the feeding/dividing trophozoite stage and the quiescent/transmission cyst stage (Figs. 11.1 and 11.2). The trophozoite inhabits the lumen of the large bowel of its host where it feeds and divides by binary fission (Fig. 11.2). In the intestinal environment the primary food source of the parasite is bacteria, although it will also ingest other organisms and food particles it encounters. In response to unknown stimuli the organism undergoes a transformation to produce the infective cyst form that is then passed into the environment in stool. Transmission to a new host occurs through ingestion of fecally contaminated food or water. The ingested cysts are resistant to stomach acids and pass unscathed into the small bowel. Emergence of the amoebae from cysts occurs as they leave the small bowel. The parasites then colonize the mucosal surface of the large bowel, completing the cycle

(Figs. 11.1 and 11.2). For a review of the organism's basic biology see Clark *et al.* (2000).

Most of the proteins implicated in causing disease are part of the normal nutrient-gathering machinery of the amoeba. The surface of the amoeba has carbohydrate-binding lectin proteins, the best known of which binds specifically to galactose and *N*-galactosamine-containing structures on the surface of the bacterial cells. The bound bacteria are taken into food vacuoles where vesicles containing a variety of digestive enzymes fuse and release their contents, which include cysteine proteinases, pore-forming peptides, lipases, etc. This cocktail kills and digests the bacteria in the vacuoles and the breakdown products are taken into the

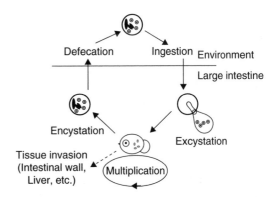

Figure 11.1 Life cycle of *Entamoeba histolytica*. The two life cycle stages (cyst and trophozoite) and where they are found are indicated. The dashed arrow indicates the pathway to disease, which is not a required part of the life cycle and, in fact, is a dead end for transmission of the infection.

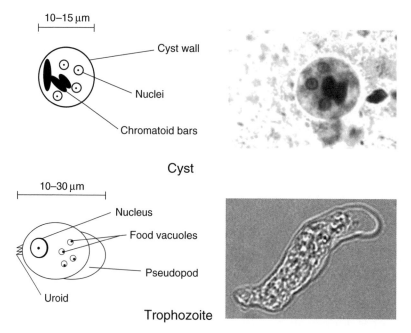

Figure 11.2 Morphological features of *Entamoeba histolytica*. The main morphological features usually visible in the cyst and trophozoite stages are indicated in diagrammatic form. A stained cyst (iron-hematoxylin) is shown for comparison – with this stain the cyst wall is only visible as a clear ring around the cyst. Chromatoid bars are transient features of the cyst and consist of crystalline ribosomes. A live trophozoite photographed under DIC optics is also shown. The cytoplasm is filled with small food vacuoles, obscuring the nucleus. The pseudopod is clear and forms at the front edge of the moving cell. At the rear of the cell is the uroid, which is where undigested waste is expelled from the cell. Photographs courtesy of John Williams, London School of Hygiene and Tropical Medicine.

amoebic cytoplasm for the parasite's use. The system goes awry when the lectin binds instead to the surface of mucosal epithelial cells. The mucosa cannot be taken into a food vacuole but the lytic vesicles still fuse with the bound lectin regions of the surface and dump their contents outside the *Entamoeba* cell. This initiates mucosal tissue necrosis and attracts host immune cells to the site. *E. histolytica* can lyse neutrophils and the contents of these immune cells are released, which also contributes to tissue destruction. Eventually, amoebae may penetrate the mucosa itself and can be transported through the portal veins to the liver, initiating a liver abscess. For a review of pathogenesis see Espinosa-Cantellano and Martinez-Palomo (2000).

The mechanism by which *Entamoeba* ingests and digests bacteria seems to be common to all species. Why only *E. histolytica* can cause disease in humans is therefore unclear – the human intestine can be host to at least five other species of *Entamoeba*, none of which cause any symptoms. In fact,

E. dispar is so similar that it was only recognized as a distinct species recently (Diamond and Clark 1993) and is considered a sibling species, yet it has never been shown to cause disease.

Origins of the genome project

The importance of *E. histolytica* as a pathogen was the primary reason for sequencing its genome. A second and distinct reason was its evolutionary interest. *Entamoeba* was long thought to represent a "primitive" eukaryote as it lacked many of the cellular characteristics normally associated with the eukaryotic state. It has a nucleus but appears to lack other cell organelles such as mitochondria, rough endoplasmic reticulum, Golgi apparatus, peroxisomes, and centrioles, at least at the morphological level (Clark *et al.* 2000). However, when phylogenetic trees of small subunit ribosomal RNA sequences were produced, the *Entamoeba* branch appeared to arise relatively late in the day, diverging more recently than several lineages with classical mitochondria and more "typical" eukaryotic cell organization. These phylogenetic trees implied that the primitive appearance of *Entamoeba* was deceiving and that it was actually a highly derived organism where many common eukaryotic morphological features had been lost (Clark *et al.* 2000; Simpson and Patterson, Chapter 1, this book).

More recent phylogenetic trees have suggested that the slime-mold *Dictyostelium*, which has mitochondria and essentially all typical eukaryotic features, is a free-living organism specifically related to *Entamoeba* (see Simpson and Patterson, Chapter 1 and Schaap, Chapter 15, this book). This reinforces the conclusion that *Entamoeba* is a derived rather than a primitive organism as well as suggesting a useful genetic "partner" for future analyses of *Entamoeba* genes.

Representatives of many *Entamoeba* research groups met in October 1998 to plan the genome project. This working group made several important decisions, including the strain to be sequenced, HM-1:IMSS, which was selected for sequencing as it was the strain already being used for most genetic research on *E. histolytica*. It was decided to revive a vial of the strain that had been cryopreserved in the early 1970s, shortly after it was isolated. In this way it was hoped that any genome changes that might have arisen during long-term *in vitro* cultivation would be minimized. Preliminary sequencing started in 2000 with the full genome project starting in 2001 at both The Institute for Genomic Research (USA) and the Sanger Institute (GB).

What we knew at the start

One of the technical innovations in the original application for funding was the idea of using shotgun sequencing of small fragments of DNA to obtain the whole genome sequence of *E. histolytica*. At that time the approach had only been used for bacterial genomes as eukaryotic genomes were thought to be too complex to assemble from hundreds of thousands of individual sequence reads. The reason for proposing this "radical" approach was simply that we did not know much about the *E. histolytica* genome before the project started. Although it was thought that the chromosome

number was 14 and the genome size was estimated to be 20 Mb (Willhoeft and Tannich 1999), our ability to separate individual chromosomes for purification was limited and no one had reported success in generating libraries of large insert clones–two approaches that were being used as first steps in the analysis of other eukaryotic genomes at the time. As a result, a whole genome shotgun was actually a necessity rather than an "experiment". *E. histolytica* was thus to be a test case, but in the interim the shotgun approach for eukaryotes has become the norm.

As for the genome structure itself, relatively little was known as only a few dozen targeted genes had been sequenced up to that point in time. Nevertheless, it was clear that the genome was very A+T rich, that the ribosomal RNA genes (rDNA) were carried on an abundant circular DNA of 25 kb in size (Bhattacharya *et al.* 2000), that allelic chromosomes varied substantially in length (although the basis for this was unknown; Willhoeft and Tannich 1999), and that a number of Long and Short Interspersed Nuclear Elements (LINEs and SINEs; derived from retrotransposon-like mobile genetic elements) were present in the chromosomes (Bhattacharya *et al.* 2003; van Dellen *et al.* 2002). These potential difficulties were thought to be of little consequence at the time but came back to haunt the project at a later date (see section Assembly, below).

Genome data

The following sections summarize some of the major findings of the genome project. The work was published recently (Loftus *et al.* 2005) but space constraints in that publication meant that many aspects of the data analysis could not be included. Where such results are referred to below they are attributed to the person responsible as "unpublished analyses". It is anticipated that these data will eventually be published separately.

Problems encountered

Although we knew going into the project that the rDNA circles were abundant, we thought that

Table 11.1 Unusual features of the *E. histolytica* genome

20% of DNA is ribosomal RNA genes
10% of DNA is transfer RNA genes
6% of DNA is transposable element related
Large family of trans-membrane kinases
No folate metabolism
Many genes acquired by lateral gene transfer
No telomere sequences

---

there was a straightforward method for removing them from the DNA preparations prior to library construction. A single site for the enzyme *Ppo*I had been shown to exist in the circle and nowhere else in the genome. By linearizing the circles with the enzyme they could be separated by size from the rest of the genomic DNA by simple gel electrophoresis. Unfortunately, when the approach was tried it was discovered that this led to a substantial drop in the amount of genomic DNA recovered for library construction. The approach was abandoned, with the consequence that as many as 20% of the sequence reads were redundant as they consisted entirely of rDNA sequences (Loftus *et al.* 2005). This large percentage of nuclear DNA being made up of rDNA alone is actually comparable to that in *Dictyostelium* (Eichinger *et al.* 2005) and the free-living amoeba *Naegleria* (Clark and Cross 1987), which also have extrachromosomal rDNA.

In a similar vein, but totally unexpectedly, it was discovered during sequencing that over 10% of the sequence reads contained transfer RNA (tRNA) genes that were organized into tandem arrays (see below). Again most of these reads were redundant and so in total almost a third of the sequences generated proved to be "wasted" as they had to be excluded from the assembly process – an expensive waste of resources (Table 11.1).

Attempts were once again made, at the sequencing centers, to generate large insert libraries with a view to improving the assembly of the genome into chromosome–sized, contiguous sequences. Despite the expertise of those involved these efforts were largely unsuccessful, the largest inserts used in the assembly process being about 10 kb in size. Efforts to overcome this are still ongoing. The high $A + T$ content of the genome (75%) is thought to be at least partly responsible

for the difficulties encountered but may not be the only factor. This biased base composition is comparable to that in *Plasmodium* (80%; Gardner *et al.* 2002) and *Dictyostelium* (77%; Eichinger *et al.* 2005) but much higher than in the three trypanosomatid genomes available (40–54%; El-Sayed *et al.* 2005)

Repetitive DNA

The rDNA circles had been sequenced and described in the literature prior to the genome project. The 25 kb sequence encodes two copies of a complete rRNA transcription unit (small subunit, large subunit and 5.8S rRNAs but not the 5S RNA) plus several families of repeated sequences. No evidence of a chromosomal rDNA copy exists and these circles are self-replicating (Bhattacharya *et al.* 2000).

In contrast, the tRNA genes were not known in any detail before the genome project began – only one entry in GenBank identified tRNAs as being encoded (GenBank Accession Number AF265348). A study of polymorphic sequences in *E. histolytica* (Zaki and Clark 2001) stumbled onto the tRNA gene organization at about the time the genome sequencing started – short tandem repeat (STR)-containing sequences were isolated and proved to be linked to tRNA genes and organized into tandem arrays. In the final analysis (Clark *et al.* 2005), there have proven to be 25 distinct tandem arrays each encoding between one and five tRNAs per unit. In addition to tRNA genes, three arrays also encode the 5S RNA of *E. histolytica* and one unidentified small RNA is encoded also. STRs are present between almost all adjacent tRNA genes, although not all are polymorphic among strains. The copy number of array units varies between about 80 and 250 per cell with an average size of close to 1 kb. As these tRNAs and 5S RNA are not encoded elsewhere in the genome the arrayed genes must be functional and transcribed, but it is possible that they also have a structural role (see section Transcription, translation, and replication below).

Other types of tandemly repeated sequences are rare in the *E. histolytica* genome. No clear microsatellite sequences were identified and no other high-copy-number repeats could be identified. This is surprising as it means that no telomere repeats were identified (Clark CG, unpublished analyses).

Dispersed repeated DNAs are present and in some cases abundant. In particular, the SINEs and LINEs are present in many copies. The exact number is still unclear as they interfere with the assembly process (see below). Three families of non-long terminal repeat retrotransposons (EhRLE1–3) of the R4 clade, totaling about 100 full-length copies of a 4.8-kb sequence, have been identified (Bhattacharya A and Bhattacharya S, unpublished analyses). At least 80 full length (ca 550 bp each) and 180 partial sequences of the related EhSINE1 are present (Ackers JP, unpublished analyses). A second SINE (EhSINE2) is present in lower numbers. In total, complete and partial LINEs and SINEs make up about 6% of the total genome (Bhattacharya *et al.* 2003; Loftus *et al.* 2005; Mandal *et al.* 2004; van Dellen *et al.* 2002). In contrast to retrotransposons, DNA transposons are rare in the genome (Pritham *et al.* 2005).

Assembly

Assembling the raw sequence data generated in this project was a mammoth task. Not only did data from two sites need to be integrated (and initially the files were incompatible!) but, as mentioned above, almost a third of the sequences had to be excluded (170 000 of 596 500 were rRNA or tRNA related). Even after omitting unassembled single reads and those less than 2 kb in length, the *E. histolytica* genome at this point in time is still in 888 pieces. It is likely that the LINEs are part of the problem, as at over 4 kb in length they are too large to be spanned by most of the clones used in the sequencing and it is then impossible to know which pieces should be joined together.

While this assembly likely contains 99% + of the genes, it does mean that we have no real insight into whole chromosome organization and cannot compare chromosome structures or search for centromeres, etc. Fortunately, additional funding has been secured to try and complete the genome assembly and this work is ongoing at TIGR at the time of writing (October 2005).

Transcription, translation, and replication

Despite still not having a clear picture of chromosomal architecture we do have a much better idea of how the DNA is transcribed and replicated. RNA polymerase II transcription in *E. histolytica* is known to be resistant to alpha amanitin inhibition (Lioutas and Tannich 1995). Resistance to alpha amanitin resides in the largest subunit of RNA polymerase II and is dependent on mutations at specific residues. The *E. histolytica* sequence diverges from the consensus seen in most other eukaryotes at these residues (Gilchrist CA, unpublished analyses). In *Giardia* and *Trichomonas* transcription is also alpha-amanitin resistant but *E. histolytica* shows little similarity to the variants seen in those organisms (Seshadri *et al.* 2003). Ten out of the 12 RNA polymerase core proteins known in *S. cerevisiae* can be identified by homology in *E. histolytica* as can a number of other transcription factor subunits (Gilchrist CA, unpublished analyses).

Originally thought to have few introns in its genes, the *E. histolytica* genome sequence reveals that 25% of genes are in fact predicted to have introns and 6% to have more than one (Loftus *et al.* 2005). Many components of the splicing machinery necessary for their removal from mRNA precursors can be identified by homology searches (Gilchrist CA, unpublished analyses). Indirect evidence for the capping of mRNAs in *E. histolytica* (Ramos *et al.* 1997) is confirmed by the identification of genes involved in RNA capping and cap methylation. Only a few of the proteins involved in mRNA polyadenylation can be identified by homology searches, and of those that are seen several differ from the ones identified in the *T. brucei* genome using a similar approach (Hendriks *et al.* 2003).

Ribosomal protein genes are well conserved and almost all can be identified in the genome, only L41 being an exception. No mitochondrial-type ribosomal protein genes were identified, in keeping with the apparent absence of a mitochondrial genome in this organism (see below). Most translation elongation and release factors can be identified, suggesting that cytoplasmic translation in *E. histolytica* differs little from other eukaryotes (Bhattacharya S and Bhattacharya A, unpublished analyses).

The genome sequence contains almost 200 cell-cycle-related genes based on homology to other eukaryotes but a number of those involved in chromosome segregation, and cytokinesis in particular, were conspicuously absent (Lohia A, unpublished analyses). As *E. histolytica* has a closed mitosis and no condensation of chromosomes occurs this may not be too surprising. One remarkable absentee, however, is telomerase. No evidence for this enzyme could be detected and no evidence for the telomere-like repeat sequences normally found at the ends of eukaryotic chromosomes was found in the random shotgun sequences. How the ends of chromosomes in *E. histolytica* are replicated is therefore unclear. It is possible that the tRNA arrays may be a functional replacement – their repetitive nature and high redundancy makes them a strong candidate (Clark *et al.* 2005) but there is no direct evidence for this at present.

Genome coding capacity

Metabolism

The known metabolic pathways of the organism are shown in the genome paper (Loftus *et al.* 2005) but they are too numerous and complex to cover in detail in a review of this type. I will therefore select a small number of interesting observations that have been made to date, in particular those that differ from more typical eukaryotes.

Many biosynthetic pathways appear to be totally absent from the organism, perhaps not surprising given the parasitic nature of the organism and the fact that it will be able to scavenge many nutrients from ingested bacteria. Among those missing (Anderson and Loftus 2005) are many amino acid biosynthetic pathways, those for fatty acids and sterols, as well as those for purines and pyrimidines. The latter was a surprise as it had previously been thought to have this ability and to use the standard pathway (Reeves 1984). Either preformed pyrimidines are an essential nutrient for the organism or an alternative synthetic pathway exists that remains to be identified.

The only pathways for amino acid synthesis that appear to be present are those for cysteine and serine. Cysteine is the major intracellular thiol in *E. histolytica* where it compensates for the absence

of glutathione. In contrast, there are numerous pathways for amino acid degradation and some of these may be involved in energy generation through conversion of amino acids into pyruvate and 2-oxobutanoate, from which ATP can be generated via pyruvate:ferredoxin oxidoreductase and acetyl-CoA synthase (ADP forming).

Another remarkable finding is the complete absence of folate metabolism – neither folate transporters nor any enzyme requiring folate as a cofactor was found. *Trichomonas* is the only other organism known to lack folate metabolism. Folate is usually an essential cofactor needed for thymidylate synthesis and methionine recycling.

A major lipid in *E. histolytica* is ceramide aminoethylphosphonate, yet none of the enzymes for synthesizing the aminoethylphosphonate headgroup could be identified, suggesting a novel pathway. Although *E. histolytica* cannot synthesize lipids *de novo* it can extend pre-existing fatty acids acquired from its food organisms. This modification involves an unusual acetyl-CoA carboxylase of a type previously only found in *Giardia* and fatty acid elongases of a type normally found only in plants.

A lot of uncertainty still exists regarding the details of the metabolism of *E. histolytica*. Part of the uncertainty lies in the identification of only partial pathways in several cases. Another problem is that the identity of enzymes based on sequence homology may be misleading. For example, Hofer and Duchêne (2005) have noted that a single gene in *E. histolytica* shares similar levels of identity to delta-levulinate synthase (involved in heme metabolism) and serine palmitoyl transferase (involved in sphingolipid metabolism). As there is no other evidence for heme metabolism in *E. histolytica* the latter is likely to be the correct assignment. Likewise, a single gene has similarity to dihydropyrimidinase, dihydroorotase, and allantoinase. All cleave amide bonds, but the substrate specificity of the *E. histolytica* enzyme is unknown. This uncertainty can only be resolved by experimental work, using recombinant enzymes to investigate the biochemical function of the proteins involved. The understanding of metabolic pathways of *E. histolytica* may have been greatly improved by the genome sequence but it should be viewed with caution until such time as experimental confirmation is forthcoming.

Transporters

A total of 174 transporters were found in the genome of *E. histolytica* (Loftus *et al.* 2005). Surprisingly, no proteins homologous to the hexose transporters normally present in eukaryotes were found, despite the fact that glucose is known to be taken up and forms the major energy source for the organism. Instead, monosaccharide transporters of a prokaryotic type (glucose/ribose porters) were found and these appear to have replaced the standard hexose transporters in this eukaryote. Functional verification of their function has not yet been performed.

Environmental sensing and signaling

One of the remarkable findings of the genome analysis has been the large family of transmembrane serine/threonine protein kinases (Loftus *et al.* 2005). The 90 members fall into three families depending on the types of extracellular domain they contain and a number are expressed at the same time (Beck *et al.* 2005). Complementing these protein kinases are over 100 protein phosphatases. The extracellular ligands and the intracellular downstream effector molecules are completely unknown at present, but they indicate a sophisticated and discriminatory ability to sense the environment. Cytosolic signal transduction-related genes are also abundant (Nozaki T, unpublished analyses).

Oxygen detoxification

E. histolytica lacks catalase and glutathione-dependent enzymes normally associate with oxygen-related detoxification (Fahey *et al.* 1984). It contains a single copy of an iron-containing superoxide dismutase (Tannich *et al.* 1991) to remove superoxide ion and several peroxiredoxins capable of removing both hydrogen peroxide and organic peroxides, including phospholipid hydroperoxides that would otherwise damage the cell membranes. Peroxiredoxins depend on thioredoxin and thioredoxin reductase and candidate genes for both of these have been found. Other proteins that may protect the organism from intracellular peroxide have also been identified, including genes encoding rubrerythrin and flavoprotein A, which may detoxify nitric oxide (Bruchhaus I, unpublished analyses). The redundancy of anti-oxidant pathways underscores the sensitivity of this amitochondrial organism to the presence of oxygen. Experimental evidence suggests that only low levels of oxygen can be tolerated and in the absence of the above pathways it would be even lower.

The absence of glutathione from *E. histolytica* has been known for many years (Fahey *et al.* 1984) and as mentioned earlier it is thought that cysteine acts as a functional replacement in this organism. Conflicting reports of the presence of trypanothione exist (Ariyanayagam and Fairlamb 1999; Ondarza *et al.* 2005). The putative trypanothione reductase gene reported from strain HK-9 (Tamayo *et al.* 2005) has no homologue in the genome sequence (strain HM-1:IMSS), nor in the (admittedly incomplete) genomes of four other *Entamoeba* species.

Pathogenesis

Genes involved in pathogenesis had received a great deal of attention prior to the start of the genome project and it is therefore not surprising that few major discoveries came out of the project. The first step in tissue invasion involves the galactose/N-acetyl galactosamine specific lectin, which consists of three subunits, two of which are disulfide bonded. The number of genes for each subunit detected in the genome sequence matched that predicted from previous experimental work, with the exception of the identification of a highly divergent gene related to the heavy subunit of the lectin (Mann BJ, unpublished analyses).

Cysteine proteases are responsible for the cytopathic effect seen *in vitro* and are thought to degrade extracellular matrix proteins during tissue invasion. At least 30 genes belonging to three distinct families (A, B, and C) have been identified (Bruchhaus I, unpublished analyses; Bruchhaus *et al.* 2003). Unusually, most genes in group B have glycosyl phosphatidylinositol (GPI-) anchors while those in group C all have transmembrane domains. Surprisingly, very few of these genes appear to be transcribed under normal laboratory culture conditions and their roles during the normal life cycle of the organism remain unclear.

The three isoforms of pore-forming peptides (amoebapores) have been well characterized (Leippe and Herbst 2004) and shown to be linked to virulence. In addition, 16 other genes for saposin-like proteins (a group of lipid-interacting proteins that includes the amoebapores) have been identified (Bruhn and Leippe 2001), but these apparently lack pore-forming activity. Their function to date is unknown.

Evolution

Lateral gene transfer (LGT)

Single gene studies had already hinted at the presence of genes in the *E. histolytica* genome that had been acquired from bacteria by LGT (Field *et al.* 2000; Nixon *et al.* 2002). Some, but not all, of these were also present in other anaerobic eukaryotes, primarily *Giardia* and *Trichomonas*. With the full complement of *E. histolytica* genes available it became possible to evaluate the contribution of such genes to the genetic makeup of the organism. A conservative search for examples of prokaryote to eukaryote LGT estimated that 96 genes had been acquired specifically by the *Entamoeba* lineage after its divergence from other known eukaryotes (Loftus *et al.* 2005). This number represents 1% of all genes identified. It is important to note that it excludes a number of previously described examples of LGT in *E. histolytica* as those genes are also found in *Giardia* and/or *Trichomonas*. To place this in context, only 18 examples of LGT were detected in the *Dictyostelium* genome, less than 0.15% of the predicted genes (Eichinger *et al.* 2005).

The majority of these 96 genes encode metabolic enzymes (see Andersson, Chapter 7, this book), with the rest being hypothetical or unclassified proteins. LGT appears to have had a significant impact on carbohydrate and amino acid metabolism, with several genes expanding the range of

substrates usable by the amoeba – for example the degradative enzymes aspartase and tryptophanase which channel amino acids into energy generation, as mentioned above, are among the 96. As a bacteriovorous organism, *E. histolytica* is constantly being exposed to bacterial DNA and so the high level of LGT is perhaps not surprising, but this on its own is not enough to explain the high percentage of LGT genes as *Dictyostelium* has a similar exposure.

The mitosome, a mitochondrial remnant

The relatively late branch point of *E. histolytica* in phylogenetic trees implies secondary loss rather than primitive absence of the missing classical eukaryotic features of the cell. One of the "missing" organelles is the mitochondrion, but it is now clear that a subcellular compartment derived from the mitochondrion persists in *E. histolytica*. This mitochondrial remnant has been named the mitosome. These organelles are abundant (ca. 250 per cell) and tiny (<0.5 μm in diameter) (Léon-Avila and Tovar 2004). There is no genome and no evidence for any energy generation function for the organelle (Table 11.2). Earlier biochemical analyses had already shown the absence of tricarboxylic acid cycle enzymes and the absence of heme implied that no cytochromes were present either (Reeves 1984). At the start of the genome project the only protein known to be located in the mitosome was the mitochondrial-type chaperonin cpn60 (Mai *et al.* 1999; Tovar *et al.* 1999), and two other genes encoding proteins likely to be targeted to the mitosome had also been identified – a second chaperone, mt-hsp70 (Bakatselou *et al.* 2000), and the enzyme pyridine nucleotide transhydrogenase (Clark and Roger 1995), which transfers reducing equivalents between NAD and NADP. It was anticipated that the genome project would provide answers regarding the function of the mitosome.

Surprisingly, the genome project revealed relatively few additional genes encoding proteins likely to be targeted to the mitosome. One gene encoding a transporter of the mitochondrial type was found. Although phylogenetic comparisons do not show a strong affinity with specific

Table 11.2 "Mitochondrial" features and the *E. histolytica* genome – a list of features typical of mitochondria in the majority of eukaryotes is given along with an indication of whether they are thought to be present in the *E. histolytica* mitosome (van der Giezen *et al.* 2005)

Missing:
Mitochondrial DNA
Cytochromes
TCA cycle
NADH dehydrogenase
Electron transport
Energy generation
Present:
ATP transporter
Pyridine nucleotide transhydrogenase
3 chaperonin proteins (Cpn60, Cpn10, mt-Hsp70)
Mitochondrial-type protein import
Iron-sulfur center synthesis?

subgroups of transporter it has recently been proven experimentally that this is an ADP/ATP transporter (Chan *et al.* 2005). This is consistent with the presence of mt-hsp70 as that chaperone consumes ATP. The only mitochondrial metabolic pathway for which evidence has been found is iron-sulfur cluster synthesis, as genes for IscS and IscU are present but even this is not unambiguously a mitosomal pathway. IscS and IscU are among the 96 LGT genes but do not cluster with mitochondrial homologues, rather they are most closely related to sequences in *Helicobacter* and *Campylobacter* (van der Giezen *et al.*, 2004). Whether the encoded proteins are actually located in the mitosomes will need to be determined experimentally.

It is likely that iron-sulfur cluster assembly is the major, if not the only, remaining mitochondrial function of the *E. histolytica* mitosome, as no other candidate role has been uncovered. It has been shown that this is the role of the mitosome in *Giardia* (Tovar *et al.* 2003). Indeed, it is the only known universally conserved pathway of mitochondrion-derived organelles – mitosomes and hydrogenosomes – and may be the only reason that the compartment has been retained in mitosomal eukaryotes.

It has been shown experimentally that the protein importation pathway of the *E. histolytica*

mitosome is functionally homologous to that of mitochondria (Mai *et al.* 1999; Tovar *et al.* 1999). It is therefore surprising that little evidence for the multiprotein complexes responsibly for importing proteins across the outer and inner membranes of the organelle (TOMs and TIMs, respectively) has been uncovered. Only one tentative protein identification has been made (TOM40) and no evidence for the processing peptidase responsible for the known removal of amino terminal leader sequences has been detected (van der Giezen *et al.* 2005). It is clear that a full understanding of the mitosome will require substantial experimental investigation.

Comparative genomics

At the same time as the *E. histolytica* genome sequencing was under way, pilot studies of the genomes of four other *Entamoeba* species were also undertaken. The species chosen were largely a matter of necessity as not all species can be grown in quantity in bacteria-free culture. Nevertheless, the four species investigated represent a good cross section of the genus. *E. dispar* is a very close relative of *E. histolytica* yet is not a pathogen; *E. moshkovskii* is a secondarily free-living species that occasionally infects humans; *E. terrapinae* is a commensal species infecting turtles; *E. invadens* causes an invasive disease in lizards and snakes that resembles amoebiasis in humans and this species is also the model system for studying *Entamoeba* cyst formation. In terms of genetic distance from *E. histolytica*, the order from closest to most divergent is *E. dispar*, *E. moshkovskii*, *E. terrapinae*, and *E. invadens* (Silberman *et al.* 1999).

From the relatively superficial comparisons made so far it is clear that most of the peculiarities of *E. histolytica* are shared by other species in the genus. For example most of the 96 LGT genes are also found in the other species, implying that their entry into the lineage is quite ancient (Clark CG, unpublished analyses). The tRNA genes are also abundant and in tandem arrays, although the specific organization of the array units varies among species (Tawari B and Clark CG, unpublished). As mentioned earlier, retrotransposons are abundant in *E. histolytica* and *E. dispar* but not in the other species, while the reverse is true of DNA transposons (Pritham *et al*. 2005). One curiosity is that although *E. histolytica* lacks evidence for a ribonucleotide reductase, this gene is present in *E. moshkovskii* and *E. invadens*, at least, implying that its loss from *E. histolytica* is a comparatively recent event (Loftus *et al*. 2005). Further analyses await the completion of further sequencing of *E. dispar* and *E. invadens* that is currently ongoing.

What we still do not know and future directions

The holes in our knowledge of the *E. histolytica* genome will be clear to anyone who has read this far. We still have little understanding of the overall organization of a chromosome in this organism due to the fragmentary nature of the "final" assembly obtained from the shotgun sequencing. Funding has been obtained to allow additional attempts at completion to be undertaken. The approach will involve, in part, HAPPY mapping (a method for genome mapping based on the analysis of approximately HAPloid DNA samples using the PolYmerase chain reaction), which has been used successfully by the *Cryptosporidium* and *Dictyostelium* genome projects. If successful, this will allow the drawing together of the genome fragments into complete chromosomes. It is also hoped that large-insert libraries will, finally, be successful.

Most of the other holes in our understanding will require extensive experimental investigation. The mitosome function, for example, may require proteomic analysis to identify the proteins involved in its structure and function. This assumes that sufficiently pure preparations can be made, of course. The expression patterns and roles of the large families of transmembrane kinases and cysteine proteinases are already being investigated using subgenomic microarray approaches. Ultimately it is to be hoped that whole genome microarrays will become available as tools to investigate not only patterns of gene expression, but also interstrain variability in gene content and organization.

Expansion of the genome surveys of *E. invadens* and *E. dispar* has already been mentioned. The current plan is to increase coverage of the genome to at least $10 \times$ from the current $2 \times$ (approximately). This should allow the identification of > 99% of the genes and investigation of synteny between species, among other benefits.

The ultimate goal of the genome project must be to reveal important aspects of the biology of the organism that are relevant to its role as a major pathogen. At the moment we understand quite a bit about the molecules involved in the process but very little about why *E. histolytica* invades tissue in some people but not in others. The availability of the genome sequence has already started to change the way in which amoebiasis research is performed and this will only continue into the future.

Acknowledgements

The *E. histolytica* genome project was supported by grants from the Wellcome Trust (064057) and the National Institute of Allergy and Infectious Diseases (R01-AI46516). The comparative genomics sequence surveys were supported by NIAID and the Biotechnology and Biological Sciences Research Council (3/G16197). I also thank my colleagues involved in the analysis of the genome data, on whose work I have drawn in this review.

References

Anderson, I. J. and Loftus, B. J. (2005). *Entamoeba histolytica*: observations on metabolism based on the genome sequence. *Exp. Parasitol*, **110**, 173–177.

Ariyanayagam, M. R. and Fairlamb, A. H. (1999). *Entamoeba histolytica* lacks trypanothione metabolism. *Mol. Biochem. Parasitol.*, **103**, 61–69.

Bakatselou, C., Kidgell, C., and Clark, C. G. (2000). A mitochondrial-type hsp70 gene of *Entamoeba histolytica*. *Mol. Biochem. Parasitol.*, **110**, 177–182.

Beck, D. L., Boettner, D. R., Dragulev, B., Ready, K., Nozaki, T., and Petri, W. A., Jr. (2005). Identification and gene expression analysis of a large family of transmembrane kinases related to the Gal/GalNAc lectin in *Entamoeba histolytica*. *Eukaryot. Cell*, **4**, 722–732.

Bhattacharya, A., Bhattacharya, S., and Ackers, J. P. (2003). Nontranslated polyadenylated RNAs from *Entamoeba histolytica*. *Trends Parasitol.*, **19**, 286–289.

Bhattacharya, A., Satish, S., Bagchi, A., and Bhattacharya, S. (2000). The genome of *Entamoeba histolytica*. *Int. J. Parasitol.*, **30**, 401–410.

Bruchhaus, I., Loftus, B. J., Hall, N., and Tannich, E. (2003). The intestinal protozoan parasite *Entamoeba histolytica* contains 20 cysteine protease genes, of which only a small subset is expressed during in vitro cultivation. *Eukaryot. Cell*, **2**, 501–509.

Bruhn, H. and Leippe, M. (2001). Novel putative saposin-like proteins of *Entamoeba histolytica* different from amoebapores. *Biochim. Biophys. Acta*, **1514**, 14–20.

Chan, K. W., Slotboom, D. J., Cox, S., Embley, T. M., Fabre, O., van der Giezen, M., Harding, M., Horner, D. S., Kunji, E. R., León-Avila, G., and Tovar, J. (2005). A novel ADP/ATP transporter in the mitosome of the microaerophilic human parasite *Entamoeba histolytica*. *Curr. Biol.*, **15**, 737–742.

Clark, C. G., Ali, I. K. M., Zaki, M., Loftus, B. J., and Hall, N. (2006). Unique organization of tRNA genes in the protistan parasite *Entamoeba histolytica*. *Mol. Biochem. Parasitol*, **146**, 24–29.

Clark, C. G. and Cross, G. A. M. (1987). rRNA genes of *Naegleria gruberi* are carried exclusively on a 14-kilo-base-pair plasmid. *Mol. Cell. Biol.*, **7**, 3027–3031.

Clark, C. G., Espinosa Cantellano, M., and Bhattacharya, A. (2000). *Entamoeba histolytica*: an overview of the biology of the organism. In J. I. Ravdin, ed. *Amebiasis*. Imperial College Press, London, pp. 1–45.

Clark, C. G. and Roger, A. J. (1995). Direct evidence for secondary loss of mitochondria in *Entamoeba histolytica*. *Proc. Natl. Acad. Sci. USA*, **92**, 6518–6521.

Diamond, L. S. and Clark, C. G. (1993). A redescription of *Entamoeba histolytica* Schaudinn, 1903 (Emended Walker, 1911) separating it from *Entamoeba dispar* Brumpt, 1925. *J. Eukaryot. Microbiol.*, **40**, 340–344.

Eichinger, L., Pachebat, J. A., Glockner, G., Rajandream, M. A., Sucgang, R., Berriman, M., Song, J., Olsen, R., Szafranski, K., Xu, Q., Tunggal, B., Kummerfeld, S., Madera, M., Konfortov, B. A., Rivero, F., Bankier, A. T., Lehmann, R., Hamlin, N., Davies, R., Gaudet, P., Fey, P., Pilcher, K., Chen, G., Saunders, D., Sodergren, E., Davis, P., Kerhornou, A., Nie, X., Hall, N., Anjard, C., Hemphill, L., Bason, N., Farbrother, P., Desany, B., Just, E., Morio, T., Rost, R., Churcher, C., Cooper, J., Haydock, S., van Driessche, N., Cronin, A., Goodhead, I., Muzny, D., Mourier, T., Pain, A., Lu, M., Harper, D., Lindsay, R., Hauser, H., James, K., Quiles, M., Madan Babu, M., Saito, T., Buchrieser, C., Wardroper, A., Felder, M., Thangavelu, M., Johnson, D., Knights, A., Loulseged, H., Mungall, K., Oliver, K., Price, C., Quail, M. A., Urushihara, H., Hernandez, J., Rabbinowitsch, E., Steffen, D., Sanders, M., Ma, J., Kohara, Y., Sharp, S., Simmonds, M., Spiegler, S., Tivey, A., Sugano, S., White, B., Walker, D., Woodward, J., Winckler, T., Tanaka, Y., Shaulsky, G., Schleicher, M., Weinstock, G.,

Rosenthal, A., Cox, E. C., Chisholm, R. L., Gibbs, R., Loomis, W. F., Platzer, M., Kay, R. R., Williams, J., Dear, P. H., Noegel, A. A., Barrell, B., and Kuspa, A. (2005). The genome of the social amoeba *Dictyostelium discoideum*. *Nature*, **435**, 43–57.

El-Sayed, N. M., Myler, P. J., Blandin, G., Berriman, M., Crabtree, J., Aggarwal, G., Caler, E., Renauld, H., Worthey, E. A., Hertz-Fowler, C., Ghedin, E., Peacock, C., Bartholomeu, D. C., Haas, B. J., Tran, A. N., Wortman, J. R., Alsmark, U. C., Angiuoli, S., Anupama, A., Badger, J., Bringaud, F., Cadag, E., Carlton, J. M., Cerqueira, G. C., Creasy, T., Delcher, A. L., Djikeng, A., Embley, T. M., Hauser, C., Ivens, A. C., Kummerfeld, S. K., Pereira-Leal, J. B., Nilsson, D., Peterson, J., Salzberg, S. L., Shallom, J., Silva, J. C., Sundaram, J., Westenberger, S., White, O., Melville, S. E., Donelson, J. E., Andersson, B., Stuart, K. D., and Hall, N. (2005). Comparative genomics of trypanosomatid parasitic protozoa. *Science*, **309**, 404–409.

Espinosa-Cantellano, M. and Martínez-Palomo, A. (2000). Pathogenesis of intestinal amebiasis: from molecules to disease. *Clin. Microbiol. Rev.*, **13**, 318–331.

Fahey, R. C., Newton, G. L., Arrick, B., Overdank-Bogart, T., and Aley, S. B. (1984). *Entamoeba histolytica*: a eukaryote without glutathione metabolism. *Science*, **224**, 70–72.

Field, J., Rosenthal, B., and Samuelson, J. (2000). Early lateral transfer of genes encoding malic enzyme, acetyl-CoA synthetase and alcohol dehydrogenases from anaerobic prokaryotes to *Entamoeba histolytica*. *Mol. Microbiol.*, **38**, 446–455.

Gardner, M. J., Hall, N., Fung, E., White, O., Berriman, M., Hyman, R. W., Carlton, J. M., Pain, A., Nelson, K. E., Bowman, S., Paulsen, I. T., James, K., Eisen, J. A., Rutherford, K., Salzberg, S. L., Craig, A., Kyes, S., Chan, M. S., Nene, V., Shallom, S. J., Suh, B., Peterson, J., Angiuoli, S., Pertea, M., Allen, J., Selengut, J., Haft, D., Mather, M. W., Vaidya, A. B., Martin, D. M., Fairlamb, A. H., Fraunholz, M. J., Roos, D. S., Ralph, S. A., McFadden, G. I., Cummings, L. M., Subramanian, G. M., Mungall, C., Venter, J. C., Carucci, D. J., Hoffman, S. L., Newbold, C., Davis, R. W., Fraser, C. M., and Barrell, B. (2002).. Genome sequence of the human malaria parasite *Plasmodium falciparum*. *Nature*, **419**, 498–511.

Hendriks, E. F., Abdul-Razak, A., and Matthews, K. (2003). tbCPSF30 depletion by RNA interference disrupts polycistronic RNA processing in Trypanosoma brucei. *J. Biol. Chem.*, **278**, 26870–26878.

Hofer, M. and Duchêne, M. (2005). *Entamoeba histolytica*:construction and applications of subgenomic databases. *Exp. Parasitol*, **110**, 178–183.

Leippe, M. and Herbst, R. (2004). Ancient weapons for attack and defense: the pore-forming polypeptides of pathogenic enteric and free-living amoeboid protozoa. *J. Eukaryot. Microbiol.*, **51**, 516–521.

Léon-Avila, G. and Tovar, J. (2004). Mitosomes of *Entamoeba histolytica* are abundant mitochondrion-related remnant organelles that lack a detectable organellar genome. *Microbiology*, **150**, 1245–1250.

Lioutas, C. and Tannich, E. (1995). Transcription of protein-coding genes in *Entamoeba histolytica* is insensitive to high concentrations of α-amanitin. *Mol. Biochem. Parasitol.*, **73**, 259–261.

Loftus, B., Anderson, I., Davies, R., Alsmark, U. C., Samuelson, J., Amedeo, P., Roncaglia, P., Berriman, M., Hirt, R. P., Mann, B. J., Nozaki, T., Suh, B., Pop, M., Duchene, M., Ackers, J., Tannich, E., Leippe, M., Hofer, M., Bruchhaus, I., Willhoeft, U., Bhattacharya, A., Chillingworth, T., Churcher, C., Hance, Z., Harris, B., Harris, D., Jagels, K., Moule, S., Mungall, K., Ormond, D., Squares, R., Whitehead, S., Quail, M. A., Rabbinowitsch, E., Norbertczak, H., Price, C., Wang, Z., Guillen, N., Gilchrist, C., Stroup, S. E., Bhattacharya, S., Lohia, A., Foster, P. G., Sicheritz-Ponten, T., Weber, C., Singh, U., Mukherjee, C., El-Sayed, N. M., Petri, W. A., Jr., Clark, C. G., Embley, T. M., Barrell, B., Fraser, C. M., and Hall, N. (2005).. The genome of the protist parasite *Entamoeba histolytica*. *Nature*, **433**, 865–868.

Mai, Z., Ghosh, S., Frisardi, M., Rosenthal, B., Rogers, R., and Samuelson, J. (1999). Hsp60 is targeted to a cryptic mitochondrion-derived organelle ("Crypton") in the microaerophilic protozoan parasite *Entamoeba histolytica*. *Mol. Cell. Biol.*, **19**, 2198–2205.

Mandal, P. K., Bagchi, A., Bhattacharya, A., and Bhattacharya, S. (2004). An *Entamoeba histolytica* LINE/SINE pair inserts at common target sites cleaved by the restriction enzyme-like LINE-encoded endonuclease. *Eukaryot. Cell*, **3**, 170–179.

Nixon, J. E.J., Wang, A., Field, J., Morrison, H. G., McArthur, A. G., Sogin, M. L., Loftus, B. J., and Samuelson, J. (2002). Evidence for lateral transfer of genes encoding ferredoxins, nitroreductases, NADH oxidase, and alcohol dehydrogenase 3 from anaerobic prokaryotes to *Giardia lamblia* and *Entamoeba histolytica*. *Eukaryot. Cell*, **1**, 181–190.

Ondarza, R. N., Hurtado, G., Iturbe, A., Hernandez, E., Tamayo, E., and Woolery, M. (2005). Identification of trypanothione from the human pathogen *Entamoeba histolytica*, by mass spectrometry and chemical analysis. *Biotech. Appl. Biochem.*, **42**, 175–181.

Pritham, E. J., Feschotte, C., and Wessler, S. R. (2005). Unexpected diversity and differential amplification of transposable elements in four species of *Entamoeba* protozoans. *Mol. Biol. Evol.*, **22**, 1751–1763 .

Ramos, M. A., Mercado, G. C., Salgado, L. M., Sanchez-Lopez, R., Stock, R. P., Lizardi, P. M., and Alagón, A.

(1997). *Entamoeba histolytica* contains a gene encoding a homologue to the 54kDa subunit of the signal recognition particle. *Mol. Biochem. Parasitol.*, **88**, 225–235.

Reeves, R. E. (1984). Metabolism of *Entamoeba histolytica* Schaudinn, 1903. *Adv. Parasitol.*, **23**, 105–142.

Seshadri, V., McArthur, A. G., Sogin, M. L., and Adam, R. D. (2003). *Giardia lamblia* RNA polymerase II: amanitin-resistant transcription. *J. Biol. Chem.*, **278**, 27804–27810.

Silberman, J. D., Clark, C. G., Diamond, L. S., and Sogin, M. L. (1999). Phylogeny of the genera *Entamoeba* and *Endolimax* as deduced from small subunit ribosomal RNA gene sequence analysis. *Mol. Biol. Evol.*, **16**, 1740–1751.

Tamayo, E. M., Iturbe, A., Hernandez, E., Hurtado, G., Gutierrez, X. M. L., Rosales, J. L., Woolery, M., and Ondarza, R. N. (2005). Trypanothione reductase from the human parasite *Entamoeba histolytica*: a new drug target. *Biotech. Appl. Biochem.*, **41**, 105–115.

Tannich, E., Bruchhaus, I., Walter, R. D., and Horstmann, R. D. (1991). Pathogenic and nonpathogenic *Entamoeba histolytica*: identification and molecular cloning of an iron-containing superoxide dismutase. *Mol. Biochem. Parasitol.*, **49**, 61–72.

Tovar, J., Fischer, A., and Clark, C. G. (1999). The mitosome, a novel organelle related to mitochondria in the amitochondriate parasite *Entamoeba histolytica*. *Mol. Microbiol.*, **32**, 1013–1021.

Tovar, J., Leon-Avila, G., Sanchez, L. B., Sutak, R., Tachezy, J., van der Giezen, M., Hernandez, M., Muller, M., and Lucocq, J. M. (2003). Mitochondrial remnant organelles of *Giardia* function in iron-sulphur protein maturation. *Nature*, **426**, 172–176.

Van Dellen, K., Field, J., Wang, Z., Loftus, B., and Samuelson, J. (2002). LINEs and SINE-like elements of the protist *Entamoeba histolytica*. *Gene*, **297**, 229–239.

van der Giezen, M., Cox, S., and Tovar, J. (2004). The iron-sulfur cluster assembly genes iscS and iscU of *Entamoeba histolytica* were acquired by horizontal gene transfer. *BMC Evol. Biol.*, **4**, 7.

van der Giezen, M., Tovar, J., and Clark, C. G. (2005). Mitochondrion-derived organelles in protists and fungi. *Int. Rev. Cytol.*, **244**, 175–225.

WHO (1998). *The World Health Report 1998. Life in the 21st Century: a Vision for All*. World Health Organization, Geneva, Switzerland.

Willhoeft, U. and Tannich, E. (1999). The electrophoretic karyotype of *Entamoeba histolytica*. *Mol. Biochem. Parasitol.*, **99**, 41–53.

Zaki, M. and Clark, C. G. (2001). Isolation and characterization of polymorphic DNA from *Entamoeba histolytica*. *J. Clin. Microbiol.*, **39**, 897–905.

CHAPTER 12

Genome reduction in microsporidia

Patrick J. Keeling

Canadian Institute for Advanced Research, Botany Department, University of British Columbia, Vancouver, Canada

Introduction to microsporidia

Microsporidia are a highly diverse group of obligate intracellular parasites, closely related to fungi, with a number of unique characteristics which have simultaneously sparked interest in the cell, molecular, and evolutionary biology of the group, and at the same time hindered our understanding because of the unusual nature of these parasites on many levels. With the exception of their complex infection apparatus, they are relatively simple cells that have lost or reduced a variety of structures. Microsporidian genomes are also unusual and reduced, some ranking among the smallest of any eukaryotic nuclear genome. The complete sequence of one microsporidian genome is known, that of *Encephalitozoon cuniculi*, and there are genome sequence surveys for several other species and one expressed sequence tag project. Altogether these data paint an unusual picture of microsporidian genomes as a place where some generally poorly conserved characteristics become more highly selected while other normally important constraints are relaxed.

As intracellular parasites, microsporidia can only grow and divide by penetrating another eukaryotic cell and taking advantage of it to varying degrees. The life cycle of microsporidia varies considerably due to the diversity of the group – over 1200 species have been described up to the last checklist in 1999 (Sprague and Becnel 1999), but this is certainly a small proportion of microsporidian diversity since they have not been looked for in most of their potential host range. However, the basic principals behind their infection mechanism are conserved (Keohane and Weiss 1999; Vávra and Larsson 1999). The infections stage of the life cycle is the spore (Fig. 12.1), which is bounded by a thick, rigid spore wall and is the only stage of microsporidia that can survive outside the host. The spore is largely considered to be dormant until triggered to germinate. The environmental cues that trigger germination are known in some species, but at best only partially, and a single species can form different types of spore which may respond to different environmental factors (Vávra and Larsson 1999). Whatever the cues are, the spore seems to import water,

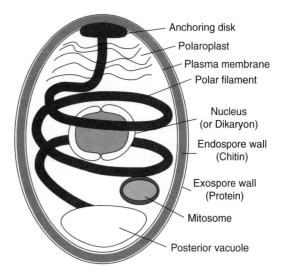

Figure 12.1 Schematic of microsporidian spore showing major structures associated with infection as described in the text.

Labels (clockwise): Anchoring disk · Polaroplast · Plasma membrane · Polar filament · Nucleus (or Dikaryon) · Endospore wall (Chitin) · Exospore wall (Protein) · Mitosome · Posterior vacuole

leading to an increase in osmotic pressure and an expansion of a posterior vacuole. This increased pressure leads to a rupturing of the thinnest portion of the spore wall at the apex, and the subsequent expulsion of a structure called the polar filament (Keohane and Weiss 1999). Within the spore the polar filament is coiled inside the cytoplasm, but when it is ejected it everts and becomes a tube through which the contents of the spore are forced. This expulsion of the polar tube is extremely rapid and if it hits or pierces another cell, the spore contents are injected directly into that cell, leading to infection if the target cell is a suitable host for the parasite (Keohane and Weiss 1999).

This unique form of infection has proved quite successful, as microsporidia are known to infect all phyla of animals, and it is not inconceivable that all animal species are infected by at least one microsporidiam. In addition, a small number are known to infect protist species, namely certain ciliates and apicomplexans (Vivier 1975). The latter are especially interesting since the apicomplexan hosts, archaegregarines, are themselves animal parasites with intracellular life stages, so these microsporidia are hyperparasites. The microsporidia that infect archaegregarines are also of interest because they are a particular class, known as metchnikovellids, with a less developed polar tube and infection mechanism, and are believed to be the earliest offshoots of the microsporidian lineage (Vivier 1975).

Deception and controversy in the evolutionary history of microsporidia

As has been the case with many highly adapted parasites, the evolutionary origin of microsporidia has been difficult to determine because they are difficult to compare with other eukaryotes. Early attempts to classify them placed them with other spore-forming parasites, spore-forming bacteria, and fungi, or simply left them on their own (Kudo 1947; Levine et al. 1980; Lom and Vávra 1962; Nägeli 1857). A more recent and influential hypothesis sprang from the observation that microsporidia lacked mitochondria and suggested that they, and three other protist groups collectively

called Archezoa, might be very ancient lineages of eukaryotes that arose prior to the endosymbiosis that gave rise to mitochondria (Cavalier-Smith 1983; Cavalier-Smith 1998). This was supported by the absence of visible mitochondria (Vávra 1965) and by their apparently prokaryotic-sized ribosomes (Ishihara and Hayashi 1968). The first molecular data from microsporidia provided what seemed to be compelling support for the Archezoa hypothesis for microsporidia, by showing that they were among the first branches on the eukaryotic tree based on small subunit ribosomal RNA (SSU rRNA), and also a 5.8s – LSU rRNA fusion otherwise only found in prokaryotes (Vossbrinck et al. 1987; Vossbrinck and Woese 1986). Initially, protein-coding genes backed up this phylogenetic position (Brown and Doolittle 1995; Kamaishi et al. 1996a; Kamaishi et al. 1996b), but eventually another conclusion began to emerge. Phylogenies based on α- and β-tubulin genes showed microsporidia branching with or within the fungi, not with the other "ancient" lineages, diplomonads and parabasalia (Edlind et al. 1996; Keeling and Doolittle 1996). Other protein genes soon began to show similar phylogenies (Fast et al. 1999; Hirt et al. 1999) and analyses of rRNAs and some proteins began to discredit early conclusions, showing they were more likely based on the fast rate of evolution seen in many microsporidian gene sequences (Hirt et al. 1999; Van de Peer et al. 2000).

The recent complete genome of the vertebrate parasite *Encephalitozoon cuniculi* (Katinka et al. 2001) has all but ended this controversy, and a recent analysis of 99 protein phylogenies from that genome also showed a strong correlation between the level of conservation of a gene and whether it supported the relationship between fungi and microsporidia (Thomarat et al. 2004). At present there is no reason to doubt that microsporidia are close relatives of fungi, but it is still not certain if they actually are fungi or simply a sister group to fungi. Analysis of α- and β-tubulin genes supports the latter (Keeling 2003; Keeling et al. 2000), but there are difficulties with the tubulins because microsporidia and all fungi except chytrids have relatively divergent tubulin gene sequences, hypothesized to be linked to the loss of all 9 + 2 microtubule structures in these taxa (Keeling et al.

2000). This might be a shared derived character, but the accelerated rates of tubulin evolution could generate misleading phylogenies. Indeed, another analysis of elongation factor-1α (EF-1α) genes has suggested microsporidia do not branch within fungi (Tanabe *et al.* 2002), but this study is also difficult to interpret because microsporidian EF-1α genes have been demonstrated to be evolving differently from other eukaryotes in a covarion fashion, making the evolution of this microsporidian gene hard to interpret (Inagaki *et al.* 2004).

At the same time as the ancient origin of microsporidia was being questioned directly by phylogenies, the idea that they were primitively amitochondriate also came under attack. Although most mitochondria retain a small genome (Gray *et al.* 1999), most mitochondrial proteins in all eukaryotes are encoded by nuclear genes and the protein products are post-translationally targeted to the organelle. In "amitochondriate" eukaryotes, it was reasoned, a mitochondriate ancestry, or perhaps even a nearly unrecognizable cryptic mitochondrion might be revealed by finding such genes in the nuclear genome (Clark and Roger 1995). The first direct evidence for a mitochondrion microsporidia was the discovery of a nuclear-encoded mitochondrion-derived HSP70 in several species (Germot *et al.* 1997; Hirt *et al.* 1997; Peyretaillade *et al.* 1998b; Arisue *et al.* 2002). Other genes were subsequently found (Fast and Keeling 2001) and the organelle itself was discovered in *Trachipleistophora hominis* by localizing the HSP70 to small, double-membrane bound structures with no other obvious visible characteristics (Williams *et al.* 2002). Once again, the complete sequence of the *E. cuniculi* genome revealed substantial evidence for the presence of mitochondria, and gave the first complete picture of the proteome of a cryptic mitochondrion in an "amitochondriate" (Katinka *et al.* 2001). Only 22 protein-coding genes were identified as being targeted to the organelle (and several of these are closely related copies of one another), which was concluded to most likely have been retained for the manufacture of iron sulfur clusters essential for several cytosolic proteins (Katinka *et al.* 2001; Lill and Muhlenhoff 2005).

Microsporidian genomes – *Encephalitozoon cuniculi*

The "transition" of microsporidia (in our minds) from ancient and primitive protists to fungi has a substantial impact on how we think about such cells; the complete genome sequence of *E. cuniculi* was a critical force in various aspects of that transition, and microsporidian genomes in general are also a model for how this transition changes our thinking. Microsporidian genomes have attracted some attention since they were first examined because they were quickly found to be unusually small. Eukaryotic genome sizes vary tremendously (Fig. 12.2), but the first microsporidian karyotypes showed many of them to be within the size range of bacterial genomes, although they are in form like other eukaryotes (Biderre *et al.* 1999; Biderre *et al.* 1995; Biderre *et al.* 1994; Streett 1994). Indeed, the smallest known microsporidian genome, that of *E. intestinalis*, is now known to be only 2.3 Mpb (Peyretaillade *et al.* 1998a), or just over half that of *E. coli* K12. Other microsporidian genomes span the range from this to as much as 19.5 Mbp, which is only marginally smaller than many protist genomes, but overall microsporidian genomes rank among the smallest of any eukaryotic group (Keeling and Fast 2002; Méténier and Vivarès 2001). As early branching eukaryotes, we would be tempted to consider microsporidian genomes poorly developed and primitively small compared to other eukaryotes, but clearly this is not true. Ancestors of microsporidia probably had relatively unremarkable genomes, like most fungi today: microsporidian genomes are the product of severe reduction, most likely related to their adaptation to intracellular parasitism.

The completion of the *E. cuniculi* genome (Katinka *et al.* 2001) was a landmark for microsporidian research: as eluded to in the discussion of microsporidian evolutionary origins, the impact of this data on microsporidian molecular biology, cell biology, and infection research was immediate and profound, although one imagines the benefits of complete genomic data from microsporidia will only increase as these fields continue to develop. The *E. cuniculi* genome is also of great interest for

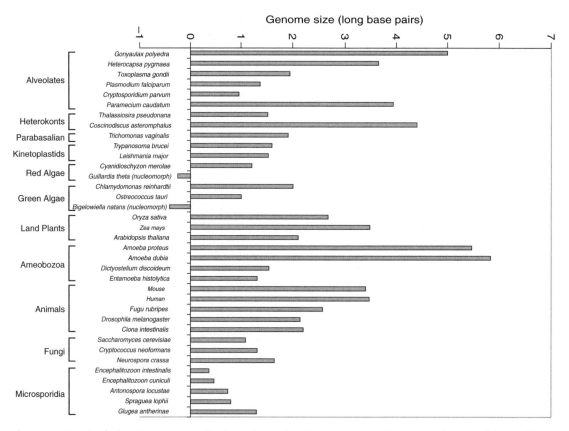

Figure 12.2 Sample of eukaryotic genome sizes (log base pair) organized by phylogenetic relationship. Note the range of sizes and the lack of correlation between size and evolutionary group. References to genome size data may be found in Keeling *et al.* (2005) and Lynch and Conery (2003).

eukaryotic genomics in general, however, since it marks an extreme end of the spectrum of genome architecture. It falls at the small end of the scale even for microsporidia, with only 2.9 Mbp on 11 chromosomes, ranging from 217 to 315 Kbp each – so what can it tell us about how genomes become so small? There are two main ways to shrink a genome: genes can be deleted, or genes can be forced into less space. *E. cuniculi* has done both.

The overall characteristics of the organism and its genome have been reviewed in detail elsewhere (Vivarès *et al.* 2002; Vivarès and Méténier 2001), but briefly each chromosome has telomeric repeats and subtelomeric regions consisting of repeats containing the rRNA operons. Between these repeats are dense protein-coding regions, but in the initial annotation only 1997 protein-coding

genes were identified. A few more will certainly be found, but there is little space unaccounted for, so this number will not grow significantly. This reduction of gene number is the most important contributor to the reduction in genome size: if one increased the coding potential of the *E. cuniculi* genome to that of yeast but kept its gene density, the genome would increase by about 300%. The loss of genes is also relatively easy to understand in the case of microsporidia: as intracellular parasites they have developed a dependence on their host cells for a large number of nutrients and all of their energy. For example the *E. cuniculi* genome lacks genes for proteins to synthesize many small molecules such as nucleotides, fatty acids, and most amino acids, all of which must be provided by the host.

The compaction of the *E. cuniculi* genome is less significant in terms of overall genome size than is gene loss: if *E. cuniculi* had the number of genes it has today but the gene density of yeast, it would only increase the genome size by 140%. However, the effects of compaction on the genome are still very interesting because of their extremity. The gene density is 0.97 per Kbp (compared to yeast at about 0.5 genes per Kbp), with few repeats and few large non-coding regions and no identifiable transposons. Naturally, genes are separated by very short intergenic regions (on average 129 bp), only 13 spliceosomal introns have been found so far (Biderre *et al.* 1998; Katinka *et al.* 2001), and these are generally short (23–52 bp). Perhaps most interestingly, the genes themselves are shorter on average than homologs in yeast. Katinka and coworkers (Katinka *et al.* 2001) put forward the hypothesis that the reduced protein complement of microsporidia has led to a reduction in the network of interactions, which in turn allows proteins to loose interaction domains leading to smaller proteins. This would suggest that the deletions should be localized to particular parts of the protein, and indeed they have shown that the ends are often truncated.

Comparative genomics

The *E. cuniculi* genome surely represents many of the main characteristics of microsporidian genomes in general, but we must nonetheless remember that this is only one species of a group with vast diversity, and its genome ranks among the smallest and therefore perhaps most extreme. In addition, a single genome can say little about certain patterns of change over time, so there are many questions that will only become apparent with comparative genomics of microsporidia. In addition to the *E. cuniculi* genome, there are small sequence surveys from *Vittaforma corineae* (Mittleider *et al.* 2002) and *Spraguea lophii* (Hinkle *et al.* 1997), as well as a larger survey (Slamovits *et al.* 2004a) and an ongoing complete genome project (http://gmod.mbl.edu/perl/site/antonospora01?page = intro) from *Antonospora locustae* (formerly *Nosema locustae* (Slamovits *et al.* 2004b)). These are very likely just the beginning: a

four-fold coverage survey is about to commence for *Enterocytozoon beineusi* and other smaller projects are underway for other species, so the coming years should see the generation of at least samples of genomes from a reasonable diversity of microsporidia. What can we expect from these data? In general we would predict that the metabolic and overall proteomic diversity will be much greater than previously imagined. Taking the mitochondrion as an example, the small *A. locustae* survey already revealed another two proteins in addition to those already characterized in *E. cuniculi* (Williams and Keeling 2005). Similarly, *S. lophii* was found to encode a reverse transcriptase (Hinkle *et al.* 1997), while *E. cuniculi* has no evidence of transposons (Katinka *et al.* 2001), and two interesting enzymes, catalase and photolyase, have been found in *A. locustae* but not *E. cuniculi* (Fast *et al.* 2003; Slamovits and Keeling 2004), and the former appears to have been acquired by lateral gene transfer from a bacterium (Fast *et al.* 2003). With good samples of many genomes, individual but nevertheless important cases like these should abound, and each has the potential to reveal additional functional diversity and therefore better inform our generalizations.

As far as the genome as a whole is concerned, comparisons between *E. cuniculi* and *A. locustae* have already revealed one surprising characteristic. Over time, the conservation of gene order is lost, mostly due to many short-range events and a few longer-range events which eventually randomize the genome. *A. locustae* and *E. cuniculi* are very distantly related species of microsporidia (Slamovits *et al.* 2004b), so it is surprising that they share a fairly high level of gene order conservation. Comparing 94 pairs of genes it was found that 13% of pairs were present in both genomes, while almost 26% of pairs were within five genes of one another (Slamovits *et al.* 2004a). It is impossible to quantify our expectations for gene order conservation without knowing how long ago two species diverged, which we cannot know for microsporidia, but if we compare this level of conservation with other fungi, the contrast is obvious. The closely related yeasts *Saccharomyces cerevisiae* and *Candida albicans* share only 9% of gene pairs and *S. cerevisiae* and the fission yeast

Schizosaccharomyces pombe share no gene order conservation. For the microsporidian genomes to be evolving at the same rate as these ascomycetes they would have to have evolved around the time that *Saccharomyces* and *Candida* diverged, which does not fit with the distribution or host range of microsporidia (Slamovits *et al.* 2004a). More likely, their genomes are evolving more slowly. It has been suggested that this is because compaction has reduced the number of potentially harmless breakpoints in the genome (Hurst *et al.* 2002; Slamovits *et al.* 2004a), which are needed to reorganize without ill effect. Indeed, there is a negative correlation between conservation of gene pairs and the length of their intergenic regions (Hurst *et al.* 2002; Slamovits *et al.* 2004a).

The conservation in genome order is potentially a general feature of compacted genomes, but no other comparison of such genomes is available as of yet, so it remains to be seen if other highly reduced genomes behave similarly. In the meantime, however, the conserved arrangement of microsporidian chromosomes has proved to be of use in detecting highly divergent genes. Using the map of the *E. cuniculi* genome, the two polar tube proteins from *A. locustae* were identified when the sequence conservation between these proteins was too low to allow detection by conventional means. Localization and biochemical analysis confirmed the identity of the two divergent sequences (Polonais *et al.* 2005), demonstrating one practical importance of comparative genomics.

Functional implications of genome structure

As we have seen, there is reason to believe that the structure of microsporidian genomes has an effect on how they evolve over time, but what about function in the short term? There are several characteristics of genome structure that suggest this may be the case. For example there is a strong correlation between the orientation of genes and the size of the intergenic regions between them (Keeling and Slamovits 2004). In *E. cuniculi*, convergent genes have the shortest intergenic spaces at an average of 128 bp, while divergent genes are on average separated by 184 and genes in the same

direction are intermediate at 156 bp. The obvious suggestion is that the 5′ control elements require more space than do 3′ elements and the regions upstream of genes are accordingly less susceptible to reduction without harm (but see below). If space in the genome is indeed used so frugally as that, then the control regions of genes must be well adapted to their immediate environment, which would further contribute to the slow pace of rearrangements discussed above since control elements may be overlapping or at least affecting one another, so moving a gene might be more likely to affect its expression even if it were successfully transposed or inverted.

With genes packed so close to one another, one might expect gene expression to be somewhat odd as well, but an analysis of expressed sequence tags (ESTs) shows it to be very odd indeed. Of 871 cDNAs with identifiable genes, 97 cDNAs from 70 different loci encoded part or all of more than one gene, and several encoded part or all of three or more genes (Williams *et al.* 2005). These cDNAs are not likely to represent operons since the genes encoded on them are not generally in the same strand. Instead, multigene transcripts have been suggested to be the result of compaction once more, where intergenic spaces have reduced to such an extent that control regions begin to overlap with protein-coding sequence of other genes. The overlapping transcription of adjacent genes that results is known from other genomes in a small number of genes, but in general has been shown to be incompatible with expression of both genes (Peterson and Myers 1993; Prescott and Proudfoot 2002). The deleterious effects of such overlaps can be classified as transcriptional collision when transcripts of two convergent genes overlap or promoter occlusion when the transcript of one gene reads through the promoter of an adjacent gene (in either direction). Both of these situations, thought to be problematic in other genomes (Peterson and Myers 1993; Prescott and Proudfoot 2002), occur regularly in microsporidian cDNAs (Fig. 12.3).

The question remains, however, are these unusual characteristics of microsporidian transcription due to the compacted nature of the genome? If they are, then other compacted nuclear genomes might be expected to show similar signs of over-lapping transcription. There are other genomes with small genomes or high gene densities (Abrahamsen *et al.* 2004; Courties *et al.* 1998; Matsuzaki *et al.* 2004), but the most extreme examples are nucleomorphs. Nucleomorphs are relict nuclei of algal endosymbionts found in cryptomonads and chlorarachniophytes (McFadden 2001; Kawach *et al.*, Chapter 13, this book). Their genomes are eukaryotic in form but encode only a few hundred kilobase pairs and are by far the smallest nuclear genomes known (McFadden *et al.* 1997). The cryptomonad host is thought to be a member of the chromalveolates and its endosymbiont is derived form a red alga (Fast *et al.* 2001; Douglas *et al.*, 2001), while the chlorarachniophytes are members of the cercozoa and have a green alga-derived endosymbiont (Keeling 2001; Van de Peer *et al.* 1996), so the nuclear reduction characteristic of both arose independently. The first cDNAs sequenced from these extremely gene-dense genomes were from the chlorarachniophyte nucleomorph, and indeed these revealed multiple gene sequences (Gilson and McFadden 1996). However, since only a

handful of cDNAs were examined the possibility of operons or processing could not be ruled out. Nevertheless, analysis of cDNAs corresponding to about 10% of genes in both genomes (Williams *et al.* 2005) has revealed that overlapping transcriptions is prevalent in nucleomorph genomes, and indeed at a much higher level than that observed in microsporidia (almost 100% of transcripts from the cryptomonad nucleomorph encoded more than one gene or gene fragment and one transcript encoded at least part of four genes). Overall, there does appear to be a correlation between compacted genomes and multigene transcripts (which also extends to the less compact genome of the apicomplexan *Cryptosporidium*), suggesting that compaction may have led to this as hypothesized above. In some cases this is inevitable (e.g. cases where the coding sequence of two genes physically overlap), but the extent of the overlapping transcriptions in both microsporidia and nucleomorphs is still surprising. Whatever the cause may be, overlapping transcription also adds another interesting constraint to the evolution of such genomes since promoters and/or termination signals from many genes will be embedded within

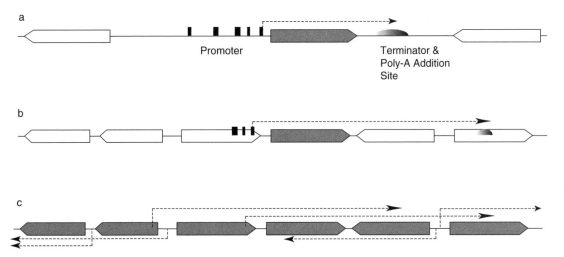

Figure 12.3 Schematic of potential effects of genome compaction on gene expression. (a) In a "typical" eukaryotic genome control elements such as promoter, terminator, and poly-A addition signals are normally found in intergenic regions. When a gene is expressed (the shaded gene) transcripts (dashed line) generally do not overlap with transcripts of adjacent genes. (b) In microsporidian genomes (and other gene-dense genomes), it appears that some control elements are now encoded within adjacent genes, so that transcripts from an expressed gene may begin within the upstream gene and end within or beyond downstream genes. If we plot expression of all genes in this hypothetical region of such a genome (c), several possible effects of overlapping expression are illustrated, including transcriptional collision and promoter occlusion, and it is easy to see why determining the expression profile of such a genome presents an unusual challenge.

the coding sequence of adjacent genes. Obviously if these genes are rearranged with respect to one another it has a strong probability of disrupting whatever genes are separated from their control regions.

The future of microsporidian genomics?

The complete sequence of the *E. cuniculi* genome has been of importance in many ways already, but we hope that the above sections demonstrate the potential value of comparative genomics, and therefore complete sequences from several additional taxa. In addition, *E. cuniculi* is not the only microsporidian of medical or commercial importance, so there are many good reasons to be optimistic that microsporidian genomics has an interesting future. Other model systems of practical importance where genome projects would be feasible and informative could include both medically important parasites (e.g. *Enterocytozoon beineusi*, *Encephalitozoon hellem*, *Vittaforma corneae*), commercially important parasites (e.g. *Nosema bombycis*, *Antonospora locustae*, as well as many fish parasites and species that infect pest insects), and parasites where we might learn something about factors that determine host range (e.g. *Brachiola algerae*) or complex life cycles including multiple hosts (e.g. *Ambylospora* species).

The small size of these genomes is of course one reason to study them in itself, but it also makes relatively broad-sampling for comparative genomics much more tractable that it would be for many other eukaryotic groups where much more sequencing would be required. Presently, most comparisons within microsporidia are between two organisms, but more whole genome data or even large-scale surveys will improve this greatly by adding much-needed "dimension" to the data. Such diversity in sampling will also reveal many characteristics about known data that are not really representative of the group as a whole, since our present sampling only begins to scratch the surface of the known biodiversity of microsporidia.

Another dimension to this work will be added by considering microsporidian genomes in the context of other reduced genomes more carefully.

Similarities have already been drawn between microsporidian genomes and those of nucleomorphs (Katinka *et al.* 2001; Williams *et al.* 2005), but true generalities common to compacted genomes will emerge with multiple samples of both, as well as other eukaryotes with small genomes such as the green alga *Ostreococcus* (Courties *et al.* 1998) and the less-reduced genomes such as *Cryptosporidium* (Abrahamsen *et al.* 2004). The commonalities and differences in these systems have the potential to tell us much about how their genomes got to be the way they are and why, as well as how their current condition affects genome function.

Acknowledgements

I am indebted to Claudio Slamovits and Bryony Williams for many discussions on genome compaction and microsporidian genomics. Work in my lab on microsporidian genomics is supported by the Canadian Institutes of Health Research and Burroughs – Wellcome Fund. PJK is a Fellow of the Canadian Institute for Advanced Research and a new investigator of the CIHR and the Michael Smith Foundation for Health Research.

References

Abrahamsen, M. S., Templeton, T. J., Enomoto, S., Abrahante, J. E., Zhu, G., Lancto, C. A., Deng, M., Liu, C., Widmer, G., Tzipori, S., Buck, G. A., Xu, P., Bankier, A,T., Dear, P. H., Konfortov, B. A., Spriggs, H. F., Iyer, L., Anantharaman, V., Aravind, L., and Kapur, V. (2004). Complete genome sequence of the apicomplexan, *Cryptosporidium parvum*. *Science*, **304**, 441–445.

Arisue, N., Sanchez, L. B., Weiss, L. M., Muller, M., and Hashimoto, T. (2002). Mitochondrial-type hsp70 genes of the amitochondriate protists, *Giardia intestinalis*, *Entamoeba histolytica* and two microsporidians. *Parasitol. Int.*, **51**, 9–16.

Biderre, C., Mathis, A., Deplazes, P., Weber, R., Méténier, G., and Vivarès, C. P. (1999). Molecular karyotype diversity in the microsporidian *Encephalitozoon cuniculi*. *Parasitology*, **118**, 439–445.

Biderre, C., Méténier, G., and Vivarès, C. P. (1998). A small spliceosomal-type intron occurs in a ribosomal protein gene of the microsporidian *Encephalitozoon cuniculi*. *Mol. Biochem. Parasitol.*, **94**, 283–286.

Biderre, C., Pagès, M., Méténier, G., Canning, E. U., and
Vivarès, C. P. (1995). Evidence for the smallest nuclear
genome (2.9 Mb) in the microsporidium *Encephalitozoon cuniculi*. *Mol. Biochem. Parasitol.*, **74**, 229–231.

Biderre, C., Pagès, M., Méténier, G., David, D., Bata, J.,
Prensier, G., and Vivarès, C. P. (1994). On small genomes in eukaryotic organisms: molecular karyotypes
of two microsporidian species (Protozoa) parasites of
vertebrates. *C. R. Acad. Sci. III*, **317**, 399–404.

Brown, J. R. and Doolittle, W. F. (1995). Root of the universal tree of life based on ancient aminoacyl-tRNA
synthetase gene duplications. *Proc. Natl. Acad. Sci. U S
A*, **92**, 2441–2445.

Cavalier-Smith, T. (1983). A 6-kingdom classification and
a unified phylogeny. In H. E. A. Schenk and W. S.
Schwemmler, eds. *Endocytobiology II: Intracellular space
as oligogenetic*. Walter de Gruyter, Berlin, New York,
pp. 1027–1034.

Cavalier-Smith, T. (1998). A revised six-kingdom system
of life. *Biol. Rev. Camb. Philos. Soc.*, **73**, 203–266.

Clark, C. G. and Roger, A. J. (1995). Direct evidence for
secondary loss of mitochondria in *Entamoeba histolytica*.
Proc. Natl. Acad. Sci. U S A, **92**, 6518–6521.

Courties, C., Perasso, R., Chretiennot-Dinet, M. J., Gouy,
M., Guillou, L., and Troussellier, M. (1998). Phyloge-
netic analysis and genome size of *Ostreococcus tauri*
(Chlorophyta, Prasinophyceae). *J. Phycol.*, **34**, 844–849.

Douglas, S., Zauner, S., Fraunholz, M., Beaton, M., Penny,
S., Deng, LT., Wu, X., Reith, M., Cavalier-Smith, T.,
Maier, U. G. (2001). The highly reduced genome of an
enslaved algal nucleus. *Nature*, **410**, 1091–109.

Edlind, T. D., Li, J., Visversvara, G. S., Vodkin, M. H.,
McLaughlin, G. L., and Katiyar, S. K. (1996). Phylogenetic analysis of the b-tubulin sequences from amitochondriate protozoa. *Mol. Phylogenet. Evol.*, **5**,
359–367.

Fast, N. M. and Keeling, P. J. (2001). Alpha and beta
subunits of pyruvate dehydrogenase E1 from the
microsporidian *Nosema locustae*: Mitochondrion-
derived carbon metabolism in microsporidia. *Mol.
Biochem. Parasitol.*, **177**, 201–209.

Fast, N. M., Kissinger, J. C., Roos, D. S., and Keeling, P. J.
(2001). Nuclear-encoded, plastid-targeted genes suggest a single common origin for apicomplexan and
dinoflagellate plastids. *Mol. Biol. Evol.*, **18**, 418–426.

Fast, N. M., Law, J. S., Williams, B. A., and Keeling, P. J.
(2003). Bacterial catalase in the microsporidian *Nosema
locustae*: implications for microsporidian metabolism
and genome evolution. *Eukaryot. Cell*, **2**, 1069–1075.

Fast, N. M., Logsdon, J. M., Jr., and Doolittle, W. F. (1999).
Phylogenetic analysis of the TATA box binding protein
(TBP) gene from *Nosema locustae*: evidence for a
microsporidia-fungi relationship and spliceosomal
intron loss. *Mol. Biol. Evol.*, **16**, 1415–1419.

Germot, A., Philippe, H., and Le Guyader, H. (1997).
Evidence for loss of mitochondria in Microsporidia
from a mitochondrial- type HSP70 in *Nosema locustae*.
Mol. Biochem. Parasitol., **87**, 159–168.

Gilson, P. R. and McFadden, G. I. (1996). The miniaturized nuclear genome of a eukaryotic endosymbiont
contains genes that overlap, genes that are cotranscribed, and the smallest known spliceosomal introns.
Proc. Natl. Acad. Sci. U S A, **93**, 7737–7742.

Gray, M. W., Burger, G., and Lang, B. F. (1999). Mitochondrial evolution. *Science*, **283**, 1476–1481.

Hinkle, G., Morrison, H. G., and Sogin, M. L. (1997).
Genes coding for reverse transcriptase, DNA-directed
RNA polymerase, and chitin synthase from the
microsporidian *Spraguea lophii*. *Biol. Bull.*, **193**, 250–251.

Hirt, R. P., Healy, B., Vossbrinck, C. R., Canning, E. U.,
and Embley, T. M. (1997). A mitochondrial Hsp70
orthologue in *Vairimorpha necatrix*: molecular evidence
that microsporidia once contained mitochondria. *Curr.
Biol.*, **7**, 995–998.

Hirt, R. P., Logsdon, J. M., Jr., Healy, B., Dorey, M. W.,
Doolittle, W. F., and Embley, T. M. (1999). Microsporidia are related to Fungi: evidence from the largest
subunit of RNA polymerase II and other proteins. *Proc.
Natl. Acad. Sci. U S A*, **96**, 580–585.

Hurst, L. D., Williams, E. J., and Pal, C. (2002). Natural
selection promotes the conservation of linkage of
co-expressed genes. *Trends Genet.*, **18**, 604–606.

Inagaki, Y., Susko, E., Fast, N. M., and Roger, A. J. (2004).
Covarion shifts cause a long-branch attraction artifact
that unites microsporidia and archaebacteria in EF-1α
phylogenies. *Mol. Biol. Evol.*, **21**, 1340–1349.

Ishihara, R. and Hayashi, Y. (1968). Some properties of
ribosomes from the sporoplasm of *Nosema bombycis*. *J.
Invert. Pathol.*, **11**, 377–385.

Kamaishi, T., Hashimoto, T., Nakamura, Y., Masuda, Y.,
Nakamura, F., Okamoto, K., Shimizu, M., and
Hasegawa, M. (1996a). Complete nucleotide sequences
of the genes encoding translation elongation factors 1 alpha and 2 from a microsporidian parasite,
Glugea plecoglossi: implications for the deepest branching
of eukaryotes. *J. Biochem.* (Tokyo), **120**,1095–1103.

Kamaishi, T., Hashimoto, T., Nakamura, Y., Nakamura,
F., Murata, S., Okada, N., Okamoto, K.-I., Shimzu, M.,
and Hasegawa, M. (1996b). Protein phylogeny of
translation elongation factor EF-1a suggests Microsporidians are extremely ancient eukaryotes. *J. Mol.
Evol.*, **42**, 257–263.

Katinka, M. D., Duprat, S., Cornillot, E., Méténier, G.,
Thomarat, F., Prenier, G., Barbe, V., Peyretaillade, E.,

Brottier, P., Wincker, P., Delbac, F., El Alaoui, H., Peyret, P., Saurin, W., Gouy, M., Weissenbach, J., and Vivarès, C. P. (2001). Genome sequence and gene compaction of the eukaryote parasite *Encephalitozoon cuniculi*. *Nature*, **414**, 450–453.

Keeling, P. J. (2001). Foraminifera and Cercozoa are related in actin phylogeny: two orphans find a home? *Mol. Biol. Evol.*, **18**, 1551–1557.

Keeling, P. J. (2003). Congruent evidence from alpha-tubulin and beta-tubulin gene phylogenies for a zygomycete origin of microsporidia. *Fungal. Genet. Biol.*, **38**, 298–309.

Keeling, P. J. and Doolittle, W. F. (1996). Alpha-tubulin from early-diverging eukaryotic lineages and the evolution of the tubulin family. *Mol. Biol. Evol.*, **13**, 1297–1305.

Keeling, P. J. and Fast, N. M. (2002). Microsporidia: biology and evolution of highly reduced intracellular parasites. *Annu. Rev. Microbiol.*, **56**, 93–116.

Keeling, P. J., Fast, N. M., Law, J. S., Williams, B. A. P., and Slamovits, C. H. (2005). Comparative genomics of microsporidia. *Folia Parasitol.* (Praha), **52**, 8–14.

Keeling, P. J., Luker, M. A., and Palmer, J. D. (2000). Evidence from beta-tubulin phylogeny that microsporidia evolved from within the fungi. *Mol. Biol. Evol.*, **17**, 23–31.

Keeling, P. J. and Slamovits, C. H. (2004). Simplicity and complexity of microsporidian genomes. *Eukaryot. Cell*, **3**, 1363–1369.

Keohane, E. M. and Weiss, L. M. (1999). The structure, function, and composition of the microsporidian polar tube. In M. Wittner and L. M. Weiss, eds. *The microsporidia and microsporidiosis*. American Society for Microbiology Press, Washington, D. C., pp. 196–224.

Kudo, R. R. (1947). *Protozoology*. Charles C. Thomas, Springfield, Il.

Levine, N. D., Corliss, J. O., Cox, F. E., Deroux, G., Grain, J., Honigberg, B. M., Leedale, G. F., Loeblich, A. R. D., Lom, J., Lynn, D., Merinfeld, E. G., Page, F. C., Poljansky, G., Sprague, V., Vavra, J., and Wallace, F. G. (1980). A newly revised classification of the protozoa. *J. Protozool.*, **27**, 37–58.

Lill, R. and Muhlenhoff, U. (2005). Iron-sulfur-protein biogenesis in eukaryotes. *Trends Biochem. Sci.*, **30**, 133–141.

Lom, J. and Vávra, J. (1962). A proposal to the classification within the subphylum Cnidospora. *System. Zool.*, **11**, 172–175.

Lynch, M. and Conery, J. S. (2003). The origins of genome complexity. *Science*, **302**, 1401–1404.

Matsuzaki, M., Misumi, O., Shin, I. T., Maruyama, S., Takahara, M., Miyagishima, S. Y., Mori, T., Nishida, K., Yagisawa, F., Yoshida, Y., Nishimura, Y., Nakao, S.,

Kobayashi, T., Momoyama, Y., Higashiyama, T., Minoda, A., Sano, M., Nomoto, H., Oishi, K., Hayashi, H., Ohta, F., Nishizaka, S., Haga, S., Miura, S., Morishita, T., Kabeya, Y., Terasawa, K., Suzuki, Y., Ishii, Y., Asakawa, S., Takano, H., Ohta, N., Kuroiwa, H., Tanaka, K., Shimizu, N., Sugano, S., Sato, N., Nozaki, H., Ogasawara, N., Kohara, Y., and Kuroiwa, T. (2004). Genome sequence of the ultrasmall unicellular red alga *Cyanidioschyzon merolae* 10D. *Nature*, **428**, 653–657.

McFadden, G. I. (2001). Primary and secondary endosymbiosis and the origin of plastids. *J. Phycol.*, **37**, 951–959.

McFadden, G. I., Gilson, P. R., Douglas, S. E., Cavalier-Smith, T., Hofmann, C. J., and Maier, U. G. (1997). Bonsai genomics: sequencing the smallest eukaryotic genomes. *Trends Genet.*, **13**, 46–49.

Méténier, G. and Vivarès, C. P. (2001). Molecular characteristics and physiology of microsporidia. *Microbes Infect.*, **3**, 407–415.

Mittleider, D., Green, L. C., Mann, V. H., Michael, S. F., Didier, E. S., and Brindley, P. J. (2002). Sequence survey of the genome of the opportunistic microsporidian pathogen, *Vittaforma corneae*. *J. Eukaryot. Microbiol.*, **49**, 393–401.

Nägeli, K. (1857). Über die neue Krankheit der Seidenraupe und verwandte Organismen. *Bot. Zeitung.*, **15**, 760–761.

Peterson, J. A. and Myers, A. M. (1993). Functional analysis of mRNA 3′ end formation signals in the convergent and overlapping transcription units of the *S. cerevisiae* genes RHO1 and MRP2. *Nucleic Acids Res.*, **21**, 5500–5508.

Peyretaillade, E., Biderre, C., Peyret, P., Duffieux, F., Méténier, G., Gouy, M., Michot, B., and Vivarès, C. P. (1998a). Microsporidian *Encephalitozoon cuniculia* unicellular eukaryote with an unusual chromosomal dispersion of ribosomal genes and a LSU rRNA reduced to the universal core. *Nucleic Acids Res.*, **26**, 3513–3520.

Peyretaillade, E., Broussolle, V., Peyret, P., Metenier, G., Gouy, M., and Vivares, C. P. (1998b). Microsporidia, amitochondrial protists, possess a 70-kDa heat shock protein gene of mitochondrial evolutionary origin. *Mol. Biol. Evol.*, **15**, 683–689.

Polonais, V., Prensier, G., Méténier, G., Vivarès, C. P., and Delbac, F. (2005). Microsporidian polar tube proteins: highly divergent but closely linked genes encode PTP1 and PTP2 in members of the evolutionarily distant *Antonospora* and *Encephalitozoon* groups. *Fungal. Genet. Biol.*, **42**, 791–803.

Prescott, E. M. and Proudfoot, N. J. (2002). Transcriptional collision between convergent genes in budding yeast. *Proc. Natl. Acad. Sci. U S A*, **99**, 8796–8801.

Slamovits, C. H., Fast, N. M., Law, J. S., and Keeling, P. J. (2004a). Genome compaction and stability in microsporidian intracellular parasites. *Curr. Biol.*, **14**, 891–896.

Slamovits, C. H. and Keeling, P. J. (2004). Class II photolyase in a microsporidian intracellular parasite. *J. Mol. Biol.*, **341**, 713–721.

Slamovits, C. H., Williams, B. A. P., and Keeling, P. J. (2004b). Transfer of *Nosema locustae* (Microsporidia) to *Antonospora locustae* n. com. based on molecular and ultrastructural data. *J. Eukaryot. Microbiol.*, **51**, 207–213.

Sprague, V. and Becnel, J. J. (1999). Checklist of available generic names for microsporidia with type species and type hosts. In M. Wittner and L. M. Weiss, eds. *The microsporidia and microsporidiosis.* American Society for Microbiology Press, Washingon, D. C., pp. 517–530.

Streett, D. A. (1994). Analysis of *Nosema locustae* (Microsporidia: Nosematidae) chromosomal DNA with pulsed-field gel electrophoresis. *J. Invert. Pathol.*, **63**, 301–303.

Tanabe, Y., Watanabe, M., and Sugiyama, J. (2002). Are Microsporidia really related to Fungi: a reappraisal based on additional gene sequences from basal fungi. *Mycological Res.*, **106**, 1380–1391.

Thomarat, F., Vivarès, C. P., and Gouy, M. (2004). Phylogenetic analysis of the complete genome sequence of *Encephalitozoon cuniculi* supports the fungal origin of microsporidia and reveals a high frequency of fast-evolving genes. *J. Mol. Evol.*, **59**, 780–791.

Van de Peer, Y., Ben Ali, A., and Meyer, A. (2000). Microsporidia: accumulating molecular evidence that a group of amitochondriate and suspectedly primitive eukaryotes are just curious fungi. *Gene*, **246**, 1–8.

Van de Peer, Y., Rensing, S. A., and Maier, U.-G. (1996). Substitution rate calibration of small subunit ribosomal RNA identifies chlorarachniophyte endosymbionts as remnants of green algae. *Proc. Natl. Acad. Sci. U S A*, **93**, 7732–7736.

Vávra, J. (1965). Étude au microscope électronique de la morphologie et du développement de quelques microsporidies. *C. R. Acad. Sci.*, **261**, 3467–3470.

Vávra, J. and Larsson, J. I. R. (1999). Structure of the Microsporidia. In M. Wittner and L. M. Weiss, eds. *The microsporidia and microsporidiosis.* American Society for Microbiology Press, Washington, D. C., pp. 7–84.

Vivarès, C. P., Gouy, M., Thomarat, F., and Méténier, G. (2002). Functional and evolutionary analysis of a eukaryotic parasitic genome. *Curr. Opin. Microbiol.*, **5**, 499–505.

Vivarès, C. P. and Méténier, G. (2001). The microsporidian *Encephalitozoon*. *Bioessays*, **23**, 194–202.

Vivier, E. (1975). The microsporidia of the protozoa. *Protistology*, **11**, 345–361.

Vossbrinck, C. R., Maddox, J. V., Friedman, S., Debrunner-Vossbrinck, B. A., and Woese, C. R. (1987). Ribosomal RNA sequence suggests microsporidia are extremely ancient eukaryotes. *Nature*, **326**, 411–414.

Vossbrinck, C. R. and Woese, C. R. (1986). Eukaryotic ribosomes that lack a 5.8S RNA. *Nature*, **320**, 287–288.

Williams, B. A., Hirt, R. P., Lucocq, J. M., and Embley, T. M. (2002). A mitochondrial remnant in the microsporidian *Trachipleistophora hominis*. *Nature*, **418**, 865–869.

Williams, B. A. and Keeling, P. J. (2005). Microsporidian mitochondrial proteins: expression in *Antonospora locustae* spores and identification of genes coding for two further proteins. *J. Eukaryot. Microbiol.*, **52**, 271–276.

Williams, B. A., Slamovits, C. H., Patron, N. J., Fast, N. M., and Keeling, P. J. (2005). A high frequency of overlapping gene expression in compacted eukaryotic genomes. *Proc. Natl. Acad. Sci. U S A*, **102**, 10936–10941.

CHAPTER 13

Nucleomorphs: remnant nuclear genomes

Oliver Kawach, Maik S. Sommer, Sven B. Gould, Christine Voß, Stefan Zauner, and Uwe-G. Maier

Philipps University of Marburg, Cell Biology, Marburg, Germany

Introduction

Many organisms acquired their plastids by the engulfment and intracellular reduction of a eukaryotic phototroph. This process is called secondary endosymbiosis. Nature provides several stages of reduction of the secondary endosymbiont, which ultimately creates complex plastids surrounded by three or four membranes. In two cases, the chlorarachniophytes and the cryptophytes, a remnant cytoplasm as well as a pygmy cell nucleus, a nucleomorph, are maintained from the secondary endosymbiont. Although chlorarachniophytes and cryptophytes evolved independently, analyses of partially and fully-sequenced nucleomorph genomes have revealed several unusual derived traits. These traits include similar patterns of reduction, subtelomeric rDNAs, and equal chromosome numbers. However, other characteristics, such as gene content and intron density, highlight the different origin and evolution of nucleomorph genomes. In this chapter we will compare the characteristics of these stream-lined genomes, thereby highlighting the evolution and the reduction strategy of nucleomorphs.

In the second half of the last century, microscopists noticed that plastids, the organelles for photosynthesis, differ not only in their pigment composition but also in the number of surrounding membranes. Consequently, it was speculated that plastids surrounded by two membranes evolved from a cyanobacteria-like cell (primary endosymbiosis) (Margulis 1970; Cavalier-Smith 1982, 2002a), whereas plastids with additional membranes may have originated from endosymbiotic eukaryotic algae (secondary endosymbiosis) (Gibbs 1981), which were subsequently reduced to so-called "complex plastids". Today, for most scientists these explanations for the origin of plastids are beyond question.

Nature provides various groups of organisms that evolved through secondary endosymbiosis (Stoebe and Maier 2002; Cavalier-Smith 2002a; Hjorth *et al.* 2004): in two, the euglenopyhtes and the peridinin-containing dinoflagellates (see Hackett and Bhattacharya, Chapter 3, this book), the plastid is surrounded by three membranes, whereas stramenopiles (heterokonts), haptophytes, sporozoa, cryptomonads, and chlorarachniophytes harbor plastids covered by four membranes (Fig. 13.1).

Clear evidence for the evolution of plastids surrounded by more than two membranes via secondary endosymbiosis came from early work on cryptomonads, where a second nucleus, the nucleomorph, was detected. In cryptomonads this remnant of the engulfed eukaryotic symbiont is located within a separated small cytoplasm, the periplastidal compartment (Greenwood 1974). Some years later, a nucleomorph with a very similar subcellular localization was also detected in phototrophic amoeba, the chlorarachniophytes (Hibberd and Noris 1984) (Fig.13.2).

Intensive research has enabled the reconstruction of the phylogeny of the algal groups with plastids surrounded by more than two membranes. The impressive result of this international effort was the

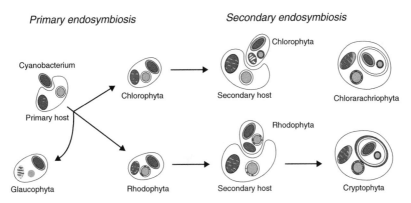

Primary endosymbiosis   *Secondary endosymbiosis*

Figure 13.1 Primary and secondary endosymbiois and the evolution of cryptophytes and chlorarachniophytes. In primary endosymbiosis, a cyanobacteria-like cell was engulfed and reduced to a plastid by a heterotrophic eukaryote. Such organelles are surrounded by a double membrane and found in glaucophytes, red and green algae, as well as in land plants. By capturing a green alga as a secondary endosymbiont the chlorarachniophytes evolved. In the case of cryptophytes, the secondary endosymbiont was a red alga. In both groups, reduction of the secondary endosymbiont is not complete, as indicated by a remnant cell nucleus of the secondary endosymbiont, the nucleomorph.

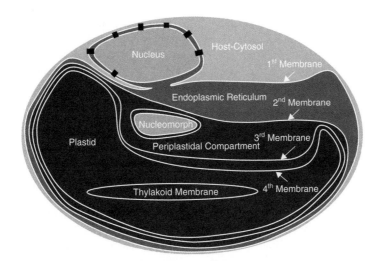

Figure 13.2 Scheme of a cryptomonad cell. Organelles are indicated, including their surrounding membranes. The four plastid membranes are shown and marked by "1st Membrane" (outermost membrane) to "4th Membrane" (inner membrane of the plastid envelope).

finding that secondary plastids evolved poly-phyletically, from either engulfed green or red algae (Van de Peer *et al.* 1996; Martin *et al.* 1998; Palmer and Delwiche 1996). It was suggested by Cavalier-Smith (1999) that the groups with a green algal symbiont (euglenophytes and chlorarachniophytes) may have originated monophyletically (Cabozoa). Additionally, a monophyletic origin was postulated for secondary evolved organisms with a red algal symbiont (heterokonts, haptophytes, peridinin-containing dinoflagellates, sporozoa, and cryptomonads). Therefore the organisms with a secondary endosymbiont of red origin were grouped together into the Chromalveolata (Cavalier-Smith 1999). Indeed, some phylogenetic investigations support the chromalveolata hypothesis (Fast *et al.* 2001; Harper and Keeling 2003; Patron *et al.* 2004), whereas no explicit molecular evidence exist for the Cabozoa concept. However, the nucleomorph-containing groups, the chlorarachniophytes and the cryptophytes, are definitely polyphyletic, as the chlorarachniophyte plastid is derived from a green and the cryptophyte from a red alga (Van de Peer *et al.* 1996). Thus, research on the nucleomorph genomes is a marvelous opportunity to study parallel genome evolution in phylogenetically distantly related, but morphologically similar, organisms.

Karyotyping of nucleomorph genomes

Early studies demonstrated that the nucleomorphs of cryptophytes harbor DNA (Ludwig and Gibbs 1985; Hansmann *et al.* 1986; Hansmann 1988). Therefore, cryptophytes and chlorarachniophytes accommodate two genomes of prokaryotic origin (those of plastids and mitochondria) and, in addition, two of eukaryotic origin (within the cell nucleus and the nucleomorph). This made it difficult to assign isolated DNAs to a specific genome. A major breakthrough in the assignment of DNA to a specific genome came from a method that involves purifying nucleomorphs of the cryptophyte *Pyrenomonas salina* (now *Rhodomonas salina*, Hansmann and Eschbach 1990). Separation of the nucleomorph genome by pulse field gel electrophoresis revealed three small chromosomes with a total length of approximately 660 kb (Eschbach *et al.* 1991; Maier *et al.* 1991). Even smaller genomes were found by studying nucleomorphs from other cryptophytes (Rensing *et al.* 1994) and, in parallel, the nucleomorph genome of one chlorarachniophyte, *Bigelowiella natans* (McFadden *et al.* 1994), where a eukaryotic genome of 380 kb was detected–the smallest eukaryotic genome investigated so far.

Further karyotyping of the nucleomorph genomes led to the first surprising parallelism between the nucleomorphs of chlorarachniophytes and cryptophytes: all are comprised of three chromosomes with individual lengths in the range of 100 kb (Eschbach *et al.* 1991; McFadden *et al.* 1994). Douglas *et al.* (2001) offered a likely explanation for the identical chromosome number in nucleomorphs. The key was that microscopists were not able to detect higher order condensation of the nucleomorph chromosomes in cryptophytes. Therefore it is likely that the chromosomes are compacted only into 30-nm chromatin threads. This would lead, if packed into three individual chromosomes, to a maximal length of 1.5 μm per

chromosome, the same width as a nucleomorph. Thus, extra folding would be unnecessary for mitotic segregation. As a higher fragmentation into smaller chromosomes may increase the risk of mitotic loss, a chromosome number of three seems to be the optimum for maintaining the coding capacity of the nucleomorphs (Douglas *et al.* 2001; Cavalier-Smith 2002b). However, as some cryptophytic nucleomorph genomes are twice the size of that of *B. natans* (Rensing *et al.* 1994), additional models are necessary to explain the conserved three chormosome arrangement of the nucleomorph genomes.

Genomics

Secondary endosymbiosis led either to plastids surrounded by three or four membranes, or to the intermediate forms in chlorarachniophytes and cryptophytes with a nucleomorph embedded in a remnant cytoplasm (periplastidal compartment) (Maier *et al.* 2000a; Cavalier-Smith 2003). Secondarily evolved algae, with the exception of cryptophytes and chlorarachniophytes, have managed to eliminate all traces of their secondary cell nucleus. Therefore, studying the functions and parallel evolution of nucleomorph genomes together with their coding capacity became of major interest (McFadden *et al.* 1997a). Consequently, sequencing projects on the nucleomorph genomes of one cryptophyte (*Guillardia theta*) (Zauner *et al.* 2000; Douglas *et al.* 2001) and a chlorarachniophyte (*Bigelowiella natans*) (Gilson and McFadden 1996, 2002) were initiated. The nucleomorph genome sequencing project on the chlorarachniophyte *B. natans* was recently finished (G. McFadden, personal communication) and that of the cryptophyte *Guillardia theta* was published in 2001 (Douglas *et al.* 2001) (Table 13.1).

Initial analyses of the nucleomorph genomes generated surprising results as subtelomeric rRNA-operons were detected in all nucleomorph chromosomes of the chlorarachniophyte and the cryptophyte (Gilson *et al.* 1997; Zauner *et al.* 2000). Moreover, mapping and sequencing of these regions showed that they are located close to the telomeres in each chromosome (Gilson and McFadden 1996; Zauner *et al.* 2000; Douglas *et al.*

Table 13.1 Comparison of the nucleomorphs of *Guillardia theta* and *Bigellowiella natans*

	Guillardia theta	*Bigellowiella natans*
Number of chromosomes	3	3
Size (kilobases)	551	380
A/T content (protein coding genes)	77%	73%
Number of genes	531	320 (predicted)
Gene density (gene/bp)	1/977	1/1141

Data from Douglas *et al.* (2001), Gilson and McFadden (2002).

2001). This parallelism between the nucleomorph genomes does not exist in the orientation of the rRNA-operons in cryptophytes and chlorarachniophytes, indicating the different origin of the nucleomorphs, but their similar genomic evolution. The fact that each nucleomorph chromosome end harbors in its subtelomeric regions rRNA genes led to the speculation that these regions could be hot spots for recombination and responsible for telomere maintenance (Douglas *et al.* 2001). This could explain why telomerase-encoding genes were not detected in the nucleomorph genomes. The intracellular parasite *Encephalitozoon cuniculi* shows a similar rRNA cistron localization in the subtelomeric region of each of its small chromosomes (Katinka *et al.* 2001). However, in other small chromosomes as detected in *Giardia lamblia*, *Saccharomyces cerevisiae*, or *Plasmodium falciparum* (see Huang and Kissinger, Chapter 8, this book) (Gardner *et al.* 2002) rRNA genes are not always located at the chromosome ends (Cavalier-Smith 2003). Nevertheless, it was speculated that in all of these subtelomeric sequences repeats are present, which could be hot spots for recombination, too (Cavalier-Smith 2003).

Because all analyzed nucleomorph chromosomes harbor rRNA genes at both ends, the chromosomes can be divided into a single copy and the terminal repeated telomere/rRNA region. Single copy regions are enriched in A/T (65–80%) in cryptophytes and chlorarachniophytes (Douglas *et al.* 2001; Gilson and McFadden 2002), which can be regarded as a characteristic feature of sexually isolated genomes and could reflect Muller's ratchet and a mutational hyperdrive (Gilson and McFadden 2002). If so, one would expect that spacers and introns are more A/T rich, which is

true for both nucleomorph genomes (up to 90%). The terminal regions, on the other hand, are less A/T rich (50–65%), which may be caused predominantly by structural constrains of the rRNAs.

The nucleomorph genomes are miniaturized but optimized in their coding capacity. Gene density is similar to bacterial genomes, with one gene in 977 bp in *G. theta* (Douglas *et al.* 2001) and one gene in 1141 bp in the smallest chromosome of *B. natans* (Gilson and McFadden 2002). Such a density is caused by minimizing spacer regions and by reducing either the number of spliceosomal introns or their length. In chlorarachniophytes, spliceosomal introns are common (3.3 introns per gene) (Gilson and McFadden 1996). However, these introns are very small (18–20 nt) and represent the smallest spliceosomal introns detected thus far. On the other hand, cryptophytes do not encode many introns in the nucleomorph and only 17 spliceosomal introns were identified (Douglas *et al.* 2001). These small numbers of spliceosomal introns in the nucleomorph of *G. theta* can be explained by its red algal origin, introns are rarely found in red-algal protein-encoding genes (Matsuzaki *et al.* 2004). Interestingly, tRNA introns are enriched in the nucleomorph of *G. theta*, which contradicts the general streamlining tendency of nucleomorph genomes (Kawach *et al.* 2005). However, as indicated by the genome sequence of the red alga *Cyanidioschyzon merolae*, the ancestor of modern red algae and the symbiont of cryptophytes could have possessed such tRNA introns (Matsuzaki *et al.* 2004).

An extreme in genome reduction and spacer elimination is shown by overlapping genes, which were detected in both nucleomorph genomes (Gilson and McFadden 1996; Douglas *et al.* 2001).

Furthermore, efficiency can be enhanced by cotranscription of neighboring genes, which was detected in the nucleomorph of *G. theta* and *B. natans* (Gilson and McFadden 1996; Fraunholz *et al.* 1998; Kawach *et al.* 2005; Williams *et al.* 2005). However, it is not proven that these transcripts are used to synthesize different proteins or that the termination of transcription is inaccurate.

In cryptophytes most nucleomorph genes encode housekeeping functions (Douglas *et al.* 2001). These include genes for proteins maintaining the nucleomorph, for example histones (with the exception of histone H2a and H1) and factors of their acetylation and deacetylation (Douglas *et al.* 2001). Furthermore, nucleomorph genes encode components for an expression apparatus, such as subunits for the RNA-polymerases, factors possibly involved in RNA maturation, and ribosomal proteins. Maturation of the proteins is managed by chaperones (TCP-complex, Hsp-proteins; Archibald *et al.* 2001) and their degradation by the ubiquitin/proteasome-machinery, for which subunits are encoded in the nucleomorph. Factors for other main functions of a eukaryotic cytoplasm and a nucleomorph, such as the cytoskeleton (Keeling *et al.* 1999) and a cyclin/cdk for cell cycle control, were detected as well. Furthermore, genes encoding membrane proteins were identified, which probably function as carriers and transporters, thus managing the exchange of molecules between the two eukaryotic cytoplasms (Douglas *et al.* 2001).

In red algae starch is deposited in the cytoplasm (Meeuse *et al.* 1960; Viola *et al.* 2001). In cryptophytes this localization is retained, but genes encoding enzymes involved in starch synthesis and degradation are missing in the nucleomorph genome of *G. theta*. However, Douglas *et al.* (2001) noted that some nucleomorph-encoded gene products may be involved in the regulation of starch degradation. Chlorarachniophytes store carbohydrates as *β*-1,3 glucan within vesicles attached to the symbiont (McFadden *et al.* 1997b) and it was predicted that the chlorarachniophyte nucleomorph would encode only a few, if any, enzymes for carbohydrate biochemistry (Gilson and McFadden 2002).

Clearly, most of the functions mentioned are necessary to maintain a eukaryotic compartment. But a self-preserving nucleomorph cannot be the reason for its maintenance in cryptophytes and chlorarachniophytes, whereas it was eliminated in phylogenetically closely related organisms with a red-algal symbiont. The most likely explanation for its perpetuation is that the nucleomorph is essential for the maintenance and functionality of the plastid. This is supported by the finding that both nucleomorph genomes harbor genes for proteins with plastid functions.

Nucleomorphs encode plastid proteins

Because red algal plastid genomes only encode a few more plastid proteins than green algal or higher plant plastid genomes, one would expect that, similar to *Arabidopsis thaliana* (Martin *et al.* 2002), genes of plastid origin represent around 18% of the nucleomorph genome. Surprisingly, only 30 of such genes (less than 7%) have been detected in the nucleomorph genome of *G. theta* thus far (Douglas *et al.* 2001). Therefore most plastid functions must be encoded in the cell nucleus. However, because there are absolutely no hints that the nucleomorph-encoded plastid proteins have orthologs in the cell nucleus, the nucleomorph of cryptophytes has to be maintained until all the genes for plastid functions have successfully been transferred to the cell nucleus. This conclusion is also likely for the nucleomorph of chlorarachniophytes. Interestingly, most of the nucleomorph-encoded plastid proteins have regulatory functions, and only few of them are structural components (Fraunholz *et al.* 1998; Wastl *et al.* 1999, 2000; Maier *et al.* 2000b, Hjorth *et al.* 2005). Thus, the nucleomorph still has intimate control over its plastid.

Feeding the plastid and nucleomorph with their functions

As mentioned above, most functions of the secondary endosymbiont are encoded in the cell nucleus. This enables the host to control the secondary endosymbiont, but requires the evolution of an efficient transport of nuclear-encoded proteins into the secondary symbiont (Cavalier-Smith 1999). In both, cryptophytes and chlorarachniophytes, nuclear-encoded proteins with a plastid destination harbor a bipartite topogenic signal, which is composed of an N-terminal peptide sequence followed by a transit peptide (Wastl and Maier 2000; Rogers *et al.* 2004). However, the outermost plastid membrane (membrane one) of the complex plastid is morphologically different in both groups of algae (Maier *et al.* 2000a; Cavalier-Smith 2002b). Whereas in cryptomonads this membrane is continuous with the envelope of the cell nucleus and therefore studded with ribosomes, the first membrane of the chlorarachniophyte plastid is not fused with other internal host membranes and not covered with ribosomes. Thus, slightly different import mechanisms can be expected for nuclear-encoded plastid proteins in the two groups. Work on cryptomonads showed that transport of plastid proteins across the outermost membrane is managed by the N-terminal part of the topogenic signal, the signal peptide, which might be cleaved off after transit (Wastl and Maier 2000). Such a processing leads to a preprotein with an N-terminal transit peptide located in the lumen between the two outermost membranes. Consequently, the transit peptide may be used as a signal for the transport of plastid-located proteins across the second membrane, as described recently (Cavalier-Smith 2003; Gould *et al.* 2005). As nuclear-encoded proteins with a final destination within the periplastidal compartment and the plastid have to traverse the same two outermost membranes, it will be of interest to study the relevant topogenic signals of both groups of proteins. For this purpose, the complete nucleomorph sequence of *G. theta* is extremely helpful, as *in silico* analyses and cell biological knowledge make it easy to identify essential proteins for the periplastidal compartment, including its nucleomorph, that are not encoded in the nucleomorph genome. EST projects that are underway will show if the predictions are correct and which signals are needed for proper protein transport to the symbiont's cytoplasm.

As mentioned above, chlorarachniophytes contain nuclear-encoded plastid proteins with a bipartite signal sequence (Archibald *et al.* 2003), but harbor no ribosomes attached to the outermost plastid membrane. Thus, the import mechanism into the plastid of chlorarachniophytes should differ from that of cryptophytes, and it can be suggested that these proteins are synthesized at the ER and pass via vesicles to the outermost plastid membrane.

Small proteins in small genomes

In respect to their size, nucleomorph genomes are the minimalists among eukaryotes (McFadden *et al.* 1997a; Gilson *et al.* 1997). This fact, together with their sexual isolation, may cause the rapid evolution of their proteins and rRNAs. Another striking feature of nucleomorph proteins is that even housekeeping proteins are sometimes smaller than their homologs (Cavalier-Smith 2002b). This could be of major advantage for the "postgenomic era", in which not only domains for important functions have to be determined but also the crystal structure of the proteins. Proteins probably will be crystallized more successfully when protein structures are reduced.

Comparative genomics

From the results of further genome studies of green algae (e.g. *Chlamydomonas reinhardtii*, Kathir *et al.* 2003) and the red algae *Cyandioschyzon merolae* (Matsuzaki *et al.* 2004), a more sophisticated comparison between the nucleomorph genomes and their free-living relatives will be possible. Such analyses will show whether important characters such as intron distribution or cotranscription, which is also found in small chromosomes of other eukaryotes (Iwabe and Miyata 2001), are caused by streamlining the nucleomorph genomes or already exist in the free-living relatives of the symbiont. Moreover, the genome projects allow the determination of proteomes, for example of the plastid. This is an important task, not only to construct metabolic maps or to predict functions (Ralph *et al.* 2004) but also to reconstruct the phylogeny of the corresponding genes and thus the mechanisms of cell evolution. Additionally, it should become possible to assign the number of nuclear-encoded plastid proteins, as well as some cytosolic and mitochondrial, which variants evolved by dual gene transfer (from the cyanobacterial progenitor of the plastid into the first, and subsequently into the second, nucleus) (Deane *et al.* 2000), by horizontal gene transfer from an distantly unrelated organism, or by new targeting of existing proteins of cytosolic or organellar origin (for example Schnarrenberger and Martin 2002; Martin *et al.* 2002; Fast *et al.* 2001; Andersson and Roger 2002; Timmis *et al.* 2004).

But knowledge beyond the plastid proteome can be studied. Genomic sequence from the diatom *Thalassiosira pseudonana* was recently published (Armbrust *et al.* 2004), and this, together with data from closely related organisms such as api-complexa (Gardner 2002), dinoflagellates, and cryptomonads, are contributing to a better under-standing of the host phylogeny and its cell biology. Analysis of these sequences will reveal whether organisms with secondary red algal plastids use the same mechanisms for plastidal protein import (McFadden and van Dooren 2004). Moreover, phylogenetic concepts can be rigorously tested, as for example the chromalveolata hypothesis, which postulates a monophyletic origin of all organisms with a secondary red algal plastid (Cavalier-Smith 1999).

Conclusions

Streamlining of a eukaryotic genome, as it is found in the nucleomorphs of cryptophytes and chlorar-achniophytes, is unique in respect to its degree of miniaturization. This comprises several oppo-rtunities to understand genome evolution on a molecular level. As genome projects are underway or already available, a comprehensive study of parallel genome reduction will soon be possible, which will indicate the structural, and perhaps functional, restrictions in genome reduction. Moreover, as nucleomorph genomes are shown to harbor only a minimum of genetic ballast, these genomes highlight at least a small set of functions essential for a eukaryotic cytoplasmic compartment.

Acknowledgements

We are supported by the Deutsche Forschungs-gemeinschaft.

References

Andersson, J. O. and Roger, A. J. (2002). A cyanobacterial gene in nonphotosynthetic protests–an early chloroplast acquisition in eukaryotes? *Curr. Biol.*, **12**, 115–119.

Archibald, J., Cavalier-Smith, T., Maier, U. G., and Dou-glas, S. (2001). Molecular chaperones encoded by a reduced nucleus–the cryptomonad nucleomorph. *J. Mol. Evol.*, **52**, 490–501.

Archibald, J. M., Rogers, M. B., Toop, M., Ishida, K., and Keeling, P. J. (2003). Lateral gene transfer and the evolution of plastid-targeted proteins in the secondary plastid-containing alga *Bigelowiella natans*. *Proc. Natl. Acad. Sci. U S A*, **100**, 7678–7683.

Armbrust, E. V., Berges, J. A., Bowler, C., *et al.* (2004). The genome of the diatom *Thalassiosira pseudonana*: ecology, evolution, and metabolism. *Science*, **306**, 79–86.

Cavalier-Smith, T. (1982). The origins of plastids. *Biol. J. Linn. Soc.*, **17**, 289–306.

Cavalier-Smith, T. (1999). Principles of protein and lipid targeting in secondary symbiogenesis: euglenoid, dinoflagellate, and sporozoan plastid origins and the eukaryote family tree. *J. Eukaryot. Microbiol.*, **46**, 347–366.

Cavalier-Smith, T. (2002a). Chloroplast evolution: sec-ondary symbiogenesis and multiple losses. *Curr. Biol.*, **12**, 62–64.

Cavalier-Smith, T. (2002b). Nucleomorphs: enslaved algal nuclei. *Curr. Opin. Microbiol.*, **5**, 612–619.

Cavalier-Smith, T. (2003). Genomic reduction and evo-lution of novel genetic membranes and protein-targeting machinery in eukaryote-eukaryote chimaeras (meta-algae). *Phil. Trans. Roy. Soc. Lond. B, Biol. Sci.*, **358**, 109–133.

Deane, J. A., Fraunholz, M., Su, V., Maier, U. G., Martin, W., Durnford, D. G., and McFadden, G. I. (2000). Evi-dence for nucleomorph to host nucleus gene transfer: light-harvesting complex proteins from cryptomonads and chlorarachniophytes. *Protist*, **151**, 239–252.

Douglas, S., Zauner, S., Fraunholz, M., *et al.* (2001). The highly reduced genome of an enslaved algal nucleus. *Nature*, **410**, 1091–1096.

Eschbach, S., Hofmann, C. J., Maier, U. G., Sitte, P., and Hansmann, P. (1991). A eukaryotic genome of 660 kb: electrophoretic karyotype of nucleomorph and cell nucleus of the cryptomonad alga, *Pyrenomonas salina*. *Nucleic Acids Res.*, **19**, 1779–1781.

Fast, N. M., Kissinger, J. C., Roos, D. S., and Keeling, P. J. (2001). Nuclear-encoded, plastid-targeted genes suggest a single common origin for apicomplexan and dinoflagellate plastids. *Mol. Biol. Evol.*, **18**, 418–426.

Fraunholz, M. J., Moerschel, E., and Maier, U. G. (1998). The chloroplast division protein FtsZ is encoded by a nucleomorph gene in cryptomonads. *Mol. Gen. Genet.*, **260**, 207–211.

Gardner, M. J., Hall, N., Fung, E., *et al.* (2002). Genome sequence of the human malaria parasite *Plasmodium falciparum*. *Nature*, **419**, 498–511.

Gibbs, S. P. (1981). The chloroplasts of some algal groups may have evolved from endosymbiotic eukaryotic algae. *Ann. New York Acad. Sci.*, **361**, 193–208.

Gilson, P. R., Maier, U. G., and McFadden, G. I. (1997). Size isn't everything: lessons in genetic miniaturisation from nucleomorphs. *Curr. Op. Genet. Dev.*, **7**, 800–806.

Gilson, P. R. and McFadden, G. I. (1996). The miniaturized nuclear genome of eukaryotic endosymbiont contains genes that overlap, genes that are cotranscribed, and the smallest known spliceosomal introns. *Proc. Natl. Acad. Sci. USA*, **93**, 7737–7742.

Gilson, P. R. and McFadden, G. I. (2002). Jam packed genomes – a preliminary, comparative analysis of nucleomorphs. *Genetica*,**115**, 13–28.

Gould, S., Sommer, M., Hadfi, K., Zauner, S., and Maier, U.-G. (2005) Protein targeting into the complex plastid of cryptophytes. *J. Mol. Evol.*, in press.

Greenwood, A. D. (1974). The Cryptophyta in relation to phylogeny and photosynthesis. In J. Sanders and D. Goodchild, eds. *Electron microscopy*. Australian Academy of Sciences, Canberra, pp. 566–567.

Hansmann, P. (1988). Ultrastructural localization of RNA in cryptomonads. *Protoplasma*, **146**, 81–88.

Hansmann, P. and Eschbach, S. (1990). Isolation and preliminary characterization of the nucleus and the nucleomorph of a cryptomonad, *Pyrenomonas salina*. *Eur. J. Cell Biol.*, **52**, 373–378.

Hansmann, P., Falk, H., Scheer, U., and Sitte, P. (1986). Ultrastructural localization of DNA in two Cryptomonas species by use of a monoclonal DNA antibody. *Eur. J. Cell Biol.*, **42**, 152–160.

Harper, J. T. and Keeling, P. J. (2003). Nucleus-encoded, plastid-targeted glyceraldehyde-3-phosphate dehydrogenase (GAPDH) indicates a single origin for chromalveolate plastids. *Mol. Biol. Evol.*, **20**, 1730–1735.

Hibberd, D. J. and Norris, R. E. (1984). Cytology and ultrastructure of *Chlorarachnion reptans* (Chlorarchaniophyta division nova, Chlorarachniophyceae classis nova). *J. Phycol.*, **20**, 310–330.

Hjorth, E., Hadfi, K., Gould, S. B., Kawach, O., Sommer, M. S., Zauner, S., and Maier, U. G. (2004). Zero, one, two, three, and perhaps four. Endosymbiosis and the gain and loss of plastids. *Endocyt. Cell Res.*, **15**, 459–468.

Hjorth, E., Hadfi, K., Zauner, S., and Maier, U. G. (2005). Unique genetic compartmentalization of the SUF system in cryptophytes and characterization of a SufD mutant in *Arabidopsis thaliana*. *FEBS Letters*, **579**, 1129–1135.

Iwabe, N. and Miyata, T. (2001). Overlapping genes in parasitic protist *Giardia lamblia*. *Gene*, **280**, 163–167.

Kathir, P., LaVoie, M., Brazelton, W. J., Haas, N. A., Lefebvre, P. A., and Silflow, C. D. (2003). Molecular map of the *Chlamydomonas reinhardtii* nuclear genome. *Eukaryot. Cell*, **2**, 362–379.

Katinka, M. D., Duprat, S., Cornillot, E., *et al.* (2001). Genome sequence and gene compaction of the eukaryote parasite *Encephalitozoon cuniculi*. *Nature*, **414**, 450–453.

Kawach, O., Voß, C., Wolff, J., Hadfi, K., Maier, U. G., and Zauner, S. (2005). Unique tRNA introns of an enslaved algal cell. *Mol. Biol. Evol.*, **22**, 1694–1701.

Keeling, P. J., Deane, J. A., Hink-Schauer, C., Douglas, S. E., Maier, U. G., and McFadden, G. I. (1999). The secondary endosymbiont of the cryptomonad *Guillardia theta* contains alpha-, beta-, and gamma-tubulin genes. *Mol. Biol. Evol.*, **16**, 1308–1313.

Maier, U. G., Douglas, S. E., and Cavalier-Smith, T. (2000a). The nucleomorph genomes of cryptophytes and chlorarachniophytes. *Protist*, **151**, 103–109.

Maier, U. G., Fraunholz, M., Zauner, S., Penny, S., and Douglas, S. (2000b). A nucleomorph-encoded CbbX and the phylogeny of RuBisCo regulators. *Mol. Biol. Evol.*, **17**, 576–583.

Maier, U. G., Hofmann, C. J., Eschbach, S., Wolters, J., and Igloi, G. L. (1991). Demonstration of nucleomorph-encoded eukaryotic small subunit ribosomal RNA in cryptomonads. *Mol. Gen. Genet.*, **230**, 155–160.

Margulis, L. (1970). *Origin of eukaryotic cells*. Yale University Press, New Haven.

Martin, W., Rujan, T., Richly, E., *et al.* (2002). Evolutionary analysis of *Arabidopsis*, cyanobacterial, and chloroplast genomes reveals plastid phylogeny and thousands of cyanobacterial genes in the nucleus. *Proc. Natl. Acad. Sci. U S A*, **99**, 12246–12251.

Martin, W., Stoebe, B., Goremykin, V., Hapsmann, S., Hasegawa, M., and Kowallik, K. V. (1998). Gene transfer to the nucleus and the evolution of chloroplasts. *Nature*, **393**, 162–165.

Matsuzaki, M., Misumi, O., Shin, I. T., *et al.* (2004). Genome sequence of the ultrasmall unicellular red alga *Cyanidioschyzon merolae* 10D. *Nature*,**428**, 653–657.

McFadden, G. I., Gilson, P. R., Douglas, S. E., Cavalier-Smith, T., Hofmann, C. J., and Maier, U. G. (1997a). Bonsai genomics: sequencing the smallest eukaryotic genomes. *Trends Genet.*, **13**, 46–49.

McFadden, G. I., Gilson, P. R., Hofmann, C. J., Adcock, G. J., and Maier, U. G. (1994). Evidence that an amoeba acquired a chloroplast by retaining part of an engulfed eukaryotic alga. *Proc. Natl. Acad. Sci. U S A*, **91**, 3690–3694.

McFadden, G. I., Gilson, P., and Sims, I. (1997b). Preliminary characterization of carbohydrate stores from chlorarachniophytes (Division: Chlorarachniophyta). *Phycol. Res.*, **45**, 145–151.

McFadden, G. I. and van Dooren, G. G. (2004). Evolution: red algal genome affirms a common origin of all plastids. *Curr. Biol.*, **14**, 514–516.

Meeuse, B. J.D., Andries, M., Wood, J. A., and Wood, M. (1960). Floridean starch. *J. Exp. Bot.*, **11**, 129–140.

Palmer, J. D. and Delwiche, C. F. (1996). Second-hand chloroplasts and the case of the disappearing nucleus. *Proc. Natl. Acad. Sci. U S A*, **93**, 7432–7435.

Patron, N. J., Rogers, M. B., and Keeling, P. J. (2004). Gene replacement of fructose-1,6-bisphosphate aldolase supports the hypothesis of a single photosynthetic ancestor of chromalveolates. *Eukaryot. Cell*, **3**, 1169–1175.

Ralph, S. A., Van Dooren, G. G., Waller, R. F., *et al.* (2004). Tropical infectious diseases: Metabolic maps and functions of the *Plasmodium falciparum* apicoplast. *Nat. Rev. Microbiol.*, **2**, 203–216.

Rensing, S. A., Goddemeier, M., Hofmann, C. J., and Maier, U. G. (1994). The presence of a nucleomorph hsp70 gene is a common feature of Cryptophyta and Chlorarachniophyta. *Curr. Genet.*, **26**, 451–455.

Rogers, M. B., Archibald, J. M., Field, M. A., Li, C., Striepen, B., and Keeling, P. J. (2004). Plastid-targeting peptides from the chlorarachniophyte *Bigelowiella natans*. *J. Eukaryot. Microbiol.*, **51**, 529–535.

Schnarrenberger, C. and Martin, W. (2002). Evolution of the enzymes of the citric acid cycle and the glyoxylate cycle of higher plants. A case study of endosymbiotic gene transfer. *Eur. J. Biochem.*, **269**, 868–883.

Stoebe, B. and Maier, U. G. (2002). One, two, three: nature's tool box for building plastids. *Protoplasma*, **219**, 123–130.

Timmis, J. N., Ayliffe, M. A., Huang, C. Y., and Martin, W. (2004). Endosymbiotic gene transfer: organelle genomes forge eukaryotic chromosomes. *Nat. Rev. Genet.*, **5**, 123–135.

Van de Peer, Y., Rensing, S. A., Maier, U. G., and De Wachter, R. (1996). Substitution rate calibration of small subunit ribosomal RNA identifies chlorarachniophyte endosymbionts as remnants of green algae. *Proc. Natl. Acad. Sci. U S A*, **93**, 7732–7736.

Viola, R., Nyvall, P., and Pedersen, M. (2001). The unique features of starch metabolism in red algae. *Proc. Roy. Soc. Lond.*, **268**, 1417–1422.

Wastl, J., Fraunholz, M., Zauner, S., Douglas, S., and Maier, U. G. (1999). Ancient gene duplication and differential gene flow in plastid lineages: the GroEL/Cpn60 example. *J. Mol. Evol.*, **48**, 112–117.

Wastl, J. and Maier, U. G. (2000). Transport of proteins into cryptomonads complex plastids. *J. Biol. Chem.*, **275**, 23194–23198.

Wastl, J., Sticht, H., Maier, U. G., Rösch, P., and Hoffmann, S. (2000). Identification and characterization of a eukaryotically encoded rubredoxin in a cryptomonad alga. *FEBS Letters*, **471**, 191–196.

Williams, B. A., Slamovits, C. H., Patron, N. J., Fast, N. M., and Keeling, P. J. (2005). A high frequency of overlapping gene expression in compacted eukaryotic genomes. *Proc. Natl. Acad. Sci. U S A*, **102**, 10936–10941.

Zauner, S., Fraunholz, M., Wastl, J., *et al.* (2000). Chloroplast protein and centrosomal genes, a tRNA intron, and odd telomeres in an unusually compact eukaryotic genome, the cryptomonad nucleomorph. *Proc. Natl. Acad. Sci. U S A*, **97**, 200–205.

CHAPTER 14

Genomic insights into diatom evolution and metabolism

E. Virginia Armbrust,[1] Tatiana A. Rynearson,[2] and Bethany D. Jenkins[2,3]

[1] *Marine Molecular Biotechnology Laboratory, School of Oceanography, University of Washington, Seattle, WA, USA*
[2] *Graduate School of Oceanography, University of Rhode Island, Narragansett, RI, USA*
[3] *Department of Cell and Molecular Biology, University of Rhode Island, Kingston, RI, USA*

Introduction

Diatoms are unicellular, photosynthetic eukaryotes found throughout marine and freshwater environments (Round *et al.* 1990) and are estimated to generate up to 40% of primary production in the global ocean. The whole genome sequence of the diatom *Thalassiosira pseudonana* and the EST database from the diatom *Phaeodactylum tricornutum* have provided new insights into the evolution and metabolic capabilities of diatoms. In this chapter we highlight unusual diatom features including the targeting of nuclear-encoded proteins to the plastid, formation of the silica-based frustule, and aspects of nitrogen and carbon assimilation. We also discuss the evolutionary implications of genomic variation within individual species and examine genomic differences between diatom species. Because current genome sequencing projects focus on two of the three major diatom lineages, we suggest expanding genome sequencing to the third lineage, permitting the identification of processes in the diatom genome that have been conserved, lost, or diversified over evolutionary time.

Diatoms are members of the heterokont algae and are characterized by intricately patterned cell walls (frustules) composed of silica and organic material, the presence of chlorophylls *a* and *c*, plastids derived from a secondary endosymbiosis, and an unusual mode of cell-size diminution and restoration coupled to the sexual cycle. Diatoms are the most species-rich group of unicellular algae known, with conservative estimates of tens of thousands of different species (Mann and Droop 1996). Diatoms have historically been divided into two classes, the pennate forms and the centric forms. A recent re-evaluation of diatom phylogeny (Medlin and Kaczmarska 2004) has divided diatoms into three classes: the Coscinodiscophyceae includes radial centric diatoms, the Mediophyceae includes multipolar centrics plus some radial centric diatoms, and the Bacillariophyceae includes pennate diatoms. Members of these classes differ with regards to cell shape (centrics display radial symmetry, pennates display bilateral symmetry), presence/absence of a raphe (in pennate diatoms) required for motility, structure and arrangement of the Golgi apparatus and chloroplast pyrenoid, mode of sexual reproduction, and 18S rDNA phylogeny (Fig. 14.1). The radial centric forms likely arose sometime after the Permian–Triassic boundary, with the first fossil record occurring at 180 million years ago (Rothpletz 1896). The pennate forms are believed to have evolved from the

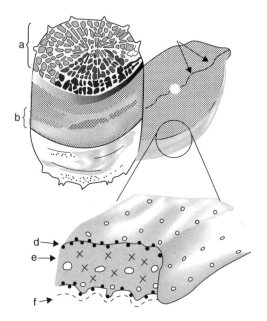

Figure 14.1 Drawing of the radial centric diatom *T. pseudonana* (upper left) and the raphid pennate diatom *Mastogloia* sp. depicting the valve (a), girdle bands (b), and raphe (c). Inset shows a cross section view of a silica frustule illustrating silica attached proteins (filled circles, d), silica embedded proteins (x marks, e), and peripherally associated proteins (dashed lines, f). Open circles are pores in the frustule.

multipolar centrics and appear in the fossil record about 70 million years ago (Moshkovitz *et al.* 1983).

Choosing the diatom species to sequence first was a challenge. None of the commonly employed criteria used to select a eukaryotic species for sequencing, such as the existence of a strong molecular infrastructure or the ability to perform forward and reverse genetics, could be used to discriminate among the different diatom species. For example when the project began in 2002, only 126 non-ribosomal nuclear gene sequences from 13 species of diatom were present in GenBank (and about half of these sequences were members of a single gene family); only 29 GenBank sequences were full-length. Chromosome number was known for relatively few diatoms, with most studies conducted prior to 1940 (reviewed in Kociolek and Stoermer 1989).

The decision to sequence the whole genome of the marine centric diatom *Thalassiosira pseudonona*, newly classified as Mediophyceae (Medlin and Kaczmarska

2004), was based primarily on ecological rather than molecular reasons, a rationale that was somewhat unusual at the time but has recently become more common (see for example, http://genome.jgi-psf.org/euk_cur1.html). A marine, rather than freshwater, diatom was chosen because, as a group, these organisms are estimated to generate about 40% of the 45 to 50 billion tonnes of organic carbon fixed annually in the sea and can generate as much as 90% of the photosynthetically-produced organic carbon that fuels coastal ecosystems (Nelson *et al.* 1995). The genus *Thalassiosira* is relatively cosmopolitan, with about 100 marine and freshwater species found in both polar and temperate environments. The species *T. pseudonana* is easy to maintain in culture with numerous isolates available from different ocean basins (see for example, http://ccmp.bigelow.org/), and preliminary estimates suggested that genome size was relatively small in this species.

Whole genome sequencing was carried out at the US Department of Energy's Joint Genome Institute (JGI) and initial analyses were conducted by an international consortium of scientists (Armbrust *et al.* 2004). With the success of the *T. pseudonana* project, the whole genomes of additional diatoms are now being sequenced by JGI. The whole genome sequence of the pennate diatom *Phaeodactylum tricornutum* (class Bacillariophyceae) will become publicly available in 2006. Approximately 12 000 expressed sequence tags (EST) for *P. tricornutum* were released publicly in 2004 (Maheswari *et al.* 2005), with an additional 80 000 EST sequences recently generated. *Phaeodactylum tricornutum* was chosen for sequencing because it is amenable to molecular manipulations such as transformation with exogenous DNA constructs containing selectable markers and reporter genes (Falciatore *et al.* 2000; Zaslavskaia *et al.* 2001). Small-scale EST projects have begun with the polar diatom *Fragillariopsis cylindrus* because of its ability to grow under extreme salinities of 100 psu and temperatures of −10°C (Mock *et al.*, 2006). *Pseudo-nitzschia multiseries* is slated to enter the sequencing queue sometime in 2006 because it is a cosmopolitan species that produces the potent neurotoxin domoic acid, the cause of amnesic shellfish poisoning (ASP) in humans if contaminated sea food is consumed (reviewed in Jeffery *et al.* 2004).

This chapter will focus primarily on findings derived from the *T. pseudonana* whole genome project. Although the chloroplast and mitochondrial genomes of *T. pseudonana* were also sequenced to completion, the focus here will be on nuclear-encoded genes. This chapter will highlight unusual diatom features uncovered in the *T. pseudonana* diatom genome project and now bolstered by comparisons with *P. tricornutum* EST sequences. We begin first with a description of the evolutionary origin of diatoms, which sets these organisms apart from the other photosynthetic eukaryotes that have been sequenced.

Genomic origins of the modern diatom: secondary endosymbiosis

Eukaryotic photosynthesis is hypothesized to have begun with a single primary endosymbiotic event in which a heterotrophic eukaryote engulfed (or was invaded by) a photosynthetic cyanobacteria (Archibald and Keeling 2004; McFadden 2001). This event generated autotrophic eukaryotes with a chloroplast bound by two membranes and a highly reduced, cyanobacterial-derived genome. Primary endosymbionts diverged into two modern-day lineages: the green algae/land plants that use chlorophyll *a* and *b*, and the red algae/glaucocystophytes that use chlorophyll *a* and phycobilin as their major photosynthetic pigments. The dominant photosynthetic eukaryotes on land are derived from a primary endosymbiosis.

Secondary endosymbiosis is hypothesized to have occurred at least three different times when heterotrophic organisms engulfed either a eukaryotic red or green alga (for details see Sommer *et al.*, Chapter 6, this book). The chromalveolate hypothesis proposes that a single secondary endosymbiotic engulfment of a red alga gave rise to all members of the chromists (heterokonts, haptophytes, and cryptomonads) and alveolates (ciliates, dinoflagellates, and apicomplexa) (Cavalier-Smith 1998), although inclusion of cryptomonads and haptophytes in this group is weakly supported (Baldauf *et al.* 2000; Harper *et al.* 2005; Van de Peer and De Wachter 1997). This hypothesis implies that the dominant photo-synthetic eukaryotes in the oceans, including diatoms, share a common ancestor because they are all derived from a secondary endosymbiosis of a red alga. The chromalveolate hypothesis also implies that non-photosynthetic organisms in these groups are derived from photosynthetic organisms and that photosynthetic capacity has been lost repeatedly over evolutionary time (described in detail in Sommer *et al.*, Chapter 6, this book).

The secondary endosymbiosis theory predicts that nuclear genes in the modern diatom have their origin from five potential sources: mitochondria or nuclear genomes of the heterotrophic host; or the mitochondria, plastid, or nuclear genome of the red algal endosymbiont. In the following sections, we provide examples of how genome and EST sequencing has provided insights into the evolution of diatoms.

Overview of *T. pseudonana* genome characteristics

The *T. pseudonana* genome was isolated from a clonal culture of a single diploid cell and was sequenced using a whole genome shotgun approach. The nuclear genome is 34 mega-base pairs (Mb). About 0.75% of the nucleotides in the genome were identified as polymorphic between chromosome pairs based on support from two or more aligning reads for each of the two alternate sequences. On average, a single nucleotide polymorphism (SNP) is expected every 150 bp, although a detailed analysis of the true distribution of SNPs has not yet been completed. Only two haplotypes were identified, consistent with descent from a single diploid founder cell.

An optically-derived, high resolution *Nhe*I restriction map of the entire genome (Lai *et al.* 2003) was used to identify and characterize the 24 diploid chromosomes that ranged in size from 0.34 to 3.3 Mb. Nine chromosomes could be separated into their respective haplotypes based on restriction site polymorphisms of the *Nhe*I sites and on the presence of insertions or inverted duplications of tens of kilo bases to a megabase in size, including rearrangements near subtelomeric regions. In other organisms, subtelomeric rearrangements are associated with diversification of gene families, generation of phenotypic diversity, promotion of disease resistance, and rapid adaptive evolution (reviewed in Mefford and Trask 2002).

The relatively high level of polymorphism observed in the *T. pseudonana* genome has been observed with other marine eukaryotic organisms, such as the ascidian *Ciona intestinalis* (~1.2%) (Dehal *et al.* 2002) and the puffer fish *Fugu rubripes* (~0.4%) (Aparicio *et al.* 2002) and is about an order of magnitude higher than what has been observed in humans (~0.08%) (International Human Genome Sequencing Consortium 2001). Nucleotide substitution rates in diatoms have been estimated for the small subunit rDNA at about 1% per 14 million years (Kooistra and Medlin 1996; Damste *et al.* 2004), which is up to three times faster than that found in metazoans (Echinodermata, Mollusca, and Actinistia–Tetrapoda). The simplest explanation for the presence of high levels of polymorphisms in the *T. pseudonana* genome (and presumably other diatom genomes as well) is that diatoms have large effective population sizes that offset the homogenizing effects of genetic drift. This possibility is supported by recent evidence that individual diatom species are composed of thousands of genetically distinct clonal lineages (Rynearson and Armbrust 2005).

Repetitive DNA sequences are relatively rare in the *T. pseudonana* genome, with only ~2% of the genome composed of transposable elements, although this percentage may increase slightly once the remaining sequence gaps are closed (slated to be completed in early 2006). Each family of transposable elements is composed of less than 100 copies, which may reflect high deletion rates and/or selection against insertion. Asexually reproducing organisms carry a reduced transposable element load (Arkhipova and Meselson 2000; Zeyl *et al.* 1996) presumably because in the absence of sexual recombination, transmission between individuals is reduced and the selective advantage of replication is eliminated (reviewed in Wright and Finnegan 2001). The frequency of sexual reproduction in diatoms is largely unconstrained, with estimates ranging from once per year to once every 40 years (Edlund and Stoermer 1997); the complete sexual cycle of *T. pseudonana* has not yet been described, although genes required for meiotic recombination are present within the genome. In contrast to the apparently rare occurrences of sexual reproduction, asexual divisions in diatoms occur on at least a daily basis under nutrient-replete conditions. Diatoms, in general, are therefore expected to possess relatively low percentages of transposable elements.

Comparative analyses

The homology-based identification of genes in *T. pseudonana* was complicated by the fact that this diatom was the first heterokont sequenced. A total of 11 242 putative nuclear protein-coding genes (longer than 50 amino acids) were identified, with just half of the predicted proteins displaying homology (E value <1 e-20) to public database proteins. Approximately 12 000 ESTs are publicly available for *P. tricornutum*, with a non-redundant EST dataset of 5108 sequences (Montsant *et al.* 2005). Using the *T. pseudonana* gene number as a reference, about one-third of the genes in *P. tricornutum* are represented by the current EST library and approximately one-third of those are similar to genes in *T. pseudonana* (E-value cutoff of 1e-30) (Montsant *et al.* 2005).

Comparative analyses of the *T. pseudonana* whole genome sequence and the *P. tricornutum* EST library have uncovered unusual combinations of features in diatoms (Montsant *et al.* 2005). For example, based on an E-value cutoff of 1 e-30, 16% of the *P. tricornutum* ESTs are similar to sequences found in *T. pseudonana* but not in the green alga *Chlamydomonas rheinhardtii* or the red alga *Cyanidioschyzon merolae* (Montsant *et al.* 2005). Unusual diatom features that will be highlighted here include the targeting of nuclear-encoded proteins to the plastid, formation of the silica-based frustule, and aspects of nitrogen and carbon assimilation. Potential differences between the two diatom species were also detected in this analysis. For example the ways in which inorganic carbon is concentrated and delivered to the carbon fixation enzyme, ribulose-1,5-bisphosphate carboxylase/oxygenase (RUBISCO), appear to differ between the two species. In the following sections we will describe how specific aspects of diatom life history and metabolism are being addressed using available genome sequence data.

Protein targeting to the plastid

The plastids in eukaryotes derived from a secondary endosymbiosis are surrounded by either three or four membranes. Diatoms have plastids bounded by four membranes. The two inner membranes originate from the plastid membranes of the primary endosymbiont and the two outer membranes originate from engulfment of the secondary endosymbiont, with the outer-most membrane contiguous with the endoplasmic reticulum (ER). In heterokonts, haptophytes, and cryptomonads, the outer-most plastid membrane is studded with ribosomes (van Dooren *et al.* 2001).

Nuclear-encoded proteins destined for the diatom plastid must cross four complex membranes. As with all ER-directed proteins, the outer-most ER-associated membrane of the plastid is cotranslationally crossed using a translocation complex that interacts with an N-terminal signal peptide (Bhaya and Grossman 1991; van Dooren *et al.* 2001). In addition to the signal sequence, a transit peptide is also required to cross the next three membranes (Apt *et al.* 2002; Kilian and Kroth 2004; Kilian and Kroth 2005; Kroth 2002). The bipartite structure of the plastid localization signal is hypothesized to have evolved through intron recombination (Kilian and Kroth 2004). Analysis of seven *P. tricornutum* nuclear genes known to encode proteins targeted to the plastid all possess introns near the 5′-most end of the gene: either within the region that encodes the ER signal sequence or the chloroplast transit sequence or else just downstream of the region encoding the entire presequence.

How nuclear-encoded plastid proteins are sorted to the plastid once they are within the ER remains unclear. Green fluorescent protein constructs can be targeted to the plastid by fusing an ER signal sequence to a minimal plastid transit targeting motif (ASAF/AFAP) (Kilian and Kroth 2005). This result indicates that once a protein enters the ER it is actively sorted to the plastid. Few obvious members of protein complexes known to direct proteins across the plastid membranes of primary endosymbionts (i.e. translocon of outer (Toc) or inner (Tic) chloroplast membranes) were identified in the *T. pseudonana* genome (Armbrust *et al.* 2004) and the few homologs that were identified are highly divergent (Reumann *et al.* 2005). Gibbs (1979) observed vesicles in the periplastidic space of heterokonts and proposed a model of vesicular transport that directs proteins from the second outermost envelope to the inner plastid membranes. More recent results indicate that plastid protein targeting in *P. tricornutum* is inhibited by treatment of cells with Brefeldin A, a drug that impairs Golgi vesicle budding and results in the accumulation of small vesicular structures near the chloroplast (Kilian and Kroth 2005). Kilian and Kroth (2005) also suggest that vesicle transport may be required for protein movement across the inner membranes bounding the plastid. Plastid targeting of nuclear-encoded proteins is an active area of research that will continue to benefit from whole genome sequences of diverse protists.

Silica metabolism

Diatoms are characterized by their intricately patterned cell walls or frustules constructed from silica and organic material (composed primarily of carbohydrates and glycoproteins). The frustule is composed of two valves connected by a series of siliceous girdle bands. The tens of thousands of species of diatoms each have uniquely structured walls that have traditionally been used to determine diatom taxonomy (Figs 14.1 and 14.2). The ability of diatoms to generate frustules made of "bio-glass" has attracted the attention of nanotechnologists because diatoms precipitate silica under physiological conditions, which contrasts sharply with the high temperatures/pressures and caustic chemicals required by human engineers (Lopez *et al.* 2005; Gordon and Parkinson 2005; Drum and Gordon 2003; Parkinson and Gordon 1999). Although they are not the only organism that can precipitate silica, diatoms are the only silica-precipitating organisms currently represented with whole genome sequence data.

Most diatoms have an absolute requirement for silicic acid, $Si(OH)_4$, although it should be noted that *P. tricornutum* does not display this requirement. Silicic acid is actively transported across a steep concentration gradient into the cytoplasm via transmembrane transporters (SITs) with no known homology to transporters in other organisms; the hypothesized function of the SITs was confirmed by injecting SIT RNA into *Xenopus laevis* oocytes and showing that germanium, a silicate analog, was actively transported into the oocytes (Hildebrand *et al.* 1997). Five SITs have been identified in the pennate diatom *Cylindrotheca fusiformis* (Hildebrand *et al.* 1997) and three new SITs were identified in the *T. pseudonana* genome. SITs from *C. fusiformis* are predicted to possess a C-terminal coiled-coil region; a comparable

C-terminal structure is absent from the *T. pseudonana* transporters suggesting that this feature may underlie differential transporter regulation in the two species (Thamatrakoln and Hildebrand 2005). Comparative analyses of SITs from additional diatoms should permit identification of important features of the transporters required for silicic acid transport.

Morphogenesis of the silica frustule occurs within a specialized vesicle known as the silica deposition vesicle. Silica precipitation is controlled by at least two components: glycoproteins known as silaffins and long-chain polyamines (Kroeger *et al.* 1999; Kroeger *et al.* 2000). In sponges, silica precipitation is controlled by silacateins, which are

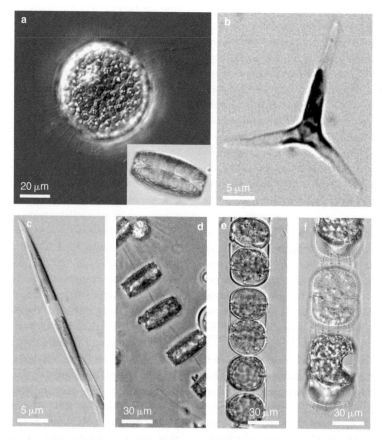

Figure 14.2 Light micrographs of (a) *Thalassiosira puntigera* (valve and girdle (inset) views), (b) *Phaeodactylum tricornutum*, (c) *Pseudo-nitzschia multiseries*, (d) *Thalassiosira rotula*, (e) *Melosira* sp., and (f) *Stephanopyxis* sp. *Phaeodactylum* and *Pseudo-nitzschia* are members of class Bacillariophyceae; *Thalassiosira* are members of class Mediophyceae; and *Melosira* and *Stephanopyxis* are members of class Coscinodiscophyceae. Structural features of diatom cells include: (a) chitin fibers and (c and e) chains connected by mucilage pads, (d) silica threads, (f) silica hooks, and (e) incomplete separation of girdle bands.

highly similar to the cathepsin L family of proteases (Cha *et al.* 1999; Shimizu *et al.* 1998); no genes encoding silacateins were detected in *T. pseudonana*. In pioneering work by Kroeger *et al.* (2002), silaffins in their native state were extracted from frustules of *C. fusiformis* and were shown to promote precipitation of silica *in vitro*. Availability of the *T. pseudonana* genome sequence facilitated identification of five additional silaffins corresponding to biochemically purified peptide sequences, despite the fact that the *T. pseudonana* proteins displayed no sequence homology to those from *C. fusiformis* (Poulsen and Kroeger 2004). Instead, silaffins from both species are rich in lysine and serine, and the lysine residues are post-translationally modified with the long-chain polyamines. The unique silica valve morphology displayed by each diatom species appears to be controlled by species-specific silaffins and long-chain polyamines (Kroeger *et al.* 2000), implying that it should be possible to experimentally manipulate valve morphology *in vivo* by using transformation to "mix and match" silaffins between species (Poulsen and Kroeger 2004).

An organic casing surrounds the silica and prevents the frustule from dissolving in water (Fig. 14.1). The protein portion of the casing is composed, in part, of calcium-binding glycoproteins known as frustulins (Kroeger *et al.* 1994). Previously identified frustulins in *C. fusiformis* (Kroeger *et al.* 1994) and the four frustulins identified in the *T. pseudonana* genome share highly conserved calcium-binding domains. A frustulin EST was one of the most redundant ESTs in the *P. tricornutum* EST collection (Montsant *et al.* 2005). A second set of extracellular glycoproteins called pleuralins is thought to be required for development of terminal girdle bands in *C. fusiformis* (Kroeger and Wetherbee 2000). No pleuralins have yet been identified in either *T. pseudonana* or *P. tricornutum*, again indicating the potential for species- or lineage-specific mechanisms of frustule formation.

Some species of diatoms can form multi-cell chains that are held together through a variety of extracellular connections, such as the incomplete separation of siliceous girdle bands, formation of silica "hooks," formation of chitin fibers (for example members of *Thalassiosira*), and secretion of mucilage pads between cells (see Fig. 14.2 for examples). Chain formation may play a role in regulating the diffusive flux of nutrients to cells (Pahlow *et al.* 1997; Karp-Boss and Jumars 1998), suggesting the potential coregulation of nutrient metabolism and aspects of frustule formation.

Nitrogen metabolism

Diatoms are capable of taking up inorganic (ammonium and nitrate) and organic (urea, amino acid) forms of nitrogen although they preferentially utilize ammonium for energetic reasons. The *T. pseudonana* genome encodes multiple transporters for ammonium and nitrate. Transporters for ammonium and nitrate in diatoms generally display higher affinities for their substrates than equivalent transporters in higher plants (Lopez *et al.* 2005). The presence of multiple transporters within an individual cell may reflect differences in substrate affinity, as has been observed with ammonium transporters in *Saccharomyces cerevisiae* (Marini *et al.* 1997). Multiple transporters may also play different roles in the uptake of nitrogen. For example the two types of ammonium transporters identified in *C. fusiformis* differ in amino acid sequence by 5% and are expressed at different levels regardless of nitrogen source, suggesting that one type may play a role in regulation rather than active transport (Hildebrand 2005).

The ammonium and nitrate transported into the cell is assimilated into biomass via the activity of glutamine synthetases: GS II is localized to the plastid and GSIII is localized to the cytosol. A third isoform of glutamine synthetase, GSI, was recently identified by *in silico* analyses, and also appears localized to the cytosol although the potential functional differences between GSI and GSIII are not yet clear (Takabayashi *et al.* 2005).

Diatoms are also able to take up nitrate in excess of immediate needs to take advantage of transient pulses of nitrate in the environment. Excess nitrate is thought to be stored in vacuoles (Lomas and Glibert 2000). Assimilated nitrogen may also be stored via asparagines, a major nitrogen storage compound in higher plants (Lea and Miflin 1980; Urquhart and Joy 1981) that is relatively inert and has a high ratio of nitrogen to carbon. Under low-carbon/high-nitrogen conditions, plants use asparagine synthetase (AS) to convert glutamine and glutamate into asparagines. A number of *P. tricornutum* ESTs and *T. pseudonana* genes display higher homology to AS genes from higher plants than to comparable genes from red and green algae (Montsant *et al.* 2005). This suggests that the asparagine synthetases in diatoms may function similarly to those in plants and that diatoms may utilize these enzymes for nitrogen storage.

One of the more unexpected findings of the *T. pseudonana* genome project was the identification of enzymes required for a complete urea cycle (Armbrust *et al.* 2004), an excretory pathway found in metazoans. Urea cycle homologs have also been identified in the *P. tricornutum* EST database (Montsant *et al.* 2005). It seems unlikely that diatoms require the urea cycle to remove "waste" urea since they can use this compound as a sole nitrogen source (Fan *et al.* 2003; Peers *et al.* 2000; Price and Harrison 1988; Oliveira and Antia 1986). The role of the urea cycle in diatom physiology remains unclear, but the cycle does appear to be actively utilized as most of the requisite genes have EST support even during exponential growth (Armbrust *et al.* 2004). The cyanobacteria *Synechococcus sp.* 6803 has three out of five key enzymes in the urea cycle and uses its partial urea cycle for arginine catabolism (Quintero *et al.* 2000). It is unknown whether other members of the chromoalveolates, such as haptophytes or the cryptomonads, also possess a complete urea cycle.

Carbon acquisition mechanisms in diatoms

Diatoms are hypothesized to generate as much organic carbon as all terrestrial rainforests combined (Mann and Droop 1996; Field *et al.* 1998) and yet the mechanisms through which diatoms fix organic carbon remain somewhat enigmatic. All diatom species examined to date take up CO_2 and HCO_3^-, the dominant form of inorganic carbon in the ocean (Colman *et al.* 2002). The first step in the fixation of CO_2 into organic material is catalyzed by ribulose-1-5-bisphosphate carboxylase (RUBISCO). This enzyme displays relatively slow kinetics, and

O_2 competes with CO_2 for the active site. The oxygenase activity of RUBISCO leads to the formation of a two-carbon compound that feeds into the photorespiratory pathway, rather than the Calvin cycle, and results in no net fixation of CO_2. Diatoms possess the Red Form I RUBISCO, which is derived from α/β purple bacteria and is characterized by a reduced affinity for O_2 and thus a reduced potential for oxygenase activity (Badger *et al.* 1998). Both subunits of the Red Form I of RUBISCO in diatoms are encoded on the plastid genome.

Diatoms utilize HCO_3^- by employing carbon concentrating mechanisms (CCMs) to convert HCO_3^- to CO_2. CCMs typically consist of HCO_3^- and CO_2 transporters and carbonic anhydrases (CA), which are enzymes that reversibly convert HCO_3^- to CO_2. The types and localization of CAs appear to differ between *T. pseudonana* and *P. tricornutum*. The only CA described in *P. tricornutum* to date was identified biochemically and found to be a Zn binding β-type CA (Satoh *et al.* 2001) localized to the plastid (Tanaka *et al.* 2005). There, the CA converts HCO_3^- to CO_2 which is then delivered directly to RUBISCO (Tanaka *et al.* 2005). A similar β-type CA has not been identified in *T. pseudonana*. Thus far, two different CAs (TWCA1 and CDCA1), originally identified biochemically in *T. weissflogii* (Lane *et al.* 2005; Lane and Morel 2000; Roberts *et al.* 1997), were identified in the *T. pseudonana* genome. TWCA1 binds zinc, but has an active site structure more similar to mammalian α-CAs than to plant β-CAs (Cox *et al.* 2000). CDCA1 binds cadmium and does not appear related to proteins identified in other organisms (Lane *et al.* 2005). Neither TWCA1 nor CDCA1 appear to possess plastid transit peptides, so these enzymes are not predicted to localize to the diatom plastid although confirmation of this possibility requires further work. Further biochemical and *in silico* analyses will also be required to confirm potential differences in the suite of CAs present in different diatom lineages (So *et al.* 2004).

The possibility of a C4-like version of a CCM has been explored in a few species of diatoms, primarily *Thalassiosira* species. In C4 photosynthesis, CO_2 is stored as a four-carbon intermediate (C4) that is transported to the plastid where it is decarboxylated to deliver CO_2 to the active site of

RUBISCO. The C4 pathway is present in about 3% of vascular plants and is believed to have evolved repeatedly via gene duplication in C3 plants during times of low atmospheric CO_2 levels (Monson 2003). Centric diatoms diversified during a time when atmospheric CO_2 concentrations were relatively low (Reinfelder *et al.* 2000) and thus it is possible they evolved a C4-like photosynthetic mechanism that differs from that of land plants and more recently evolved diatom lineages.

C4 photosynthesis occurs in plants that can divide functions between cell types: for example, one cell type uses the enzyme phosphoenolpyruvate carboxylase (PEPC) to convert HCO_3^- to malic or aspartic acid and another cell type decarboxylates the CO_2 proximal to RUBISCO via the activity of phosphoenolpyruvate kinase (PEPCK) or malic enzymes (reviewed in Westhoff and Gowik 2004). For a C4-like pathway to function in a unicellular organism the formation of C4 compounds, their subsequent decarboxylation to CO_2, and concentration of CO_2 around RUBSICO must all occur within the same cell. In *T. weissflogii*, inhibition of PEPC activity results in a decrease in photosynthetic O_2 evolution, and the storage of organic carbon intermediates prior to their fixation in the Calvin cycle (Reinfelder *et al.* 2004). This result has been taken as evidence for a C4-like pathway in *T. weissflogii* because this response is not observed in *Chlamydomonas*, an alga known to carry out C3 photosynthesis. The *T. pseudonana* genome encodes important C4 enzymes such as PEPC and PEPCK (Reinfelder *et al.* 2004). However, these enzymes participate in both C3 and C4 photosynthetic pathways so their *in silico* detection in *T. pseudonana* does not reveal C4-specific function. Furthermore, the two copies of PEPC found in *T. pseudonana* lack a conserved serine residue in a region of PEPC that contains major determinants for saturation kinetics in C4 proteins (Westhoff and Gowik 2004). It is important to note that *in silico* analyses of whole genomes generate hypotheses about potential processes carried out within a cell. Confirmation of a C4-like version of a CCM in diatoms, for example, will require: determination of the subcellular localization of the CCM enzymes and whether another decarboxylase acts in tandem with PEPCK; identification of the C4

metabolite transported from the cytosol to the chloroplast; and identification of the structural features that prevent leakage of CO_2 (Granum *et al.*, 2005). Future exploration of C4 photosynthesis in diatoms will help determine whether this ability influences species composition of phytoplankton assemblages and may explain how diatoms will cope with changes in atmospheric CO_2 concentrations.

Future considerations

Genomic information thus far available for diatoms is largely derived from the two most closely-related classes of diatoms, the Mediophyceae and the Bacillariophyceae, which represent a single monophyletic clade diverged from the more ancient Coscinodiscophyceae (Medlin and Kaczmarska 2004). Expanding genome sequencing to the Coscinodiscophyceae would provide opportunities for new insights into the evolution of diatoms through identification of processes that have been conserved, lost, and diversified in different lineages. For example most of our knowledge of silica metabolism comes from studies with representatives of the Bacillariophyceae, the most derived class of diatoms; most of our knowledge of carbon concentrating mechanisms comes from studies with representatives of the Mediophyceae. Comparative genomic analyses are beginning to highlight distinctive features of different diatom lineages and emphasize the need for caution in generalizing diatom features. For example, there is intriguing evidence that the Coscinodiscophyceae may possess unique silica-bound polypeptides that are likely involved in directing silica precipitation (Kroeger *et al.* 2000).

Four candidate genera within the Coscinodiscophyceae to consider for future sequencing projects are *Rhizosolenia, Stephanopyxis, Melosira,* and *Coscinodiscus.* Representative species from these genera have been used in laboratory studies of silica valve morphogenesis, machinery and nanopatterning (Van De Meene and Pickett-Heaps 2004; Sumper 2002; Li and Volcani 1985), nitrogen metabolism (Fisher and Cowdell 1982; Boyd and Gradmann 1999b; Carpenter *et al.* 1972; Joseph and Villareal 1998), carbon acquisition and photosynthesis (Tortell *et al.* 1997; Ming and Stephens 1985; Burkhardt *et al.* 1999), electrophysiology (Boyd and Gradmann 1999a; Gradmann and Boyd 1999), and reproductive biology (Nagai *et al.* 1995; Lin and Fu 1992; Jewson 1993). In addition, species from these genera can all be readily recognized within the marine plankton, which would facilitate examination of both laboratory isolates and field populations.

As more diatoms are slated for whole genome sequencing, even more emphasis needs to be placed on developing forward and reverse genetics in these organisms. No members of the Coscinodiscophyceae or Mediophycea and only a few members of the Bacillariophyceae can be consistently transformed with heterologous DNA constructs. Transformation efficiencies are currently relatively low (11–36 transformants per 10^7 cells), although concerted efforts are now being made to improve efficiencies (Poulsen and Kroger 2005). The complete sexual cycle of few species has been described (Chepurnov *et al.* 2004) and because diatoms are diploid, the generation of mutant lines is still in its infancy. It should also be noted that the sequenced isolates of *T. pseudonana* and *P. tricornutum* have been in culture for almost 50 years. It is to be expected that these isolates are now well-adapted to laboratory existence, but have likely lost, through accumulated mutations, attributes found in isolates more recently retrieved from nature. Thus, additional effort should be placed on obtaining and examining newly-isolated strains of *T. pseudonana* and *P. tricornutum*, in addition to the use of fresh isolates for future genome sequencing projects.

Acknowledgements

We thank P. van Dassow and K. Holterman for providing light micrographs of diatoms and M. Parker and T. Mock for providing helpful comments on the manuscript.

References

Aparicio, S., Chapman, J., Stupka, E., *et al.* (2002). Whole-genome shotgun assembly and analysis of the genome of *Fugu rubripes*. *Science*, **297**, 1301–1310.

Apt, K. E., Zaslavkaia, L., Lippmeier, J. C., *et al.* (2002). In vivo characterization of diatom multipartite plastid targeting signals. *J. Cell Sci.*, **115**, 4061–4069.

Archibald, J. M. and Keeling, P. J. (2004). The evolutionary history of plastids: a molecular phylogenetic perspective. In R. P. Hirt and D. S. Horner, ed. *Organelles, genomes and eukaryote phylogeny*, Vol. 68. CRC Press, Boca Raton, pp. 55–73.

Arkhipova, I. and Meselson, M. (2000). Transposable elements in sexual and ancient asexual taxa. *Proc. Natl. Acad. Sci., USA*, **97**, 14473–14477.

Armbrust, E. V., Berges, J. A., Bowler, C., *et al.* (2004). The genome of the diatom *Thalassiosira pseudonana*: Ecology, evolution, and metabolism. *Science*, **306**, 79–86.

Badger, M. R., Andrews, T. J., Whitney, S. M., *et al.* (1998). The diversity and coevolution of Rubisco, plastids, pyrenoids, and chloroplast-based CO_2-concentrating mechanisms in algae. *Can. J. Bot.*, **76**, 1052–1071.

Baldauf, S. L., Roger, A. J., Wenk-Siefert, I., and Doolittle, W. F. (2000). A kingdom-level phylogeny of eukaryotes based on combined protein data. *Science*, **290**, 972–977.

Bhaya, D. and Grossman, A. (1991). Targeting proteins to diatom plastids involves transport through an endoplasmic-reticulum. *Mol. Gen. Genet.*, **229**, 400–404.

Boyd, C. and Gradmann, D. (1999a). Electrophysiology of the marine diatom *Coscinodiscus wailesii*. I. Endogenous changes of membrane voltage and resistance. *J. Exp. Bot.*, **50**, 445–452.

Boyd, C. and Gradmann, D. (1999b). Electrophysiology of the marine diatom *Coscinodiscus wailesii*. III. Uptake of nitrate and ammonium. *J. Exp. Bot.*, **50**, 461–467.

Burkhardt, S., Riebesell, U., and Zondervan, I. (1999). Stable carbon isotope fractionation by marine phytoplankton in response to daylength, growth rate, and CO_2 availability. *Mar. Ecol. Prog. Ser.*, **184**, 31–41.

Carpenter, E., Remsen, C., and Schroeder, B. (1972). Comparison of laboratory and in situ measurements of urea decomposition by a marine diatom. *J. Exp. Mar. Biol. Ecol.*, **8**, 259–264.

Cavalier-Smith, T. (1998). A revised six-kingdom system of life. *Biol. Rev.*, **73**, 203–266.

Cha, J. N., Shimizu, K., Zhou, Y., *et al.* (1999). Silicatein filaments and subunits from a marine sponge direct the polymerization of silica and silicones in vitro. *Proc. Natl. Acad. Sci. USA*, **96**, 361–365.

Chepurnov, V. A., Mann, D. G., Sabbe, K., and Vyverman, W. (2004). Experimental studies on sexual reproduction in diatoms. *Int. Rev. Cytol.*, **237**, 91–154.

Colman, B., Huertas, I. E., Bhatti, S., and Dason, J. S. (2002). The diversity of inorganic carbon acquisition mechanisms in eukaryotic microalgae. *Funct. Plant Biol.*, **29**, 261–270.

Consortium, I. H.G. S. (2001). Initial sequencing and analysis of the human genome. *Nature*, **409**, 860–921.

Cox, E. H., McLendon, G. L., Morel, F. M. M., *et al.* (2000). The active site structure of Thalassiosira weissflogii carbonic anhydrase 1. *Biochemistry*, **39**, 12128–12130.

Damste, J. S. S., Muyzer, G., Abbas, B., *et al.* (2004). The rise of the rhizosolenid diatoms. *Science*, **304**, 584–587.

Dehal, P., Satou, Y., Campbell, R. K., *et al.* (2002). The draft genome of *Ciona intestinalis*: Insights into chordate and vertebrate origins. *Science*, **298**, 2157–2167.

Drum, R. W. and Gordon, R. (2003). Star Trek replicators and diatom nanotechnology. *Trends Biotech.*, **21**, 325–328.

Edlund, M. B. and Stoermer, E. F. (1997). Ecological, evolutionary, and systematic significance of diatom life histories. *J. Phycol.*, **33**, 897–918.

Falciatore, A., d'Alcala, M. R., Croot, P., and Bowler, C. (2000). Perception of environmental signals by a marine diatom. *Science*, **288**, 2363–2366.

Fan, C., Glibert, P. M., Alexander, J., and Lomas, M. W. (2003). Characterization of urease activity in three marine phytoplankton species, *Aureococcus anophagefferens*, *Prorocentrum minimum*, and *Thalassiosira weissflogii*. *Mar. Biol.*, **142**, 949–958.

Field, C. B., Behrenfeld, M. J., Randerson, J. T., and Falkowski, P. (1998). Primary production of the biosphere: integrating terrestrial and oceanic components. *Science*, **281**, 237–240.

Fisher, N. and Cowdell, R. (1982). Growth of marine planktonic diatoms on inorganic and organic nitrogen. *Marine Biol.*, **72**, 147–155.

Gibbs, S. P. (1979). Route of entry of cytoplasmically synthesized proteins into chloroplasts of algae possessing chloroplast-ER. *J. Cell Sci.*, **35**, 253–266.

Gordon, R. and Parkinson, J. (2005). Potential roles for diatomists in nanotechnology. *J. Nanosci. Nanotechnol.*, **5**, 35–40.

Gradmann, D. and Boyd, C. (1999). Electrophysiology of the marine diatom *Coscinodiscus wailesii*. II. Potassium currents. *J. Exp. Bot.*, **50**, 453–459.

Granum, E., Raven, J. A., and Leegood, R. C. How do marine diatoms fix ten billion tonnes of inorganic carbon per year? *Can. J. Bot. Can. J. Bot.*, **83**, 898–908.

Harper, J. T., Waanders, E., and Keeling, P. J. (2005). On the monophyly of chromalveolates using a six-protein phylogeny of eukaryotes. *Int. J. Syst. Evol. Microbiol.*, **55**, 487–496.

Hildebrand, M. (2005). Cloning and functional characterization of ammonium transporters from the marine diatom *Cylindrotheca fusiformis* (Bacillariophyceae). *J. Phycol.*, **41**, 105–113.

Hildebrand, M., Volcani, B. E., Gassman, W., and Schroeder, J. I. (1997). A gene family of silicon transporters. *Nature*, **385**, 688–689.

International Human Genome Sequencing Consortium. (2001). Initial sequencing and analysis of the human genome. *Nature*, 409, 860–921

Jeffery, B., Barlow, T., Moizer, K., Paul, S., and Boyle, C. (2004). Amnesic shellfish poison. *Food Chem. Toxicol.*, **42**, 545–557.

Jewson, D. H. (1993). Size and sexual reproduction of the centric diatom *Melosira varians* C. A. Ag. in Lough Neagh. *Irish Naturalists' J.*, **24**, 253–255.

Joseph, L. and Villareal, T. (1998). Nitrate reductase activity as a measure of nitrogen incorporation in *Rhizosolenia formosa* (H. Peragallo): Internal nitrate and diel effects. *J. Exp. Mar. Biol. Ecol.*, **229**, 159–176.

Karp-Boss, L. and Jumars, P. A. (1998). Motion of diatom chains in steady shear flow. *Limnol. Oceanogr.*, **43**, 1767–1773.

Kilian, O. and Kroth, P. G. (2004). Presequence acquisition during secondary endocytobiosis and the possible role of introns. *J. Mol. Evol.*, **58**, 712–721.

Kilian, O. and Kroth, P. G. (2005). Identification and characterization of a new conserved motif within the presequence of proteins targeted into complex diatom plastids. *Plant J.*, **41**, 175–183.

Kociolek, J. P. and Stoermer, E. F. (1989). Chromosome numbers in diatoms: a review. *Diatom Research*, **4**, 47–54.

Kooistra, W. H. C. F. and Medlin, L. K. (1996). Evolution of the diatoms (Bacillariophyta) IV. A reconstruction of their age from small subunit rRNA coding regions and the fossil record. *Mol. Phylo. Evol.*, **3**, 391–407.

Kroeger, N., Bergsdorf, C., and Sumper, M. (1994). A new calcium binding glycoprotein family constitutes a major diatom cell wall component. *EMBO J.*, **13**, 4676–4683.

Kroeger, N., Deutzmann, R., Bergsdorf, C., and Sumper, M. (2000). Species-specific polyamines from diatoms control silica morphology. *Proc. Natl. Acad. Sci. USA*, **97**, 14133–14138.

Kroeger, N., Deutzmann, R., and Sumper, M. (1999). Polycationic peptides from diatom biosilica that direct silica nanosphere formation. *Science*, **286**, 1129–1132.

Kroeger, N., Lorenz, S., Brunner, E., and Sumper, M. (2002). Self-assembly of highly phosphorylated silaffins and their function in biosilica morphogenesis. *Science*, **298**, 584–586.

Kroeger, N. and Wetherbee, R. (2000). Pleuralins are involved in theca differentiation in the diatom *Cylindrotheca fusiformis*. *Protist*, **151**, 263–273.

Kroth, P. G. (2002). Protein transport into secondary plastids and the evolution of primary and secondary plastids. In, Jeon, K. ed. *International Review Of Cytology – A Survey Of Cell Biology*, Vol 221. Academic Press, San Diego, pp. 191–255.

Lai, Z., Jing, J., Aston, C., *et al.* (2003). A shotgun optical map of the entire *Plasmodium falciparum* genome. *Nat. Gen.*, **23**, 309–313.

Lane, T. W. and Morel, F. M. M. (2000). Regulation of carbonic anhydrase expression by zinc, cobalt, and carbon dioxide in the marine diatom *Thalassiosira weissflogii*. *Plant Physiol.*, **123**, 345–352.

Lane, T. W., Saito, M. A., George, G. N., Pickering, I. J., Prince, R. C., and Morel, F. M. M. (2005). A cadmium enzyme from a marine diatom. *Nature*, **435**, 42.

Lea, P. and Miflin, B. (1980). Transport and metabolism of asparagine and other nitrogen compounds within the plant. In P. Stumpt and E. Conn, ed. *The Biochemistry of Plants*. Academic Press, New York, pp. 569–607.

Li, C.-W. and Volcani, B. (1985). Studies on the biochemistry and fine structure of silica shell formation in diatoms. 10. Morphogenesis of the labiate process in centric diatoms. *Protoplasma*, **124**, 147–156.

Lin, J. and Fu, J. (1992). Sexual reproduction in two species of genus Rhizosolenia. *J. Oceanography in Taiwan Strait*, **11**, 49–53.

Lomas, M. W. and Gilbert, P. M. (2000). Comparisons of nitrate uptake, storage, and reduction in marine diatoms and flagellates. *J. Phycol.*, **36**, 903–913.

Lopez, P. J., Descles, J., Allen, A. E., and Bowler, C. (2005). Prospects in diatom research. *Curr. Opin. Biotechnol.*, **16**, 180–186.

Maheswari, U., Montsant, A., Goll, J., *et al.* (2005). The Diatom EST database. *Nucleic Acids Res.*, **33**, D344–D347.

Mann, D. G. and Droop, S. J. M. (1996). Biodiversity, biogeography and conservation of diatoms. *Hydrobiologia*, **336**, 19–32.

Marini, A.-M., Soussi-Boudekou, S., Vissers, S., and Andre, B. (1997). A family of ammonium transporters in *Saccharomyces cerevisiae*. *Mol. Cell Biol.*, **17**, 4282–4293.

McFadden, G. I. (2001). Primary and secondary endosymbiosis and the origin of plastids. *J. Phycol.*, **37**, 951–959.

Medlin, L. K. and Kaczmarska, I. (2004). Evolution of the diatoms: V. Morphological and cytological support for the major clades and a taxonomic revision. *Phycologia*, **43**, 245–270.

Mefford, H. C. and Trask, B. J. (2002). The complex structure and evolution of human subtelomeres. *Nature Rev. Gen.*, **3**, 95–102.

Ming, L. and Stephens, G. (1985). Uptake of free amino acids by the diatom, *Melosira mediocris*. *Hydrobiologia*, **128**, 2.

Mock, T., Krell, A., Gloeckner, G., Kolukisaoglu, U., and

Valentin, K.. Analysis of expressed sequence tags (ESTs) from the polar diatom *Fragilariopsis cylindrus*. *J. Phycol*, **42**, 78–85.

Monson, R. K. (2003). Gene duplication, neofunctionalization, and the evolution of C-4 photosynthesis. *Int. J. Plant Sci.*, **164**, S43–S54.

Montsant, A., Jabbari, K., Maheswari, U., and Bowler, C. (2005). Comparative genomics of the pennate diatom Phaeodactylum tricornutum. *Plant Physiol.*, **137**, 500–513.

Moshkovitz, S., Ehrlich, A., and Soudry, D. (1983). Siliceous microfossils of the upper cretaceous mishash formation, Central Negev. *Israel Cret. Res.*, **4**, 73–194.

Nagai, S., Hori, Y., Manabe, T., and Imai, I. (1995). Restoration of cell size by vegetative cell enlargement in *Coscinodiscus wailesii* (Bacillariophyceae). *Phycologia*, **34**, 533–535.

Nelson, D. M., Tréguer, P., Brzezinski, M. A., Leynaert, A., and Quéguiner, B. (1995). Production and dissolution of biogenic silica in the ocean: revised global estimates, comparison with regional data and relationship to biogenic sedimentation. *Glob. Biogeochem. Cycle*, **9**, 359–372.

Oliveira, L. and Antia, N. J. (1986). Some observations on the urea-degrading enzyme of the diatom *Cyclotellacryptica* and the role of nickel in its production. *J. Plankton Res.*, **8**, 235–242.

Pahlow, M., Riebesell, U., and Wolf-Gladrow, D. A. (1997). Impact of cell shape and chain formation on nutrient acquisition by marine diatoms. *Limnol. Oceanogr.*, **42**, 1660–1672.

Parkinson, J. and Gordon, R. (1999). Beyond micromachining: the potential of diatoms. *Trends Biotechnol.*, **17**, 190–196.

Peers, G. S., Milligan, A. J., and Harrison, P. J. (2000). Assay optimization and regulation of urease activity in two marine diatoms. *J. Phycol.*, **36**, 523–528.

Poulsen, N. and Kroeger, N. (2004). Silica morphogenesis by alternative processing of silaffins in the diatom *Thalassiosira pseudonana*. *J. Biol. Chem.*, **279**, 42993–42999.

Poulsen, N. and Kroger, N. (2005). A new molecular tool for transgenic diatoms. Control of mRNA and protein biosynthesis by an inducible promoter-terminator cassette. *FEBS J.*, **272**, 3413–3423.

Price, N. M. and Harrison, P. J. (1988). Uptake of urea-C and urea-N by the coastal marine diatom *Thalassiosirapseudonana*. *Limnol. Oceanogr.*, **33**, 528–537.

Quintero, M. J., Muro-Pastor, A. M., Herrero, A., and Flores, E. (2000). Arginine catabolism in the cyanobacterium *Synechocystis* sp. strain PCC 6803 involves the urea cycle and arginase pathway. *J. Bacteriol.*, **182**, 1008–1015.

Reinfelder, J. R., Kraepiel, A. M. L., and Morel, F. M. M. (2000). Unicellular C-4 photosynthesis in a marine diatom. *Nature*, **407**, 996–999.

Reinfelder, J. R., Milligan, A. J., and Morel, F. M. M. (2004). The role of the C-4 pathway in carbon accumulation and fixation in a marine diatom. *Plant Physiol.*, **135**, 2106–2111.

Reumann, S., Inoue, K., and Keegstra, K. (2005). Evolution of the general protein import pathway of plastids (Review). *Mol. Membr. Biol.*, **22**, 73–86.

Roberts, S. B., Lane, T. W., and Morel, F. M. M. (1997). Carbonic anhydrase in the marine diatom *Thalassiosira weissflogii* (Bacillariophyceae). *J. Physiol.*, **33**, 845–850.

Rothpletz, A. (1896). über die Flysch-Fucoiden und einige andere fossile Algen, sowie über liasische, Diatomeen führende Horenschwämme. *Z. Dsch. Geol. Ges.*, **48**, 854–914.

Round, F. E., Crawford, R. M., and Mann, D. G. (1990). *The Diatoms*. Cambridge University Press, Cambridge, p. 747.

Rynearson, T. A. and Armbrust, E. V. (2005). Maintenance of clonal diversity during a spring bloom of the centric diatom *Ditylum brightwellii. Mol. Ecol.*, **14**, 1631–1640.

Satoh, D., Hiraoka, Y., Colman, B., and Matsuda, Y. (2001). Physiological and molecular biological characterization of intracellular carbonic anhydrase from the marine diatom *Phaeodactylum tricornutum. Plant Physiol.*, **126**, 1459–1470.

Shimizu, K., Cha, J., Stucky, G. D., and Morse, D. E. (1998). Silicatein α: cathepsin L-like protein in sponge biosilica. *Proc. Natl. Acad. Sci. USA*, **95**, 6234–6238.

So, A. K.-C., Espie, G. S., Williams, E. B., Shively, J. M., Heinhorst, S., and Cannon, G. C. (2004). A novel evolutionary lineage of carbonic anhydrase (ε class) is a component of the carboxysome shell. *J. Bacteriol.*, **186**, 623–630.

Sumper, M. (2002). A phase separation model for the nanopatterning of diatom biosilica. *Science*, **295**, 2430–2433.

Takabayashi, M., Wilkerson, F. P., and Robertson, D. (2005). Response of glutamine synthetase gene transcription and enzyme activity to external nitrogen sources in the diatom *Skeletonema costatum* (Bacillariophyceae). *J. Phycol.*, **41**, 84–94.

Tanaka, Y., Nakatsuma, D., Harada, H., Ishida, M., and Matsuda, Y. (2005). Localization of soluble beta-carbonic anhydrase in the marine diatom *Phaeodactylum tricornutum*. Sorting to the chloroplast and cluster formation on the girdle lamellae. *Plant Physiol.*, **138**, 207–217.

Thamatrakoln, K. and Hildebrand, M. (2005). Approaches for functional characterization of diatom silicic acid transporters. *J. Nanosci. Nanotech.*, **5**, 1–9.

Tortell, P. D., Reinfelder, J. R., and Morel, F. M. M. (1997). Active uptake of bicarbonate by diatoms. *Nature*, **390**, 243–244.

Urquhart, A. A. and Joy, K. W. (1981). Use of phloem exudate technique in the study of amino-acid-transport in pea-plants. *Plant Physiol.*, **68**, 750–754.

Van De Meene, A. and Pickett-Heaps, J. (2004). Valve morphogenesis in the centric diatom *Rhizosolenia setigera* (Bacillariophyceae, Centrales) and its taxonomic implications. *Eur. J. Phycol.*, **39**, 93–104.

Van de Peer, Y. and De Wachter, R. (1997). Evolutionary relationships among the eukaryotic crown taxa taking into account site-to-site rate variation in 18S rRNA. *J. Mol. Evol.*, **45**, 619–630.

van Dooren, G. G., Schwartzbach, S. D., Osafune, T., and McFadden, G. I. (2001). Translocation of proteins across the multiple membranes of complex plastids. *Biochim. Biophys. Acta-Mol. Cell Res.*, **1541**, 34–53.

Westhoff, P. and Gowik, U. (2004). Evolution of C4 phosphoenolpyruvate carboxylase. Genes and proteins: a case study with the genus *Flaveria. Ann. Bot. London*, **93**, 13–23.

Wright, S. and Finnegan, D. (2001). Genome evolution: Sex and the transposable element. *Curr. Biol.*, **11**, 296–299.

Zaslavskaia, L. A., Lippmeier, J. C., Shih, C., Ehrhardt, D., Grossman, A. R., and Apt, K. E. (2001). Trophic conversion of an obligate photoautotrophic organism through metabolic engineering. *Science*, **292**, 2073–2075.

Zeyl, C., Bell, G., and Green, D. (1996). Sex and the spread of retrotransposon Ty3 in experimental populations of *Saccharomyces cerevisiae. Genetics*, **143**, 1567–1577.

CHAPTER 15

The *Dictyostelium* genome – a blueprint for a multicellular protist

Pauline Schaap

School of Life Sciences, University of Dundee, UK

Introduction

The year 2005 saw the completion of the genome sequencing project of *Dictyostelium discoideum*, the first non-pathogenic protist to be completely sequenced (Eichinger *et al.* 2005). What is *Dictyostelium's* place among the protists and why is knowledge of its genome important? Compared to the morphological variety of macroscopic life forms that can be observed by the naked eye, the much greater genetic diversity of microscopic life is usually unappreciated. However, it is from these humble origins that multicellular life was independently generated many times over.

The dictyostelids are a remarkable group of organisms that can alternate long periods of life as solitary cells with episodes of participation in a multicellular assembly (Kessin 2001). This assembly displays a repertoire of behaviors, such as co-ordinated cell movement, differential cell adhesion, cell-type specialization, and organization into tissues, that bears striking similarity to the ontogeny of animal form. Their easy culture, genetic tractability, and experimental accessibility provide unique possibilities to unravel the processes that underlie these behaviors at the molecular level. In recent years, *Dictyostelium* has also become a popular system for the study of infection by bacterial pathogens (Solomon and Isberg 2000), the evolution of social behavior (Strassmann *et al.* 2000; Queller *et al.* 2003; Foster *et al.* 2004), and the evolution of developmental strategies (Alvarez-Curto *et al.* 2005). Its relatedness and partially shared life-style with the solitary amoebas allows

us to trace how pathways that originally functioned in environmental sensing became adapted for intercellular signaling and allowed the organism to evolve to increasing structural and behavioral complexity.

What is the place of *Dictyostelium* in the living world? Based on morphological and biochemical criteria, all living organisms were subdivided into five kingdoms in the past century (Margulis *et al.* 1999). All protists were classified as a single kingdom, separate from the prokaryotic monera, and the other eukaryote kingdoms of plants, animals, and fungi. The progressive use of gene or protein sequence data to reconstruct phylogenetic relationships now gives us a completely different view. There are three domains of life, the bacteria, archaea, and eukaryotes (Woese *et al.* 1990). The eukaryotes can be subdivided into at least eight major clades, each of which contains several groups of protists (Fig. 15.1). Multicellular organisms appeared independently in at least five of these major clades (Baldauf *et al.* 2000; Baldauf 2003; Simpson and Patterson, Chapter 1, this book). The dictyostelids are members of the amoebozoans, which share their clade with the animals and fungi. The amoebozoans also comprise the syncytial slime molds and the solitary lobose amoebas (Baldauf and Doolittle 1997). A recently constructed molecular phylogeny of most known dictyostelid species shows that the previous morphology-based subdivision in three genera, dictyostelids, acytostelids, and polysphondylids (Raper 1984), is incorrect. Instead most dictyostelids

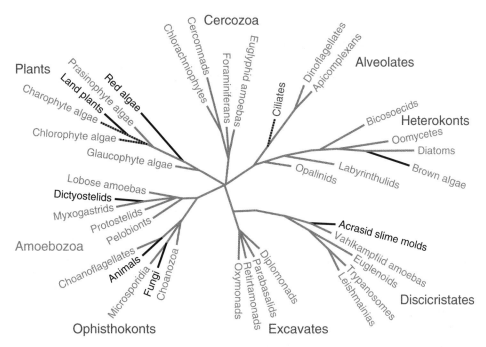

Figure 15.1 Phylogeny of the eukaryotes based on multiple protein sequences. Schematically redrawn from Baldauf, (2003), Fig. 1. The eukaryotes can be subdivided into eight major group. The unicellular protists (in grey) show the greatest genetic variety and are present in all groups. Multicellular organisms (in black) or groups with some colonial or multicellular taxa (dotted line) evolved independently from highly diverse protist ancestors.

can be placed into one of four major groups, with the acytostelids and most polysphondylids in group 2 and *D. discoideum* in the most derived group 4 (Schaap *et al.*, submitted).

Dictyostelid survival strategies

Most dictyostelids display three different responses to nutrient deprivation. The three most basal groups have retained the strategy of their solitary ancestors to encyst individually. A second strategy is the formation of sexual macrocysts. Here cells fuse to form a zygote, which then chemotactically attracts and cannibalizes a large number of its starving brethren. The zygote subsequently lays down a thick, multilayered cell wall and enters a period of dormancy (Fig. 15.2). Many species only produce macrocysts when opposite mating types are combined, but some also produce homothallic macrocysts. Microcysts and macrocysts are usually formed under conditions of high humidity and

darkness that are unfavorable for the formation of fruiting bodies (Raper 1984).

During asexual fruiting body formation, starving amoebae attract each other by secreting a chemoattractant. In the case of *D. discoideum*, the chemoattractant is cAMP, which is produced in an oscillatory manner by the aggregation center and relayed by surrounding cells (Fig. 15.3). Once aggregated, the cellular mound transforms into a slug-shaped structure, which, guided by light and warmth, moves to the top layer of the soil to find an optimal spot for spore dispersal. Meanwhile, the majority of the cells differentiate into spore-cell precursors, while a smaller proportion of cells enter the stalk-cell differentiation pathway. The two cell types first form at random, but the prestalk cells then move towards the top of the mound. Upon initiation of fruiting body formation, the prestalk cells first synthesize a central cellulose tube and then move into this tube. While doing so they differentiate into highly vacuolated stalk cells. The prespore cells follow the prestalk cells up the

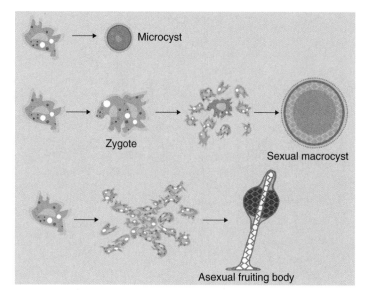

Figure 15.2 The three alternative life cycles of the dictyostelids. During microcyst formation starving amoebae encyst as individuals, as also occurs during starvation of solitary amoebozoans. Macrocyst formation is initiated by sexual fusion of two starving cells, followed by cannibalism of attracted surrounding cells and finally encystation. During asexual fruiting body formation, cells aggregate and co-operate to build a fruiting structure.

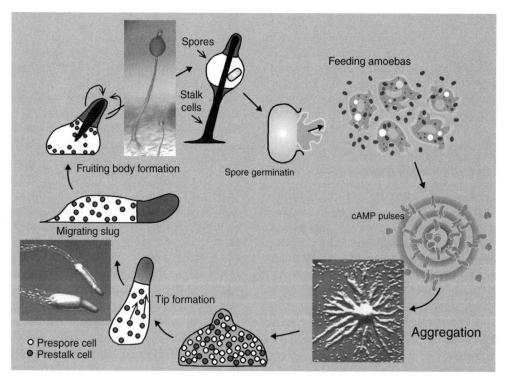

Figure 15.3 Asexual fruiting body formation in *Dictyostelium discoideum*. Co-ordinated cell movement and cell-type specialization combine to generate migrating slugs and well-proportioned fruiting structures from dispersed starving cells.

stalk. They mature into spores by loosing water and by constructing a thick individual wall from prepackaged spore coat materials.

Dictyostelium as a research tool

What attracts researchers to *D. discoideum*? Initially, probably its uncanny capacity for self-organization. Pivotal work by Raper highlighted that dictyostelids showed many of the aspects of embryonic development. He demonstrated that similar to Spemann's organizer of gastrulation, the *D. discoideum* slug has an organizer that specifies its polarity and controls morphogenesis. He also showed that, similar to vertebrate embryos, slugs can accurately regulate the proportions of their constituent cell types (Raper 1940). Sussman explored the potential of *D. discoideum* as a genetic model system in the 1950s and 1960s (Sussman 1955; Sussman and Brackenbury 1976). Konijn identified its chemoattractant as cAMP (Konijn *et al.* 1967), the well-known intermediate for hormone action in humans (Sutherland and Rall 1958) and this boosted research into cell signaling and directional cell movement. Essential molecular tools, such as gene transformation, gene disruption, and tagged mutagenesis, were developed in the 1980s and 1990s (Nellen *et al.* 1984; De Lozanne and Spudich 1987; Kuspa and Loomis 1992) and contributed enormously to its current popularity.

History of *Dictyostelium* genome sequencing

Japanese *Dictyostelium* researchers were the great forerunners in the combined efforts that eventually presented the field with a fully sequenced genome. They started in 1996 with the sequencing of whole expressed genes from different developmental stages (Morio *et al.* 1998; Urushihara 2002; Urushihara *et al.* 2004). An occidental consortium for genome sequencing was formed in 1998 between the University of Cologne, Germany, the Institute of Molecular Biotechnology in Jena, Germany, the Baylor College of Medicine in Houston, USA, and the Wellcome Trust Sanger Institute Centre in Hinxton, UK. The raw cDNA and genome sequencing data were rapidly made available to the scientific community and have been avidly screened and used by many workers both inside and outside the *Dictyostelium* field. Needless to say, both the cDNA project and the genome project have truly revolutionized *Dictyostelium* research.

The chosen genomic sequencing strategy was to perform shotgun sequencing on separated chromosomes, because *Dictyostelium's* high A + T bias made long-insert bacterial clones unstable. The PCR-based strategy of HAPPY mapping was used to assemble stretches of contiguous DNA sequence into chromosomes (Dear and Cook 1993). The sequence of the first and longest of the six chromosomes, chromosome 2, was published in 2002 (Glockner *et al.* 2002), to be followed by the sequence of the entire genome 3 years later (Eichinger *et al.* 2005). The assembled and annotated *D. discoideum* genome sequence is available at the Dictybase website (http://dictybase.org/).

Genome structure

The assembled *Dictyostelium* genome measures 34 Mb in size, subdivided over six chromosomes which range from 3.6 to 8.5 Mb. A seventh chromosome-like structure consists of about 100 copies of 88 kb of palindromic DNA that carries ribosomal RNA genes (Sucgang *et al.* 2003). A master copy of these genes is also present on chromosome 4. The overall A + T content is 77.6%, but is particularly high in introns (88%) and intergenic regions (85%). Exons have a lower (73%) A + T bias. However, this is high enough to be notable in protein sequence composition, where amino acids that are encoded by GC-rich codons are underrepresented (Eichinger *et al.* 2005).

The gene density is high, with an average of one gene per 2.5 kb. The predicted number of protein-encoding genes is 13 541, but a conservative estimate reduces this to about 10 500 functional genes (Olsen 2005). Compared to the human genome with 23 000 genes and the *Drosophila* genome with 14 000 genes this seems high, particularly when it is also compared to the 5000 genes present in the protist *Plasmodium falciparum* and the fungus *Saccharomyces cerevisiae*. Its multicellular life-style cannot fully account for the high gene number of *Dictyostelium*, since the closely related pathogen

Entamoeba histolytica, that has a strictly unicellular life-style, also has 10 000 protein-encoding genes (Loftus *et al.* 2005).

Despite its compactness, the *Dictyostelium* genome has a high content of low complexity sequence in both the intergenic and protein-encoding regions. Apart from long homopolymeric stretches, these tend to take the form of tandem repeats of three, six, nine, or higher multiples of three in exons, and therefore yield repeated sequences of amino acids. Remarkably, such repeats are common to some groups of proteins, such as the protein kinases, but do not occur at all in others.

The genome is rich in clustered transposable elements, which do not show sequence-specific targets for insertion, but primarily insert within similar elements. Non-LTR (long terminal repeat) transposons commonly insert next to transfer RNA genes, while each chromosome carries a cluster rich in *Dictyostelium* Intermediate Repeat Sequence (DIRS)-type transposons near one terminus. During mitosis the DIRS-containing ends of chromosomes circle together, suggesting that they could function as centromeres.

The ends of the *Dictyostelium* chromosomes do not carry the characteristic G + T rich repeats that mark most eukaryotic telomers. Instead, telomers probably consist of complex repetitive sequences

and partial copies of the extrachromosomal rDNA element. The *Entamoeba* genome also lacks typical telomer sequences, which seem here to be replaced by arrays of tRNA genes (Eichinger *et al.* 2005; Loftus *et al.* 2005; Eichinger and Noegel 2005).

The characteristic features of the *Dictyostelium* genome are summarized in Table 15.1.

Genome content

The genome content of an organism is a present day snap-shot that reflects both its evolutionary history and its current life-style. We know little of the evolutionary history of the dictyostelids, but their phylogenetic association with the solitary lobose amoebas makes it likely that the ancestor of the dictyostelids was a solitary amoeba. As such, it would have alternated between two developmental stages – a trophic stage were it fed on bacteria or other protists, and a dormant cyst stage, which it entered when deprived of food or under other stress conditions. Possibly, there was also a sexual stage, the precursor of the dictyostelid macrocyst. Apart from the basic housekeeping genes, the protein repertoire of such an organism would consist of: (i) a signaling system that regulated dynamic remodeling of the cytoskeleton, which allowed cells to hunt and devour prey; (ii) a sensory system to detect stress and potential predators and process this information into short-term stress responses, such as avoidance or chemical warfare; (iii) a developmental program for long-term survival of persistent stress conditions, such as starvation, drought, and extremes of temperature; and (iv) quorum- and mate-sensing mechanisms could also be essential.

The cyst stage of the solitary amoebae, is, under the name microcyst, still a prevalent survival strategy of the more basal dictyostelids. In addition, two alternative life-styles emerged – the sexual macrocyst and the asexual fruiting body. Particularly the latter alternative requires a repertoire of processes that were never part of the solitary life-style. Firstly, to be able to aggregate, the amoebae need to produce chemoattractants. Secondly, intercellular adhesion and a flexible external matrix are required to keep the

Table 15.1 Characteristics of the *D. discoideum* genome

High (A + T) bias (77.6%)
High gene density (1 gene/2.5 kb)
Rich in simple sequence repeats (> 11% of bases)
Rich in transposable elements
Contains an 88-kb extrachromosomal element consisting of ribosomal RNA genes
Complex repeat/rDNA junctions may act as telomers
DIRS clusters (*Dictyostelium* Intermediate Repeat Sequence) may act as centromeres
Many proteins contain long homopolymeric amino acid tracts
18 putative instances of genes acquired by lateral transfer
Rich in ABC transporters and polyketide synthases
Rich in actin binding proteins and cell adhesion proteins
Rich in G-protein-coupled receptors
No receptor tyrosine kinases
Poor in universally conserved transcription factors

aggregated cell mass together but capable of undergoing shape changes. Thirdly, signaling molecules should be secreted or displayed in a time- and space- controlled manner to induce cell-type-specific differentiation. Finally, cells need to co-ordinate their metabolic activities to build a central cellulosic stalk tube.

What is the gene set that made this level of co-operation possible? Since no genome of the solitary dictyostelid ancestor is or most likely will ever be available, we will for the time being have to glean this information from comparison with genomes of remotely related organisms. However, the major source of information is as yet the *D. discoideum* genome itself.

Metabolism

The most ancient organisms and many of their present day descendents, such as the plants and fungi, can synthesize all their constituents from carbon, nitrogen, and phosphorus in low molecular weight compounds or minerals. However, grazers and predators such as animals and amoebozoans, which use organic matter as food, loose a significant proportion of this metabolic potential. *Dictyostelium* appears to lack enzymes to synthesize 10 amino-acids and five vitamins, but has all the enzymes required for glycolysis and for nucleotide metabolism. Most of the lost enzymes are also missing in animals. *Dictyostelium* also lacks several enzymes of the urea cycle that sequesters toxic ammonia in animals (Payne 2005). In *Dictyostelium*, ammonia is directly secreted and has come to serve a signaling function in the regulation of cell movement (Schindler and Sussman 1977; Bonner *et al.* 1988) and cell differentiation (Wang and Schaap 1989).

In addition to its enzymatic machinery for basic metabolism, *Dictyostelium* has an unusually large number (43) of polyketide synthases, which are also found in bacteria, fungi, and plants. These enzymes catalyze the stepwise condensation of simple carboxylic acid precursors into polyketide chains, which can be further cyclized and modified to yield secondary metabolites that are active as antibiotics, antifungal, or insecticidal agents (Anjard 2005). One enzyme produces DIF-1, a

signal molecule that regulates cell-type proportions (Thompson and Kay 2000). However, the role of the other enzymes is unclear and may range from other intercellular signaling functions to chemical warfare with other soil organisms.

Transport

The *Dictyostelium* genome encodes a large number of ABC-type transporters. These proteins are used by many eukaryotes to export harmful compounds from the cell and they are a major cause for the development of chemotherapeutic resistance in human cancer and pathogen-borne diseases. ABC transporters typically consist of one or two ATP-binding cassettes and one or two sets of six transmembrane helices. The proteins with a single set of each need to dimerize to form a functional transporter. The 66 *Dictyostelium* ABC transporters can be subdivided into the seven families that have been previously characterized (Dean *et al.* 2001; Eichinger *et al.* 2005; Anjard 2005). At least 20 of the genes are expressed during growth, where they may have roles in the export of endogenously produced toxic metabolites, or to create immunity from biocidal agents that are produced by other soil inhabitants. A subset of ABC transporters with attached serine protease domain, such as TagA, TagB, and TagC, generate peptides that play crucial roles in developmental signaling. One of the peptides is SDF-2, which triggers the maturation of spore cells (Shaulsky *et al.* 1995; Anjard *et al.* 1998; Good *et al.* 2003; Anjard and Loomis 2005). As is the case with the polyketide synthase multigene family, this use of transporters may represent another example of how a cellular defense system has been recruited during dictyostelid evolution to adopt a new role in multicellular development.

Cytoskeletal proteins

Similar to other soil amoebae that feed on bacteria, *Dictyostelium* amoebae are dependent on an efficient mechanism for directed cell locomotion and capture of prey. The aggregation process of single cells and the subsequent co-ordinated cell movements that form the slug and fruiting body demand additional sophistication of the mechanisms that control cell motility. This sophistication appears to be reflected in the huge number and variety of proteins that regulate cytoskeletal function. A similar variety with a comparable domain architecture is otherwise only found in animals, although a significant amount of domain reshuffling appears to have occurred during evolution of the metazoan and amoebazoan lineages. There are in total 138 proteins with probable or established functions in modification of the actin cytoskeleton, of which 71 are thus far unique to *D. discoideum*. This is not including the actin gene family itself, which has no less than 30 members, while there are also 11 genes that encode actin-related proteins.

Small GTP-ases of the Rho family are the most common upstream regulators of actin remodeling in eukaryotes. There are 15 different members of this family in *D. discoideum*, as well as over 80 RhoGEF or RhoGAP proteins that regulate the activity of the Rho-GTPases. Overall, these proteins are more similar to those in animals and fungi than in other eukaryotes. However, also here there are some metazoan families that are missing in *Dictyostelium* and some that are unique to *Dictyostelium* (Eichinger *et al.* 2005; Rivero and Eichinger 2005).

Signaling

Receptors

The sensory system of solitary cells is predominantly involved in scanning the environment for food, mates, or potential stress situations. A hallmark of multicellular life is its complete dependence on the ability of its component cells to signal to each other. Such signals instruct cells whether to divide, differentiate, or die, and co-ordinate the coherent cell movement and growth patterns that give the organism its specific shape. To understand the evolution of multicellularity, it is therefore crucial to trace the history of intercellular signaling.

In animals, G-protein-coupled receptors, ion-channel-linked receptors, and enzyme-linked receptors are the most common proteins for primary signal detection. For the latter class, the enzyme is most commonly a protein kinase that either targets tyrosine or serine/threonine

residues, but it can also be a guanylyl cyclase that produces the second messenger cGMP. In bacteria, plants, and fungi, histidine kinases are often linked to sensory receptors.

The *Dictyostelium* genome contains an entirely unanticipated wealth of 55 G-protein coupled receptors (GPCRs). For only four of those (cAR1–4), the ligand, cAMP, and developmental roles were previously known. There are another eight cAR-like receptors and four groups of receptors that share homology with four out of the six classes of metazoan GPCRs. The largest group contains 25 smoothened/frizzled type receptors (class 5), followed by 17 metabotropic glutamate/GABA$_B$ type receptors (class 3), and two secretin type receptors (class 2) (Vassilatis *et al.* 2003; Eichinger *et al.* 2005; Hereld 2005). The four *D. discoideum* cARs are the result of recent gene duplications (Alvarez-Curto *et al.* 2005) and this may also be the case for several of the class 3 and class 5 receptors. Like the cARs, the duplicated receptors will probably still bind the same ligand. However, even bearing this in mind, there could be at least 35 ligands with unknown roles in processes such as chemotaxis, quorum sensing, mating, and development yet to be identified.

GPCRs use heterotrimeric GTP-binding proteins as intermediates for signal transduction, which consist of separate α-, β-, and γ-subunits. The *Dictyostelium* genome contains 13 G-protein α-subunits, including the unusual Gα Spalten, which also harbors a protein phosphatase domain. There is most likely only a single Gβ- and a single Gγ-subunit, both of which were identified prior to the genome project (Lilly *et al.* 1993; Zhang *et al.* 2001). The *Dictyostelium* genome contains at least seven putative Regulators of G-protein Signaling (RGS). Mammalian RGS prokins regulate the activity of heterotrimeric G-proteins by activating the intrinsic GTP-ase activity of the Gα-subunit (Ross and Wilkie 2000). Only one of these proteins, RCK1, has been studied in detail; it translocates to the plasma membrane in response to cAMP stimulation and regulates the chemotactic response (Sun and Firtel 2003).

Ion-channel-linked receptors for extracellular ligands have never been reported in *Dictyostelium* and instead of receptor guanylyl cyclases,

Dictyostelium has one receptor adenylyl cyclase that functions as an osmosensor (Saran and Schaap 2004). The receptor tyrosine kinases that play such crucial roles in cell cycle control and cell differentiation in animals (van der Geer *et al.* 1994) and other monospecific tyrosine kinases are lacking, although there are over 60 tyrosine kinase like (TKL) enzymes, that resemble both tyrosine and serine/threonine-specific protein kinases (Kimmel 2005; Eichinger *et al.* 2005). Phosphorylated tyrosine residues typically bind to SH2 domains of effector proteins in animals. There are in total 12 proteins with SH2 domains in *Dictyostelium*. Four STAT transcription factors are the best characterized *Dictyostelium* proteins with SH2 domains. However, a functional connection between a tyrosine kinase and either of the STATs has not yet been established (Araki *et al.* 2003).

There is one identified receptor (DkhA) for the spore-inducing peptide SDF-2 among the 15 members of the *Dictyostelium* histidine kinase family (Wang *et al.* 1999; Anjard 2005). There are three more family members with transmembrane helices and extracellular regions that may serve a primary receptor function. The associated histidine kinase or histidine phosphatase activity acts as the source or sink of a series of phosphoryl group transfers that, thus far, has only one physiological target in *Dictyostelium*: the receiver domain of the intracellular cAMP phosphodiesterase RegA (Thomason and Kay 2000; Shaulsky *et al.* 1998). This domain regulates cAMP hydrolyzing activity of RegA and therefore the activation state of PKA, an important regulator of multiple life cycle transitions in *Dictyostelium* (Saran *et al.* 2002).

Dictyostelium has one novel family of receptor kinases. These are the GDT kinases, so-called because two members of the family (GDT1 and GDT2) regulate the growth to development transition (Zeng *et al.* 2000; Chibalina *et al.* 2004). All nine members of the family have three transmembrane helices and a large extracellular domain with putative receptor function, while four of those harbor a functional serine/threonine kinase domain. Presently, neither the ligands nor the downstream targets for these receptors are known. The *Entamoeba* genome also harbors a large number of putative receptor-linked serine/threonine kinases. However,

these proteins have only a single transmembrane domain and share no greater similarity to the *Dictyostelium* enzymes than to other protein kinases (Loftus *et al.* 2005). Similar to *D. discoideum*, *Entamoeba* has many genes encoding G-protein coupled receptors, but no receptor tyrosine kinases, which appear to be entirely confined to the metazoans.

With so many orphan receptors present in *Dictyostelium* and its solitary cousin *Entamoeba*, this aspect of their genome sequencing projects has highlighted both the complexity and diversity of signaling in the two organisms, and our extreme ignorance of this being the case.

Downstream intermediates

Known *Dictyostelium* targets for regulation by GPCRs are the adenylyl cyclase ACA, the guanylyl cyclases sGC and GCA, the MAP kinase ERK2, phospholipase C, and the phosphatidyl inositol-3 kinase PI3K (Huang *et al.* 2003; Roelofs *et al.* 2001; Parent and Devreotes 1996). However, in the case of ERK2, activation does not require a G-protein (Maeda *et al.* 1996), and in the case of at least ACA, GCA, and sGC, the activation by G-proteins is indirect.

Most eukaryote signaling pathways incorporate at least one, but most commonly multiple, protein phosphorylation events. Not surprisingly, there is a large repertoire of putative protein kinase genes in the *Dictyostelium* genome, which encompasses many of the protein kinase families found in animals, fungi, and plants. In addition to the histidine kinases and tyrosine kinases, discussed above, there are 120 kinases that fall into either the AGC, CAMK, CMGC, or STE families of mammalian serine threonine kinases, 60 proteins that do not belong to either family, and 23 kinases with atypical catalytic domains (Eichinger *et al.* 2005; Kimmel 2005; see http://dictybase.org/GeneFamilies/ProteinKinases.html for a complete list of the proteins). Seventy-three out of the 287 protein kinase candidates have been functionally characterized, but the remaining 214 other enzymes still offer a vast reservoir of uncharted signaling potential.

Transcription factors

Regulation of gene transcription is the final goal of many signaling events and here the *Dictyostelium* genome displays a surprising frugality. Compared to yeast, the ratio of transcription factors per number of genes is three times lower in *Dictyostelium* and this reduces to six times when *Dictyostelium* is compared to metazoans. The otherwise universal family with basic helix–loop–helix DNA binding domains is missing in *Dictyostelium*, as are the forkhead and steroid receptor transcription factors. Other classes such as the homeo-domain and MADS-box containing factors, and the Myb, bZIP, GATA, CBF-A, and CBF-B type transcription factors are present, but have fairly limited numbers of representatives (Eichinger *et al.* 2005; Shaulsky and Huang 2005).

A number of important transcriptional activators were identified, such as GBF, CRTF, and CbfA, that are thus far unique to *Dictyostelium* (Schnitzler *et al.* 1994; Mu *et al.* 2001; Winckler *et al.* 2004). This suggests that the limited repertoire of known *Dictyostelium* transcription factors is possibly supplemented by entirely novel proteins.

Cell adhesion

Solitary amoebae are likely to be in need of mechanisms that allow them to adhere to their substratum, their prey, and their mates. However, for multicellular life adhesive interactions between cells and to a supportive matrix are of primary importance; firstly to maintain structural integrity and to create a physical barrier to the environment, and secondly, to organize differentiated cells into tissues and often to facilitate or participate in cell-contact-mediated signaling.

The *Dictyostelium* genome encodes a large number of putative components of cellular adhesion systems, such as α- and β-catenin, disintegrins, vinculin, formins, and α-actinin, that were previously only observed in animals. Over 60 *Dictyostelium* proteins have been found to harbor transmembrane domains and repetitive sets of domains, such as the EGF/laminin, E-set, LDL, growth factor receptor, and fibronectin domains, that are characteristic for metazoan proteins involved in cell adhesion or in cell-contact-mediated signaling (Eichinger *et al.* 2005). A number of the *Dictyostelium* proteins (AmpA, ComC, LagC, LagD) have been demonstrated to play a role in developmental signaling (Varney *et al.* 2002; Kibler *et al.* 2003).

Concluding remarks

A discussion of the evolution of the *D. discoideum* genome is premature, because the information on related genomes is so limited. Ideally, one would like to make a comparison with the closest living *Dictyostelium* relative with a solitary life-style and at least one dictyostelid with a more basal phylogenetic position. However, the genome of the amoebozoan parasite *Entamoeba histolytica* offers some scope for deduction of genome evolution in amoebozoans and comparison with their closest sister groups, the fungi and animals.

The *Dictyostelium* and *Entamoeba* genomes, with 12 000 and 10 000 protein-encoding genes, are strikingly large compared to the yeast, *Drosophila*, and human genomes, with respectively 5500, 13 000, and 22 000 genes. Particularly in view of the fact that the endoparasitic life-style of *Entamoeba* caused considerable gene loss, this leads us to suspect that a large gene repertoire is characteristic for all amoebazoans. Both *D. discoideum* and *E. histolytica* have a well-developed cell signaling system. The *D. discoideum* genome also contains a large number and variety of structural components of the cytoskeleton and signaling proteins that modulate the cytoskeleton. Although some of these have specific roles in multicellular development, the majority may serve roles in processes, such as directional cell movement and phagocytosis, that are common to most amoebozoans. Signaling and cytoskeletal remodeling are also highly evolved in the myxogastrids. These amoebozoans with a syncytial life-style display several feats that require highly co-ordinated intracellular communication, which have been likened to a primitive form of intelligence (Nakagaki *et al.* 2000), such as food seeking through a maze, synchronous division of thousands of nuclei, and construction of architecturally intricate fruiting bodies from cytoskeletal components (Stempen *et al.* 2000). In animals, behavioral responses to sensory stimuli are primarily processed by the nervous system. Here information storage and processing uses an intercellular signaling network, which unlike the intermolecular network of the amoebozoans is less dependent on an extensive gene repertoire. Unlike plants and fungi, animals are also not metabolically complex and may primarily require their modestly sized genome for their extensive cell-type specialization.

The *Dictyostelium* genome presents us with some interesting insights on how processes that are required for solitary life in the soil, such as the production and transport of secondary metabolites, have been co-opted to regulate cell-type specialization. Similarly, dictyostelid species such as *D. minutum* have co-opted food-seeking to the bacterial attractant folate as their specific mechanism for cellular aggregation (De Wit and Konijn 1983). However, the large number of orphan receptors and novel proteins with putative roles in signal processing and cell–cell interactions that are present in the *Dictyostelium* genome emphasizes how much there is yet to learn about its developmental and environmental signaling processes. For solitary amoebae such information is even more limited.

Several methods are currently in use or are being developed to functionally characterize the 12 000 *Dictyostelium* genes. These methods include: (i) the use of DNA microarrays and high throughput *in situ* hybridization for spatio-temporal profiling of gene expression patterns; (ii) the generation of a complete set of barcoded mutants to catalogue genotype–phenotype relationships; (iii) high throughput biochemical screens to establish biochemical activity; and (iv) TAP-tag and yeast two-hybrid technology to identify interacting proteins (Kuspa 2005). In addition to these technologies, comparative genomics can be a powerful tool to distinguish core regulatory processes from species-specific elaborations, to define conserved and therefore functionally relevant regions in hitherto uncharacterized proteins, to identify essential regulatory sequences in promoter regions, and to unravel the pattern of gene modifications that generated phenotypic innovations during species evolution.

For such an approach to be useful in the Amoebozoa, a robust phylogeny based on multiple protein sequences and sampling a broad range of taxa is a primary requirement. Only then can we select a representative set of species for genome sequencing and obtain meaningful information on the directionality of genome evolution.

References

Alvarez-Curto, E., Rozen, D. E., Ritchie, A. V., Fouquet, C., Baldauf, S. L., and Schaap, P. (2005). Evolutionary origin of cAMP-based chemoattraction in the social amoebae. *Proc. Natl. Acad. Sci. USA*, **102**, 6385–6390.

Anjard, C. (2005). Multigene families of *Dictyostelium*. In W. F. Loomis and A. Kuspa, eds. *Dictyosklium genomics*. Horizon Bioscience, Norfolk, UK, pp. 59–82.

Anjard, C. and Loomis, W. F. (2005). Peptide signaling during terminal differentiation of Dictyostelium. *Proc. Natl. Acad. Sci. USA*, **102**, 7607–7611.

Anjard, C., Zeng, C. J., Loomis, W. F., and Nellen, W. (1998). Signal transduction pathways leading to spore differentiation in *Dictyostelium discoideum*. *Devel. Biol.*, **193**, 146–155.

Araki, T., Tsujioka, M., Abe, T., *et al.* (2003). A STAT-regulated, stress-induced signalling pathway in *Dictyostelium*. *J. Cell Sci.*, **116**, 2907–2915.

Baldauf, S. L. (2003). The deep roots of eukaryotes. *Science*, **300**, 1703–1706.

Baldauf, S. L. and Doolittle, W. F. (1997). Origin and evolution of the slime molds (mycetozoa). *Proc. Natl. Acad. Sci. USA*, **94**, 12007–12012.

Baldauf, S. L., Roger, A. J., Wenk-Siefert, I., and Doolittle, W. F. (2000). A kingdom-level phylogeny of eukaryotes based on combined protein data. *Science*, **290**, 972–977.

Bonner, J. T., Chiang, A., Lee, J., and Suthers, H. B. (1988). The possible role of ammonia in phototaxis of migrating slugs of *Dictyostelium discoideum*. *Proc. Natl. Acad. Sci. USA*, **85**, 3885–3887.

Chibalina, M. V., Anjard, C., and Insall, R. H. (2004). Gdt2 regulates the transition of *Dictyostelium* cells from growth to differentiation. *Dev. Biol.*, **4**, 8.

Dean, M., Hamon, Y., and Chimini, G. (2001). The human ATP-binding cassette (ABC) transporter superfamily. *J. Lipid Res.*, **42**, 1007–1017.

Dear, P. H. and Cook, P. R. (1993). Happy mapping: linkage mapping using a physical analogue of meiosis. *Nucleic Acids Res.*, **21**, 13–20.

De Lozanne, A. and Spudich, J. A. (1987). Disruption of the *Dictyostelium* myosin heavy chain gene by homologous recombination. *Science*, **236**, 1086–1091.

De Wit, R. J. W. and Konijn, T. M. (1983). Identification of the acrasin of *Dictyostelium minutum* as a derivative of folic acid. *Cell Differ.*, **12**, 205–210.

Eichinger, L. and Noegel, A. (2005). Comparative genomics of *Dictyostelium discoideum* and *Entamoeba histolytica*. *Curr. Opin. Microbiol.*, **8**, 606–611.

Eichinger, L., Pachebat, J. A., Glockner, G., *et al.* (2005). The genome of the social amoeba *Dictyostelium discoideum*. *Nature*, **435**, 43–57.

Foster, K. R., Shaulsky, G., Strassmann, J. E., Queller, D. C., and Thompson, C. R. (2004). Pleiotropy as a mechanism to stabilize cooperation. *Nature*, **431**, 693–696.

Glockner, G., Eichinger, L., Szafranski, K., *et al.* (2002). Sequence and analysis of chromosome 2 of *Dictyostelium discoideum*. *Nature*, **418**, 79–85.

Good, J. R., Cabral, M., Sharma, S., *et al.* (2003). TagA, a putative serine protease/ABC transporter of *Dictyostelium* that is required for cell fate determination at the onset of development. *Development*, **130**, 2953–2965.

Hereld, D. (2005). Signal transduction via G-protein-coupled receptors, trimeric G-proteins and RGS proteins. In W. F. Loomis and A. Kuspa, eds. *Dictyostelium genomics* Horizon Bioscience, Norfolk, UK, pp. 103–123.

Huang, Y. E., Iijima, M., Parent, C. A., Funamoto, S., Firtel, R. A., and Devreotes, P. (2003). Receptor-mediated regulation of PI3Ks confines PI(3,4,5)P3 to the leading edge of chemotaxing cells. *Mol. Biol. Cell*, **14**, 1913–1922.

Kessin, R. H. (2001). *Dictyostelium: Evolution, Cell Biology and the Development of Multicellularity*. Cambridge University Press, Cambridge, UK.

Kibler, K., Svetz, J., Nguyen, T. L., Shaw, C., and Shaulsky, G. (2003). A cell-adhesion pathway regulates intercellular communication during *Dictyostelium* development. *Devel. Biol.*, **264**, 506–521.

Kimmel, A. R. (2005). The *Dictyostelium* kinome: protein kinase signaling pathways that regulate growth and development. In W. F. Loomis and A. Kuspa, eds. *Dictyostelium genomics*. Horizon Bioscience, Norfolk, UK, pp. 211–234.

Konijn, T. M., Van De Meene, J. G., Bonner, J. T., and Barkley, D. S. (1967). The acrasin activity of adenosine-3′,5′-cyclic phosphate. *Proc. Natl. Acad. Sci. USA*, **58**, 1152–1154.

Kuspa, A. (2005). Whole-genome functional analysis in *Dictyostelium*. In W. F. Loomis and A. Kuspa, eds. *Dictyostelium genomics*. Horizon Bioscience, Norfolk, UK, pp. 279–296.

Kuspa, A. and Loomis, W. F. (1992). Tagging developmental genes in *Dictyostelium* by restriction enzyme-mediated integration of plasmid DNA. *Proc. Natl. Acad. Sci. USA*, **89**, 8803–8807.

Lilly, P., Wu, L., Welker, D. L., and Devreotes, P. N. (1993). A G-protein beta-subunit is essential for *Dictyostelium* development. *Genes Devel.*, **7**, 986–995.

Loftus, B., Anderson, I., Davies, R., *et al.* (2005). The genome of the protist parasite *Entamoeba histolytica*. *Nature*, **433**, 865–868.

Maeda, M., Aubry, L., Insall, R., Gaskins, C., Devreotes, P. N., and Firtel, R. A. (1996). Seven helix chemoattractant receptors transiently stimulate mitogen- activated protein kinase in *Dictyostelium*. *J. Biol. Chem.*, **271**, 3351–3354.

Margulis, L., Schwartz, K., and Dolan, M. (1999). *Diversity of Life: The Illustrated Guide to Five Kingdoms*, 2nd edn. Jones and Bartlett, Boston.

Morio, T., Urushihara, H., Saito, T., *et al.* (1998). The *Dictyostelium* developmental cDNA project: Generation and analysis of expressed sequence tags from the first-finger stage of development. *DNA Res.*, **5**, 335–340.

Mu, X., Spanos, S. A., Shiloach, J., and Kimmel, A. (2001). CRTF is a novel transcription factor that regulates multiple stages of *Dictyostelium* development. *Development*, **128**, 2569–2579.

Nakagaki, T., Yamada, H., and Toth, A. (2000). Maze-solving by an amoeboid organism. *Nature*, **407**, 470.

Nellen, W., Silan, C., and Firtel, R. A. (1984). DNA-mediated transformation in *Dictyostelium discoideum*: Regulated expression of an actin gene fusion. *Mol. Cell. Biol.*, **4**, 2890–2988.

Olsen, R. M. (2005). How many protein encoding genes? In W. F. Loomis and A. Kuspa, eds. *Dictyostelium genomics*. Horizon Bioscience, Norfolk, UK, pp. 265–278.

Parent, C. A. and Devreotes, P. N. (1996). Molecular genetics of signal transduction in *Dictyostelium*. *Ann. Rev. Biochem.*, **65**, 411–440.

Payne, S. H. (2005) Metabolic pathways. In Loomis, W. F. and Kuspa, A. (eds.), Dictyostelium genomics. Horizon Bioscience, Norfolk, UK, pp. 41–57.

Queller, D. C., Ponte, E., Bozzaro, S., and Strassmann, J. E. (2003). Single-gene greenbeard effects in the social amoeba *Dictyostelium discoideum*. *Science*, **299**, 105–106.

Raper, K. B. (1940). Pseudoplasmodium formation and organization in *Dictyostelium discoideum*. *J. Elisha Mitch. Sci. Soc.*, **56**, 241–282.

Raper, K. B. (1984). *The Dictyostelids*. Princeton University Press, Princeton, New Jersey.

Rivero, F. and Eichinger, L. (2005). The microfilament system of *Dictyostelium discoideum*. In W. F. Loomis and A. Kuspa, eds. *Dictyostelium genomics*. Horizon Bioscience, Norfolk, UK, pp. 125–171.

Roelofs, J., Meima, M., Schaap, P., and Van Haastert, P. J. (2001). The *Dictyostelium* homologue of mammalian soluble adenylyl cyclase encodes a guanylyl cyclase. *EMBO J.*, **20**, 4341–4348.

Ross, E. M. and Wilkie, T. M. (2000). GTPase-activating proteins for heterotrimeric G proteins: regulators of G protein signaling (RGS) and RGS-like proteins. *Ann. Rev. Biochem.*, **69**, 795–827.

Saran, S., Meima, M. E., Alvarez-Curto, E., Weening, K. E., Rozen, D. E., and Schaap, P. (2002). cAMP signaling in *Dictyostelium* – complexity of cAMP synthesis, degradation and detection. *J. Muscle Res. Cell M.*, **23**, 793–802.

Saran, S. and Schaap, P. (2004). Adenylyl cyclase G is activated by an intramolecular osmosensor. *Mol. Biol. Cell*, **15**, 1479–1486.

Schaap, P., Winckler, T., Nelson, M., *et al.* Molecular phylogeny of the Dictyostelidae–evolution of form and function in social amoebas. (submitted)

Schindler, J. and Sussman, M. (1977). Ammonia determines the choice of morphogenetic pathways in *Dictyostelium discoideum*. *J. Mol. Biol.*, **116**, 161–169.

Schnitzler, G. R., Fischer, W. H., and Firtel, R. A. (1994). Cloning and characterization of the G-box binding factor, an essential component of the developmental switch between early and late development in *Dictyostelium*. *Genes Devel.*, **8**, 502–514.

Shaulsky, G., Fuller, D., and Loomis, W. F. (1998). A cAMP-phosphodiesterase controls PKA-dependent differentiation. *Development*, **125**, 691–699.

Shaulsky, G. and Huang, E. (2005). Components of the Dictyostelium gene expression regulatory machinery. In W. F. Loomis and A. Kuspa, eds. *Dictyostelium genomics*. Horizon Bioscience, Norfolk, UK, pp. 83–101.

Shaulsky, G., Kuspa, A., and Loomis, W. F. (1995). A multidrug resistance transporter/serine protease gene is required for prestalk specialization in *Dictyostelium*. *Genes Devel.*, **9**, 1111–1122.

Solomon, J. M. and Isberg, R. R. (2000). Growth of *Legionella pneumophila* in *Dictyostelium discoideum*: a novel system for genetic analysis of host-pathogen interactions. *Trends Microbiol.*, **8**, 478–80.

Stempen, H., Stephen, H., and Stephenson, S. L. (2000). *Myxomycetes : Handbook of Slime Molds*. Timber Press, Portland, Oregon.

Strassmann, J. E., Zhu, Y., and Queller, D. C. (2000). Altruism and social cheating in the social amoeba *Dictyostelium discoideum*. *Nature*, **408**, 965–967.

Sucgang, R., Chen, G., Liu, W., *et al.* (2003). Sequence and structure of the extrachromosomal palindrome encoding the ribosomal RNA genes in *Dictyostelium*. *Nucleic Acids Res.*, **31**, 2361–2368.

Sun, B. and Firtel, R. A. (2003). A regulator of G protein signaling-containing kinase is important for chemotaxis and multicellular development in *Dictyostelium*. *Mol. Biol. Cell*, **14**, 1727–1743.

Sussman, M. (1955). "Fruity" and other mutants of the cellular slime mould, *Dictyostelium discoideum*: a study of developmental aberrations. *J. Gen. Microbiol.*, **13**, 295–309.

Sussman, M. and Brackenbury, R. (1976). Biochemical and molecular-genetic aspects of cellular slime mold development. *Ann. Rev. Plant Physiol.*, **27**, 229–265.

Sutherland, E. W. and Rall, T. W. (1958). Fractionation and characterization of a cyclic adenine ribonucleotide formed by tissue particles. *J. Biol. Chemistry*, **232**, 1077–91.

Thomason, P. and Kay, R. (2000). Eukaryotic signal transduction via histidine-aspartate phosphorelay. *J. Cell Sci.*, **113**, 3141–3150.

Thompson, C. R. and Kay, R. R. (2000). The role of DIF-1 signaling in *Dictyostelium* development. *Mol. Cell*, **6**, 1509–1514.

Urushihara, H. (2002). Functional genomics of the social amoebae, *Dictyostelium discoideum*. *Mol. Cell*, **13**, 1–4.

Urushihara, H., Morio, T., Saito, T., *et al.* (2004). Analyses of cDNAs from growth and slug stages of *Dictyostelium discoideum*. *Nucleic Acids Res.*, **32**, 1647–1653.

van der Geer, P., Hunter, T., and Lindberg, R. A. (1994). Receptor protein-tyrosine kinases and their signal transduction pathways. *Ann. Rev. Cell Biol.*, **10**, 251–337.

Varney, T. R., Casademunt, E., Ho, H. N., Petty, C., Dolman, J., and Blumberg, D. D. (2002). A novel *Dictyostelium* gene encoding multiple repeats of adhesion inhibitor-like domains has effects on cell–cell and cell–substrate adhesion. *Devel. Biol.*, **243**, 226–248.

Vassilatis, D. K., Hohmann, J. G., Zeng, H., *et al.* (2003). The G protein-coupled receptor repertoires of human and mouse. *Proc. Natl. Acad. Sci. U S A*, **100**, 4903–4908.

Wang, M. and Schaap, P. (1989). Ammonia depletion and DIF trigger stalk cell differentiation in intact *Dictyostelium discoideum* slugs. *Development*, **105**, 569–574.

Wang, N., Soderbom, F., Anjard, C., Shaulsky, G., and Loomis, W. F. (1999). SDF-2 induction of terminal differentiation in *Dictyostelium discoideum* is mediated by the membrane-spanning sensor kinase DhkA. *Mol. Cell. Biol.*, **19**, 4750–4756.

Winckler, T., Iranfar, N., Beck, P., *et al.* (2004). CbfA, the C-module DNA-binding factor, plays an essential role in the initiation of *Dictyostelium discoideum* development. *Eukaryot. Cell*, **3**, 1349–1358.

Woese, C. R., Kandler, O., and Wheelis, M. L. (1990). Towards a natural system of organisms: proposal for the domains Archaea, Bacteria, and Eucarya. *Proc. Natl. Acad. Sci. USA*, **87**, 4576–9.

Zeng, C., Anjard, C., Riemann, K., Konzok, A., and Nellen, W. (2000). gdt1, a new signal transduction component for negative regulation of the growth-differentiation transition in *Dictyostelium discoideum*. *Mol. Biol. Cell*, **11**, 1631–1643.

Zhang, N., Long, Y., and Devreotes, P. N. (2001). Ggamma in *Dictyostelium*: its role in localization of Gbetagamma to the membrane is required for chemotaxis in shallow gradients. *Mol. Biol. Cell*, **12**, 3204–3213.

INDEX

Note: page numbers in *italics* refer to Figures, and those in **bold** refer to Tables.

ABC-type transporters, *Dictyostelium discoideum* 219
Acantharea **9**, 22
  EST project 90
acanthoecid choanoflagellates 13
Acetabularia species, plastid genomes 98
acetyl-CoA-synthetase, lateral gene transfer **112**
acrasids 24, 25
actin genes
  Dictyostelium discoideum 220
  Foraminifera 83
  scrambling in ciliates 70
actinophyrids 8, *18*
acytostelids 214, 215
Adenoides eludens, minicircles 56
adenylyl cyclase receptor, *Dictyostelium discoideum* 221
ADPs (actin-deviating proteins), Foraminifera 83
Aedes mosquito 34
agamonts, Foraminifera 80
  evolution 85
  Foraminifera 78
alanyl-tRNA synthetase, lateral gene transfer **112**, 117
alcohol dehydrogenase, lateral gene transfer **112**, 113
Alexandrium species
  A. tamarense
    chromosomes 52
    *cox*2 genes 58
    EST study 50, 51–2
    histones 53
    SSU rDNA *82*
  haploid DNA content **51**
Allogromia species 20
  A. laticollaris 79
  SSU rDNA *82*
Allogromiidae 78
  diversity 89
  introns 84
alpha actinin, *Dictyostelium discoideum* 222
alpha amanitin inhibition resistance, *Entamoeba histolytica* 173
alveolates 8, 10, 11, *12*, 17, 19–20, 48, *125*, 215
  plastid origin 100
alveoli, dinoflagellates 48
Ambylospora species 188

amino acid metabolism
  Dictyostelium discoideum 219
  Entamoeba histolytica 174, 176
  Tritryps 163
aminoacyl-tRNA synthetases, gene replacement 117
aminoethylphosphonate metabolism, *Entamoeba histolytica* 174
amitosis, ciliates 65–8, 73
Ammodiscus species 88
Ammonia 21
ammonium metabolism
  diatoms 207
  Dictyostelium discoideum 219
amnesic shellfish poisoning (ASP) 202
Amoebozoa 8, *12*, 13, *15*, 16, 26, 214, 215
Amphidinium species 20
  A. carterae
    EST study 50
    haploid DNA content **51**
  microcircles 56
  minicircles 55–6, 101
amylopectin metabolism, *Cryptosporidium* 132
anaerobic protists 109
  lateral gene transfer 110, **112**, 131
    metabolic adaptations 111–15
  multiple origins *110*
Ancryomonas 9, 20, 24
animals (metazoa) 8, *12*, 13
Anopheles mosquito 34
antibiotic resistance, gene transfer 123
antimalarial resistance
  association with transporter gene SNPs 42
  Plasmodium falciparum 37
antioxidant pathways, *Entamoeba histolytica* 175
Antonospora locustae genome sequencing 185, 188
Apicomplexa 8, 10, 19, 33, *110*, 125
  drug treatments 133
  gene transfer 123
    intracellular 126–8
    lateral gene transfer 116, 128–30
    retention of genes 130–3
  genome studies 126
  as hosts to microsporidia 182

Apicomplexa (*Cont.*)
  introns 52
  nuclear organization 53
  origins of LDH and MDH 130
  plastid origin 99, 100, *124*
apicoplast 19, 54–5, 100–1, *124*, 125–6
  genome 127–8
  as target for drug treatments 133
targeting signals 127, 128
apusomonads 9, 20, 24–5
Arabidopsis thaliana
    genes of cyanobacterial origin 95, 126
    genome size *184*
  gene transfer into nucleus 103
Arachne 146
arcellids 16
archaegregarines 182
Archaeplastida *see* Plantae
Archamoebae 16
Archezoa 182
arginine metabolism, diplomonads 114
Armophorea *67*
  telomere repeats **71**
armor plates, dinoflagellates 48
Arthrobacter species, β-glucurinidase gene 116
  fruiting body formation, dictyostelids 215, *216*, 222
Ashbya gossypii, dihydroorotate
        dehydrogenase gene 116
asparagine synthetases, diatoms 208
aspartase, *Entamoeba histolytica* 176
assembly
  clone-by-clone sequencing 143
  Entamoeba histolytica genome 173
  WGS sequencing 146
Astasia longa, plastid genome 100
Astrammina species 81
  A. rara 79
    DNA content 80
  A. triangularis 79
Astrolithium 21
Astrorhizida 78, 85
athalamids 78
atp genes, dinoflagellates 5, 55
AT rich regions
  Dictyostelium discoideum 217
  Entamoeba histolytica 172
  nucleomorph genomes 195
A type flavoprotein, lateral gene transfer **112**, 113

Bacillariophyceae 201
BAC libraries 141, 143–4
bacteria, gene transfer 123
bacterial ingestion and digestion, *Entamoeba histolytica*
    169, 170

Bacteroidetes, lateral gene transfer 114
base-calling programs 143
base composition, *Plasmodium* genomes 35
Bathysiphon species, ubiquitin gene 83
BES (bloodstream form VSG expression sites),
        Trypanosoma brucei 160
bicarbonate ion, uptake by diatoms 208
bicosoecids 8, 18, 25
Bigelowiella natans
    lateral gene transfer 117
    nucleomorph genome 194, **195**
  polyubiquitin gene 83
bioluminescence 48
Biomyxa 9, 25
biosynthetic pathways, *Entamoeba histolytica* 174
Blastocystis 19
Blepharisma 67
bloom formation, dinoflagellates 48
bodonids 24
bolidophyceans 19
Bolivina species 87
Brachiola algerae 188
Breviatea 25
brown algae (phaeophycaeans) 19
Bulimia marginata, RNA polymerase gene 83
Buliminida 78, 87
Bursaria truncatella 65

C4 photosynthesis 208–9
CAAT box sequences, *Plasmodium* species 41
Cabozoa 99, 193
Caecitellus 18
Cafeteria 18
calcareous Foraminifera 78, *79*
cAMP, dictyostids 215, *216*, 217, 221
Campyloacantha 15
canonical EF-1α protein 117
Capaspora 8, 13
carbon-acquisition mechanisms, diatoms 208–9
    concentrating mechanisms (CCMs), diatoms 208–9
    fixation, marine diatoms 202
carbonic anhydrases (CAs), diatoms 208
cAR-like receptors, *Dictyostelium discoideum* 220
Carpediemonas 9, 23
Carterinida 85
catalase
  acquisition by *Nosema locustae* 111
  lateral gene transfer **112**
catenins, *Dictyostelium discoideum* 222
Cbs, *Tetrahymena* 70
CDCA1 208
cell adhesion, *Dictyostelium discoideum* 221–2
centric diatoms 201

Centroheliozoa 9, 20, 25
centromere characteristics, *Plasmodium* species **37**
ceramide aminoethylphosphonate metabolism,
  Entamoeba histolytica 174
Ceratium horridum, minicircles 56, 101
cercomonads 8, 21
Cercomonas species 20, *21*
  C. longicauda, SSU rDNA *82*
  polyubiquitin gene 83
Cercozoa 3, 8, *12*, 20–1, *215*
  polyubiquitin genes 83
Cervus 15
Chaetospaeridium globosum, plastid genome **97**, 98
Chagas' disease 24
chain formation, diatoms 207
Chaos 15
"charophytes" 17
Chilodonella species 20, 64, *65, 67*
  cis-acting sequences **69**
  telomere repeats **71**
Chlamydomonas 17
Chlorarachnion 20
  cDNA library 90
chlorarachniophytes 8, 21, 22
  genome size 7
  lateral gene transfer 117
  plastid origin 99
  see also nucleomorphs
chlorophyll a/b lines 99–100
chlorophyll-c 125
chlorophytes 17
Chloroplastida *see* Viridaeplantae
chloroplasts 98
  origin *see* primary endosymbiosis
choanoflagellates 8, 13, *15*
Choanozoa *12*
cholesteric liquid crystal structure, dinoflagellate
  chromosomes 53, *54*
Chonotrichia, budding 67
Chromalveolata *12*, 26, 100, 193
  plastids 17
chromalveolate hypothesis 17, 19, 125–6, 198, 203
Chromista 17, 18
  plastid origin 100
  ciliates 64, 70
  Dictyostelium discoideum 217
  Foraminifera 80
  nucleomorphs 194
  Thalassiosira pseudonana 203
  trypanosomatids 156–7
chromosome processing, ciliates 68, 73
chromosome rearrangements, *Plasmodium* species 37
  ciliates 68–70
chromosome size, suitability of PFGE 144

chromosome spreading 140
chromosome structure
  ciliates 68–73
  dinoflagellates 52–3, *54*
  Plasmodium species 35, 37
chryosophyceans 18
chytrid fungi 13, 109
Chytriomyces 15
ciguatera fish poisoning 48
ciliates 8, 10, 20, 54, *65*, *110*, 125
  anaerobic 109
  chromalveolate hypothesis 17
  chromosomal processing 68–73
    amplification 71
    DNA deletion 68–9
    DNA unscrambling 69–70
    epigenetic regulation 72–3
    fragmentation 70
    telomere addition 70–1
  cis-acting sequences **69**
  genome content variation 71
  genome projects 71–2
  IES excision **69**
  intradomain gene transfer 117
  life cycles 64–8
  nuclear dualism 7
    origins 73–4
  protein evolution 73–4
  role in DNA unscrambling 70
  telomere repeats **71**
Ciliophrys 18
Ciona intestinalis 204
cis-acting sequences, ciliates
classification of living world 214
Clathrulina 21
closed mitosis 52
cob sequences, *Pfisteria piscicida* 58
coccolithophorids 25
Codium fragile, plastid genome 98
Coelosporidium 9, 25
coevolution, organelles and host cells 103–4
COGs (clusters of orthologous genes), Tritryps 161
Coleps 65
collodictyonids 25
Colpodea *67*
Colpodella 20
Colpodellidae 8, 19
Colponema 19
compaction
  Encephalitozoon cuniculi genome 185, 186, 187
  nucleomorph genomes 187, 195–6
comparative genomics 33, 223
  Entamoeba species 177–8
  genome sequence alignment 39
  microsporidia 185–6, 188

comparative genomics (*Cont.*)
  nucleomorphs 197–8
  Thalassiosira pseudonana 204
"complex plastids" 192
composite chromosome 12
copromyxids 9, 25
Corallochytrium 13
corals 19, 49
core Cercozoa 21
core proteome, Tritryps 161
Cornuspira antarctica, introns 84
Coscinodiscophyceae 201
  genome sequencing 209
Coscinodiscus species *18*, 209–10
cosmids 141
Cot analysis 140
Cot-based cloning and sequencing (CBCS) 146
cotranscription, nucleomorphs 196
cox genes
  Apicomplexa 126
  dinoflagellates 58
cpn60, presence in mitosome 176
Crithionina delacai 79
Cryothecomonas 8, 21
Crypthecodinium cohnii
  mitochondrial genome 58
  nuclear genome 50
cryptic mitochondrion, microsporidia 183
cryptomonads (cryptophytes) 9, 25, *110*
  genome size 7
  nucleomorphs 187
  plastids 17
   Dinophysis species 58
  plastid origin 99, 100
  as tertiary endosymbionts 102
  see also nucleomorphs
Cryptosporidium species
  1,4-α-glucan branching enzymes 130, *131*
  amylopectin metabolism 132
  C. parvum 125
   genome analysis 127, 128, 131, 139
   genome size *184*
   lateral gene transfer 129
   nucleotide biosynthesis *132*, 133
  genome mapping 147
cyanelles 95
Cyanidiales 16
Cyanidioschyzon merolae, genome analysis 95
Cyanidium species *15*
  C. caldarium, plastid genome 95, **97**, 98
cyanobacteria
  as origin of primary plastid 94–5
  sequenced genomes **97**
Cyanophora species *15*

C. paradoxa, cyanelle genome 95, **97**
Cylindrotheca fusiformis
  ammonium transporters 207
  silica metabolism 206, 207
cysteine biosynthesis, *Entamoeba histolytica* 174
cysteine proteases, *Entamoeba histolytica* 175
cyst forms
  dictyostelids 215, *216*, 218
  Entamoeba histolytica 169, *170*
cytoskeletal proteins, *Dictyostelium discoideum* 219–20

Dactylaria *15*
degenerate *ingi*/L1Tc-related retroelements (DIRE) 160
Desmidium *15*
desmids 17
Desmothoracids 8, 21
diatoms **3**, 18, 19, 201–2
  C4-like photosynthesis 209
  candidates for future sequencing 209–10
  carbon acquisition mechanisms 208–9
  genome sequencing 202–3
  nitrogen metabolism 207–8
  plastids 205
  silica metabolism 205–7
  see also Thalassiosira pseudonana
Dicer-like protein, *Tetrahymena* 73
Dictyochophyceae 18
dictyostelids 16, 214–15
  survival strategies 215, 217
Dictyostelium species **3**
  D. discoideum
   cell adhesion 222
   chromosomal processing 68
   comparative genomics 222
   genome content 218–22
   genome mapping 147
   genome sequencing projects 150, 217
   genome size *184*
   genome structure 217–**18**
   use as research tool 217
  D. minutum 223
   rDNA 172
   relationship to *Entamoeba* 171
DIF-1 production, *Dictyostelium discoideum* 219
dihydroorotate dehydrogenase genes, yeasts 116
dimorphids 9, 25
dinoflagellates 8, 10, 19, 48–9, 59, *110*, 125
  cell division 52
  evolution of nuclear organization 53–4
  haploid DNA content **51**
  histone-depauperate chromosomes 7, 53
  intracellular gene transfer 56
  kleptoplasts 57–8
  mitochondrial genome 58–9

nuclear genome
  chromosome morphology 52–3, *54*
  EST studies 50–2
  gene organization 52
  genome size 49–50
  plastid genome
    kleptoplasts 58
    peridinin plastids 55–6, 99
    tertiary plastids 57
  plastid loss 102
  plastids 54–5, 56–7
    peridinin plastids 49, 55–6, 99, 100, 101, 192
  RNA editing 56, 58–9
  tertiary endosymbiosis 102
dinokaryon DNA structure 19
Dinophysis species, plastids 58
diplomonads 9, 23, *110*
  lateral gene transfer **112**, 113, 114
  sulfide dehydrogenase 111–13
diplonemids 24
DIRE retroelements, Tritryps 160
DIRS (*Dictyostelium* intermediate repeat sequence) transposons 218
discicristates *12*, 24, *215*
Discocelis 9, 25
disintegrins, *Dictyostelium discoideum* 222
dispersed gene family 1 (DGF-1) 160, 161
  protists 7, 10, 11
DkhA 221
DNA deletion, ciliates 68–9
  epigenetic regulation 72
DNA end-repair 141
DNA extraction, Foraminifera 81
DNA fragmentation 141
  ciliates 70
DNA packaging, dinoflagellates 53
DNA unscrambling, ciliates 69–70
DRIPs *see* Ichthyosporea
drug targeting
  Apicomplexa 133
  Tritryps 165

ebriids 9, 25
E-Cbs, *Monoeuplotes* 70
editing sites, land plant plastids 98
EFL protein 117
EGF/laminin domain, *Dictyostelium discoideum* 222
Eimeria species
  E. tenella 125
  fructose-1, 6-biphosphatase gene 132–3
  genome mapping 147
elongation factor-1α (EGF-1α) genes, microsporidia 183
embryophytes *see* land plants
Emiliania huxleyi 25

Encephalitozoon cuniculi 111, 181
  comparative genomics 185–6
  genome 182, 183–5, 195
  genome size 139, *184*
Encephalitozoon hellem 188
Encephalitozoon intestinalis, genome size 183, *184*
Endodinium chattoni, chromosomes 52
endoreplication, Foraminifera 80
endosymbiosis
  origin of plastids *96*
  Paulinella chomatophora 95
  see also intracellular gene transfer (IGT); primary endosymbiosis; secondary endosymbiosis; tertiary endosymbiosis
end-repair 141
Entamoeba dispar 170
entamoebae 8, 16, 109, *110*
  comparative genomics 177–8
  intradomain gene transfer 117
  lateral gene transfer **112**
Entamoeba histolytica *3*, 16, 169–70
  chromosomal processing 68
  genome coding capacity 174–6
  genome sequencing 139, 150
    assembly 173
    origins of project 170–1
    problems 171–2
    repetitive DNA 172–3
    transcription, translation and replication 173–4
  genome size *184*, 217–18, 222
  lateral gene transfer 113–14, 128, 130–1, 176
  life cycle *169*
  morphological features *170*
  pathogenesis 169–70, 175–6
  phagotrophy 125
Enterocytozoon beineusi genome sequencing 185, 188
enteromonads 23
environmental sensing, *Entamoeba histolytica* 175
epigenetic regulation, ciliates 72–3
Epiphagus virginiana, plastid genome reduction 98
ERK2, *Dictyostelium discoideum* 221
ER signal sequence, *Thalassiosira pseudonana* 205
E-set domain, *Dictyostelium discoideum* 222
EST studies *see* expressed sequence tag studies
Euglena species
  E. gracilis
    chromosomal processing 68
    plastid genome 100
    plastids 55
euglenids 9, 24
euglenophytes
  photosynthetic loss 100
  plastids 99, 192
Euglenozoa 9, 23, 24, *110*

euglyphids 8, 21, 22
  polyubiquitin gene 83
eukaryote-to-eukaryote gene transfer 117, 124
Euplotes species *67*
  cis-acting sequences **69**
  telomere repeats **71**
eustigmatophyceans 19
Eutreptiella 23
Excavata 9, 11, *12*, 22–4, *23*, 26, *215*
Explorers Cove, studies of Foraminifera 89
expressed sequence tag (EST) studies
  diatoms 202
  dinoflagellates 50–2, 56
  Foraminifera 89–90
  microsporidia 186
expression site associated genes (ESAGs),
    Trypanosoma brucei 160, 161
extensive fragmenters, ciliates 64, *67*
  chromosome processing 68
extranuclear spindles 52
extrusomes 25
eyespot, dinoflagellates 57

fatty acid biosynthesis
  Apicomplexa 131
  Entamoeba histolytica 174
  Tritryps 164
Favella 65
FBA (fructose-1, 6-biphosphate isomerase) gene 17
Fe-hydrogenase, lateral gene transfer **112**, 114
Feulgen image analysis densitometry 140
fibronectin domain, *Dictyostelium discoideum* 222
filopodia 78
Filosa 21
filose testate amoebae 21, 22
fingerprinting 141
flow cytometry 140
flp gene, lateral gene transfer **112**
folate metabolism, lack in *Entamoeba histolytica* 174
Folliculina 65
Fonticula 9, 25
Foraminifera 3, 20, 21, 22, 78, *79*, 80
  biomedical applications 90
  EST (expressed sequence tag) studies 89–90
  genome investigation strategies 81
  introns 84
  molecular clock analyses 88
  molecular ecology 89
  molecular phylogeny 84–5, *86*, 87–8
  nuclear dualism 7
    origins 73
  protein-coding genes 83–4, 88
  reproductive strategies 80
  small subunit and large subunit ribosomal DNA 81–3

Form II rubisco gene, dinoflagellates 50
formins, *Dictyostelium discoideum* 222
fossil evidence
  age of photosynthesis 94
  Foraminifera 78, 80, 84, 88
Fragillariopsis cylindrus, EST projects 202
fragmentation of DNA, ciliates 70
fructose-1, 6-biphosphatase gene,
    Apicomplexa 132–3
fructose-1, 6-biphosphate aldolase, lateral
    gene transfer **112**
fruiting body formation, dictyostelids 215, *216*,
    218–19, 222
frustules, diatoms 201, 205–7, *206*
frustulins 207
fucoxanthin-containing dinoflagellates 57
Fugu rubripes 204
functional annotation 148
fungi 8, *12*, 13, *110*
  lateral gene transfer 116
  relationship to microsporidia 182–3
fusion hypotheses 117–19
Fusulinida 78, 85

galactose/N-acetyl galactosamine specific lectin,
    Entamoeba histolytica 175
gametocytes, malaria parasites 34
gametogenesis, Foraminifera 80, 81
gamonts, Foraminifera 80
gap closure, genome sequencing 143
GAPDH (glyceraldehyde-3-phosphate
    dehydrogenase) genes 17
G-box, *Plasmodium* species 42
GC bias, Tritryps 157
GC-content, dinoflagellates 51
GDT kinases, *Dictyostelium discoideum* 221
gene clusters, Tritryps *157–8*
gene densities
  Dictyostelium discoideum 217–18
  Encephalitozoon cuniculi 185
  nucleomorphs 95
  Tritryps 161
gene duplication 130
gene expression, *Plasmodium* species 41
gene families
  dinoflagellates 50–1
  Plasmodium species 38, **39**
gene flow, Foraminifera 89
gene order conservation, microsporidia 185–6
gene orientation, correlation with intergenic spaces,
    microsporidia 186
gene prediction 148
generative nuclei, Foraminifera 80
genetic codes 7
gene transfer 123–5

Apicomplexa, retention of genes 130–3
implications for antiapicomplexan therapy 133
see also intracellular gene transfer;
      lateral gene transfer
genome mapping 146–7
  HAPPY mapping *147–8*
  optical mapping 148
genome organizations 7
  dinoflagellates 52
  Tritryps
    gene clusters 157–*8*
    retroelements 160
    RNA genes **159**
genome processing *see* chromosome processing
genome reduction
  Apicomplexa 127, 131–2
  epiphyte plastid 98
  microsporidia 183, 184
genome scanning, ciliates 72–3
genome sequencing 139
  choice of strain or clone 139–40
  clone-by-clone strategy 140–4, *142*
  genome characteristic determination 140
  whole chromosome shotgun (WCS) strategy 144
  whole genome shotgun (WGS) strategy 144, *145*, 146
genome sequencing projects
  ciliates 71–2
  diatoms 202
    Thalassiosira pseudonana 203–4
  Dictyostelium discoideum 150, 214, 217
  Entamoeba histolytica 150, 170–8
  microsporidia 185
    Encephalitozoon cuniculi 183–5
  nucleomorphs 194–6
  Plasmodium species 33, 34–5, **36**, 42–3
  Trypanosoma cruzi 149–50
genome size 7, 139, *184*
  Apicomplexa 131
  ciliates, relationship to cell volume 71, *72*
  determination methods 140
  Dictyostelium discoideum 217
  dinoflagellates 49–50, **51**, 59
  Foraminifera 80–1
  microsporidia 181, 183
  nucleomorphs 194, 197
  Plasmodium species 35, **37**
  primary plastids 95, **97**, 98
  Thalassiosira pseudonona 203
  trypanosomatids 155
genome structure
  Dictyostelium discoideum 217–18
  Entamoeba histolytica 171
  functional implications in microsporidia 186–8
  Tritryps 155–7

Giardia species 109
  G. lamblia 23, 114, *115*, 195
    genome sequencing 139
    lateral gene transfer 128
    mitosome 177
Glaucoma species, telomere repeats **71**
glaucophytes (glaucocystophytes) 8, 16
  cyanelle genome 95, **97**
Glenodinium foliaceum, stigma 102
Globigerinella siphonifera 87
Globigerinida 78, 85, 87
Globigerinoides trilobus 80
Globobulimina species 87
globorotalids, substitution rates 88
1, 4-α-glucan branching enzymes 130, *131*
glucokinase, lateral gene transfer **112**
glucosamine-6-phosphate isomerase, lateral
    gene transfer **112**, 113
glucose metabolism, Tritryps 163–4
β-glucuronidase genes, lateral transfer 116
glutamate dehydrogenase, gene replacement 117
glutamine synthetases, diatoms 207
glutathione, absence from *Entamoeba histolytica* 175
glyceraldehyde-3-phosphate dehydrogenase
  and chromalveolate hypothesis 125
  gene replacement 117
  lateral gene transfer **112**
glycoprotein metabolism, Tritryps 164
glycostyles 26
Goniomonas 25
Gonyaulax, minicircles 101
GPI (glycosylphosphatidylinositol) metabolism,
    Tritryps 164
G-protein coupled receptors (GPCRs), *Dictyostelium
    discoideum* 220
Granuloreticulosea (heterotrophic amoebae) 8, 22
  see also Foraminifera
green algae 17, 203
  as origin of apicoplast 126
  plastid genomes **97**, 98
  secondary endosymbiosis 99, 192, 193
green primary plastid genomes **97**, 98
gregarines 19
Gromia species 20, 22, 78
  EST comparison with Foraminifera 89–90
  G. oviformis, RNA polymerase gene 83
growth factor receptor domain, *Dictyostelium
    discoideum* 222
GTP-ases, *Dictyostelium discoideum* 220
guanylyl cyclases, *Dictyostelium discoideum* 221
Guillardia theta, nucleomorph **195**
  genome sequencing 14
Gymnodinium simplex, haploid DNA content **51**
gymnophryids 8, 21

Gymnophrys species, polyubiquitin gene 83
Gymnosphaerida 9, 25–6

HAD-hydrolase gene family **39**
Haplosporidia 9, 20, 21, 22, 26
haplotype maps (HapMaps), *Plasmodium* species 42
HAPPY mapping 144, *147–8*, 150
    Dictyostelium discoideum 217
    Entamoeba histolytica 178
haptonema 25
haptophytes (prymnesiophytes) 9, 25, *110*
    plastid origin 99, 100, 192
    plastids 17
    as source of dinoflagellate plastids 1–2, 57, 58
Haynesina species, ubiquitin gene 83
helical filaments 78
Heliophyra 65
"heliozoa" 18, 25
Hemisphaerammina bradyi, SSU rDNA *82*
Heterocapsa species, haploid DNA content **51**
heterokonts *12*, 18, *215*
    plastid loss 102
    plastid origin 99, 100, 192
    as tertiary endosymbionts 102
    see also diatoms, stramenopiles
Heterolobosea 9, 23, 24, *110*
Heterotrichea *67*, 68
    extranuclear microtubules 73
heterozygous genomes, *Trypanosoma cruzi* 149–50
hexose phosphate transporters, Tritryps 163
Hidden Markov Models (HMMs) 148
histone-depauperate chromosomes 7
histone genes, nucleomorphs 196
histone H4, substitution rates in ciliates 74
histone-like proteins (HLPs)
    dinoflagellates 53
    phylogeny *55*
histones, in dinoflagellates 53
HM-1:IMSS strain, *Entamoeba histolytica* 171
Homalozoon 20
horizontal gene transfer *see* lateral gene transfer
host specificity, *Plasmodium* species 42
housekeeping functions, nucleomorph genes 196
HSP70
    presence in microsomes 176
    presence in microsporidia 183
HU proteins 53
    phylogeny *55*
hybrid-cluster protein, lateral gene transfer **112**, 113
Hydra, intradomain gene transfer 117
hydrogenosomes *110*, 114
    parabasilids 24
3-hydroxy-3-methylglutaryl coenzyme A reductase, lateral gene transfer **112**

hydroxymethyluracil (HoMeUra), presence in dinoflagellates 52
hypermastigotes, mitosis 52
hypochytrids 18

Ichthyosporea (DRIPs) 8, *12*, 13
immobilization antigen inheritance, *Paramecium* 72
ingi/RIME retroelement 160
inosine 5' monophosphate dehydrogenase (IMPDH) gene 132
internally excised sequences (IESs), ciliates 68–9, 72
intracellular gene transfer (IGT) 123, **124**
    Apicomplexa 126–8
    dinoflagellates 56
    retention of genes 131
    see also endosymbiosis; primary endosymbiosis; secondary endosymbiosis; tertiary endosymbiosis
intradomain gene transfer 117
Intramacronucleata *67*, 68
introns
    dinoflagellates 52
    Entamoeba histolytica 173
    Euglena gracilis 100
    Foraminifera 84
    nucleomorphs 195
    Paramecium and *Oxytricha* 72
Involuntinida 78, 85
iron-sulfur cluster synthesis, *Entamoeba histolytica* 177
IscS and IscU, lateral gene transfer **112**
isochore structure, *Plasmodium* species **37**
isoprenoid biosynthesis, Apicomplexa 131

Jakoba 23
jakobids 9, 23, 24
JIGSAW 148

Karenia species
    haploid DNA content **51**
    K. brevis 48, *49*
        *cox*2 genes 58
        EST study 50–1
    tertiary plastids 57, 102
Karlodinium species
    haploid DNA content **51**
    tertiary plastids 57
Karyorelictea *67*, 68
Kathablepharids 9, 25
kelp 18, 19
kinetoplast 24
kinetoplast DNA organization, Tritryps 163, 165
kinetoplastids *3*, *9*, 24
    lateral gene transfer 116
kleptoplasts, dinoflagellates 57–8
Komokiacea 9, 26

L1Tc/NARTc retroelement 160
labyrinthulids (slime-nets) 18
Lacrymaria 65
lactate dehydrogenase (LDH) 130
Lagenida 78, 85
Lander-Waterman model 143
land plants (embryophytes) 17
  plastid editing machinery 98
  plastid genomes **97**, 98
large subunit ribosomal DNA (LSU rDNA),
      Foraminifera 81
last common eukaryote ancestor 109
lateral gene transfer (LGT) 116, 119, 123–**4**, 128
  anaerobic protists 110, **112**
    adaptations 111–15
    proof 110–11
  Apicomplexa 128–30
  Entamoeba histolytica 176
  eukaryote-to-eukaryote transfer 117
  and fusion hypotheses 118, 119
  retention of genes 130–3
  Tritryps 165
LDL domain, *Dictyostelium discoideum* 222
lectin proteins, *Entamoeba histolytica* 169, 175
Lecythium, EST sequencing 90
Leeuwenhoek, A. van 10
Leishmania major
  genome sequencing 155
  genome size *184*
  genome statistics **156**
  genome structure 156
  lateral gene transfer 116
  see also Tritryps
leishmaniases 24
Lepidodinium viride, tertiary plastids 57
Le Roch, K.G. *et al.* 41
leucine aminopeptidase genes *129*
  Apicomplexa 128, 131
leucine-rich repeats, *Trichomonas vaginalis* 113
libraries, WGS sequencing 146
lichens 17
life cycles
  ciliates 64–8
  dictyostelids 215–17, *216*
  Entamoeba histolytica 169
  Foraminifera 80
  microsporidia 181
  Plasmodium species 34
    gene expression 41
light microscopy 10
linear forms, plastid genomes 99
LINEs (long interspersed nuclear elements),
      Entamoeba histolytica 172–3
Lingulodinium polyedrum

EST studies 50
  gene regulation 52
  mitochondrial genome 58
lipid metabolism, Tritryps 164
liquid crystal structure, dinoflagellate
      chromosomes 53, *54*
Litostomatea *67*
Lituolida 78, 85, 87
liver abscess formation, *Entamoeba histolytica* 169, 170
lobose amoebae 8, 13, 16
loricae 13, *15*
loss of aerobic mitochondria 109
loss of laterally transferred genes 129, 131–2
loss of plastid genomes 101, 102–3
  Apicomplexa 127
Lotharella species, polyubiquitin gene 83
Loxodes 65, *67*
luciferin-binding protein gene, dinoflagellates 50
Luffisphaera 9, 26

macrocysts, dictyostelids 215, *216*
macronucleus, ciliates 20, 64, 65
  chromosome processing 68–71
  regeneration 67–8
maize, plastid genome 99
malaria parasites 19, 33–4
  genome sequencing 33, 34–5, **36**
  see also Plasmodium species
malate deydrogenase (MDH) 130
Malawimonas 9, 23, 24
malic enzyme, lateral gene transfer **112**
mannitol cycle, *Eimeria* 133
manual curation 148
map-as-you-go strategy, genome
      sequencing 143–4, 149
mapping 146–7
  HAPPY mapping *147*–8
  optical mapping 148, *149*
mapping panel generation 147
Marginopora vertebralis 79
Massisteria 21
Massively Parallel Signature Sequencing (MPSS),
      dinoflagellates 50
Mastigamoeba species *15*
  lateral gene transfer **112**, 113
  M. invertens 25
Mastogloia species *202*
Mediophyceae 201, 202
mefloquine resistance, *Plasmodium falciparum* 37
Melosira species *206*, 209–10
membrane protein genes, nucleomorphs 196
membranes, plastids 94
merozoites, malaria parasites 34
Mesomycetozoa *see* Ichthyosporea

Mesostigma viridis, plastid genome **97**, 98
metabolic adaptation, anaerobic protists 109–10
metabolic pathways
   Dictyostelium discoideum 219
   Entamoeba histolytica 174
   Tritryps 163–5
metazoa (animals) 8, *12*, 13
metchnikovellids 182
Metopus species *67*
  telomere repeats **71**
microarray approaches 178
microcircles, *Amphidinium* 56
microcysts, dictyostelids 215, *216*, 218
micronucleus, ciliates 20, 64
  retention of deleterious mutations 73
microsporidia **3**, 13, 109, *110*, 181–2
  comparative genomics 185–6
  Encephalitozoon cuniculi genome 183–5
  evolutionary origin 182–3
  future genomic studies 188
  genome structure, functional implications
     186–8
  spore structure *181*
microtubule dynamics, reticulopodia 78
Miliammina fusca, evolution 85
Miliolida 78, 85, 87–8
mini-chromosomes, *Trypanosoma brucei* 155, 157
minicircles, dinoflagellates 55–6, 101
"minimal tiling paths" 141, *142*
Ministeria 8, 13
mitochondria 10
  absence of 16
    Entamoeba histolytica 170–1
    microsporidia 182, 183
  degeneration in Apicomplexa 132
  gene transfer into nucleus 103
  loss of 109
mitochondrial electron transport system,
    Tritryps 164
mitochondrial features, and *Entamoeba*
    histolytica genome **177**
mitochondrial genomes 95
  dinoflagellates 58–9
  Plasmodium species 42
mitosis, dinoflagellates 52
mitosomes 23, *110*
  Entamoeba histolytica 176–7, 178
mixotrophs 18, 20
  see also dinoflagellates
monocistronic transcription, *Plasmodium*
    nuclear genes 41
Monoeuplotes species 64
  M. crassus
    conserved chromosome breakage sequence 70

  telomere addition 71
    "TA" IESs 69
mosaic nuclear genomes, Apicomplexa 126–7
mosquito, as host for malaria parasites 34
MRCAs, *Plasmodium* species 42
mRNA processing, Tritryps 163
mt-hsp70, presence in mitosome 176
MT proteins, Foraminifera 90
mucin-associated surface proteins (MASPs)
    genes 160, 161
multicellular assemblies, dictyostelids
    214, 215, *216*, 223
Multicilia 9, 13, 16, 26
multigene transcripts 186, 187
multilocular Foraminifera, evolution 85
multimembranous plastids 99
multiple molecular markers 10
multiple origins, anaerobic protists *110*
MUMmer2 40
Mycetozoa 8, *12*, 13, *15*, 16, *110*
myxogastrids (myxomycetes) 8, 16, 222
Myxozoa 13

N-acetylneuraminate lyase acquisition,
    Trichomonas vaginalis 111
NADH oxidase, lateral gene transfer **112**
Naegleria species, rDNA 172
"naked" Foraminifera 78, *79*
  evolution 85
Nanochlorum eukaryotum, plastid genome 98
Nassophorea 67
Neogloboquadrina pachyderma 88, 89
Neoproterozoic origins, Foraminifera 88
Nephridiophagids 9, 26
Nephroselmis olivaceae, plastid genome **97**, 98
NEP (nuclear-encoded phage-type RNA polymerase) 98
nitrogen fixation, gene transfer 123
nitrogen metabolism, diatoms 207–8
nitrogen storage, diatoms 207
non-autonomous non-LTR retrotransposons (NARTc) 160
Nosema bombycis 188
Nosema locustae, catalase acquisition 111
Notodendrodes hyaninosphaira, SSU rDNA *82*
nuclear dualism 7
  ciliates 20
    origins 73–4
  Foraminifera 80
nuclear-encoded proteins, import into plastids 197
  Thalassiosira pseudonana 205
nuclear genome, dinoflagellates 49–54, 59
Nuclearia 15
Nucleariidae (heterotrophic amoebae) 8, 13
nuclear transfer, plastid genes 103
  Apicomplexa 126, 127, 128

dinoflagellate microcircle genes 101–2
nuclei numbers, ciliates 64
nucleomorphs **3**, 22, 192–4, *193*
  compacted genomes 187
  comparative genomics 197–8
  encoding of plastid proteins 196
  genome karyotyping 194
  genome sequencing 194–6
  protein sizes 197
nucleotide biosynthesis
  Cryptosporidium parvum 132, 133
  Entamoeba histolytica 174
  Tritryps 164
nutrient deprivation, dictyostelid responses 215, *216*
Nyctotherus species 67
  telomere repeats **71**

Ochromonas 18
Oligohymenophorea *67*
  chromosome processing 68
  cis-acting sequences **69**
  telomere repeats **71**
oomycetes (water moulds) 8, 13, 18
  gnd gene 103
  plastid loss 102
Opalina 18
opalines 19
open conoid structure 19–20
operon transfer 130
Ophisthokonts 8, *12, 13, 15*, 25, 26, *215*
optical mapping 148, *149*
Orbulina universa 80
organelles 16
orthologous genes, *Plasmodium* species 37
overlapping transcription
  microsporidia 186–7
  nucleomorphs 187, 195
oxygen detoxification, *Entamoeba histolytica* 175
oxymonads 9, 23
Oxyrrhis species 19
  O. marina, nuclear organization 54
Oxytricha species *64, 67*
  cis-acting sequences **69**
  DNA unscrambling 69
  O. trifallax, genome sequencing project 71, 72
  telomere repeats **71**

parabasal apparatus 24
parabasalids 9, 23–4, *110*
  lateral gene transfer **112**
paralogous gene families, *Plasmodium* species 38, **39**
Paramecium species *64, 65, 67*
  cis-acting sequences **69**
  P. caudatum, genome size *184*

P. putrinum, amitosis 67
P. tetraurelia
  chromosome processing 68
  genome sequencing project 71–2
  immobilization antigen inheritance 72
  "TA" IESs 69
  telomere addition 71
  telomere repeats **71**
Paramyxea 9, 26
parasitic plants 100
parasitism, adaptation to 132
pathogenesis
  Entamoeba histolytica 169–70, 175–6
  microsporidia 182
Paulinella species *21*
  P. chomatophora, endosymbiont 95
pcst gene families **39**
pelagophyceans 19
pellicle, euglenids 24
pelobionts 8, 16, 109, *110*
Peneropolis species, SSU rDNA *82*
pennate diatoms 201, *202*
PEPC (phosphoenolpyruvate carboxylase), C4
  photosynthesis 209
PEP (plastid-encoded RNA polymerase) 98
peridinin 19
peridinin chlorophyll *a*-binding protein gene,
  dinoflagellates 50
peridinin plastids, dinoflagellates 49, 55–6, 99,
  100, 101, 192
Peridinium species, tertiary plastids 57
periplastidal compartment 192, *193*
perkinsids 8, 19–20
Perkinsus species
  nuclear organization 53
  P. marinus, introns 52
peroxiredoxins, *Entamoeba histolytica* 175
pet genes, dinoflagellates 55, 56
Pexel motif, *Plasmodium* species 42
Pf-fam gene families **39**
Pfiesteria species
  haploid DNA content **51**
  P. piscicida, *cob* sequences 58
Phaeodactylum tricornutum 201, *206*
  carbonic anhydrase 208
  frustulin EST 207
  genome sequencing 202, 204
  sequenced isolates 210
  silica metabolism 205–6
  urea cycle 208
Phaeodarea 8, 21, 22
phaeophycaeans (brown algae) 19
phaeothamniophyceans 18–19
phages, presence of hydroxymethyluracil 52

Phagodinium 9, 26
phagotrophy 124–5
    anaerobic protists 111
    yeast ancestors 118
Phalansterium 9, 13, 16, 26
phosphoenolpyruvate carboxykinase, lateral gene transfer **112**
6-phosphogluconate dehydrogenase gene (*gnd*) 102–3
phospholipase C, *Dictyostelium discoideum* 221
phospholipid metabolism, Tritryps 164
photosynthesis
    C4 pathway 208–9
    origins 94–5
photosynthetic loss
    Apicomplexa 127
    euglenophytes 100
    gnd as marker 102–3
photosystem genes, dinoflagellates 56
phototaxis, dinoflagellates 49
Phred 143
Phusion 146
phycobilisomes 16
Phyllopharyngea *67*
    cis-acting sequences **69**
    telomere repeats **71**
"physical gaps" 143
physical mapping, clone-by-clone sequencing
        strategy 141
Phytomyxea *see* plasmodiophorids
Phytophthora infestans 18
PI3K, *Dictyostelium discoideum* 221
Plagiopyla 65
Planoprotostelium 15
Plantae 8, *12*, 16–17, *110*, *215*
PlasmoDB 35
plasmodiophorids (Phytomyxea) 8, 20, 21, 22
Plasmodium species 19, 33–4
    1, 4-α-glucan branching enzymes, absence 130
    comparative gene expression and regulation 41–2
    genome characteristics 35, **37–8**
    genome mapping 147
    genome sequencing 33, 34–5, **36**, 42–3, 139, 144, 146
    paralogous gene families 38, **39**
    P. falciparum 125, 195
        apicoplast 100, 101
        genome analysis 127–8, 129
        genome sequencing, choice of clone 140
        genome size *184*
        mefloquine resistance 37
        time to MRCA 42
    P. vivax, mitochondrial genomes 42
    synteny analyses 39–40
    whole genome alignment 38–40
    whole genome SNP maps 42

plastid editing machinery, land plants 98
plastid loss 101, 102–3
plastid proteins, encoding by nucleomorphs 196
plastid replacement 102
plastids
    in Apicomplexa 19
    chromalveolate 17, 18, 19
    in diatoms 205
    in dinoflagellates 19, 49, 55, 57
    euglenids 24
    import of nuclear-encoded proteins 197
        Thalassiosira pseudonana 205
    membranes 192, *193*
    origin in endosymbiosis *96*
        see also endosymbiosis
    in Plantae 16, 17
    primary 94
    sequenced genomes **97**
plastome architecture, higher plants 99
plastomes, primary plastids 95, **97–8**
pleuralins 207
Pleuronema 65
Polarella glacialis, haploid DNA content **51**
polar filament, microsporidia *181*, 182
polyadenylation, dinoflagellates 51, 52
poly-A signals, dinoflagellates 51–2
polycistronic gene clusters (PGCs), Tritryps 157, *158*, 162
Polycystinea 9, 20, 22
    EST project 90
polyketide synthases, *Dictyostelium discoideum* 219
Polykrikos species *20*
    P. schwartzii 48, *49*
polysphondylids 214, *215*
polyubiquitin genes, Foraminifera and Cercozoa 83
pore-forming peptides, *Entamoeba histolytica* 176
Porphyra purpurea, plastid genome 95, **97**
Porphyra yezoensis 95
possible estuary-associated syndrome (PEAS) 48
Postciliodesmatophora *67*, 68
Postgaardi 9, 24
post-transcriptional gene regulation, dinoflagellates 52
prasinophyte green algae 17
    as source of dinoflagellate plastids 57
Preaxostyla 23
predation, Foraminifera 88–9
primary endosymbiosis 16, 94–5, *96*, *124*, 125, 192, *193*, 202
primary plastids, genomes 95, **97–8**
primitive eukaryotes 23, 109
probable plastid hypothesis 98
Prorocentrum species, haploid DNA content **51**
Pyrocystis lunula, introns 52
prolyl-tRNA, lateral gene transfer **112**
promastigote surface antigens (PSA-2) 161
promoter elements, *Plasmodium* species 41

Prostomatea *67*
Protaspis 22
protein:DNA mass ratio, dinoflagellates 53
protein-coding genes
    Encephalitozoon cuniculi 184
    Foraminifera 83–4, 88
protein importation, *Entamoeba histolytica* mitosome 177
protein kinases
    Dictyostelium discoideum 221
    Entamoeba histolytica 175
    Tritryps 165
proteromonads 19
protostelids 16
psa, *psb* genes, dinoflagellates 56
Psammophaga simplora, endoreplication 80
pseudodendromonads 18
Pseudo-nitzschia multiseries 202, *206*
Pseudospora 9, 26
Pseudotrichomonas keilini 110
Pterocanium 21
pulsed field gel electrophoresis (PFGE) 140
    whole chromosome shotgun sequencing 144
purine biosynthesis
    Entamoeba histolytica 174
    Tritryps 164
purine salvage genes 132
Py235 gene family **39**
Pyrenomonas salina, nucleomorph genome separation 194
Pyrgo peruviana, SSU rDNA *82*
pyridine nucleotide transhydrogenase, presence in
    mitosome 176
pyrimidine biosynthesis
    Entamoeba histolytica 174
    Tritryps 164
pyrimidine salvage genes 132
Pyrocystis lunula, histone H3 53
pyruvate:ferrodoxin oxidoreductase (PFO) 114
pyst gene families **39**

quadripartite structure, plastid genomes 95, 98
Quaternary climate change, role in foraminiferan
    diversification 88

radial centric diatoms 201, *202*
Radiolaria 3, 9, 20, 22
Ramicristates 8, 13, *15*, 16, 26
Raphidophyceae 18
RCK1, *Dictyostelium discoideum* 220
rDNA circles, *Entamoeba histolytica* 171–2
receptors, *Dictyostelium discoideum* 220–1
recombination hotspot (RHS) pseudogenes 10
red algae *see* rhodophytes
redox balance, plastids 103
redox potential regulation, dinoflagellates 56

regeneration, ciliate macronuclei 67–8
regulation, *Plasmodium* genes 41–2
repeat content of genome 140
    Dictyostelium discoideum 150, 218
    dinoflagellates 50
    Entamoeba histolytica 150, 172–3
    Thalassiosira pseudonana 204
replication origin complex, Tritryps 165
reproductive strategies
    ciliates 64–8
    Foraminifera 80
Reticulomyxa species 22, 78
    evolution 85
    R. filosa 79, 81
        EST project 89–90
        introns 84
        RNA polymerase gene 83
        tubulin gene 83
        ubiquitin gene 83
reticulopodia, Foraminifera 78, *79*, 84, 88
Retortamonads 9, 23
retroelements, Tritryps 160
retrotransposable elements 130
retrotransposons, *Entamoeba histolytica* 173
Rhipidodendron 21
Rhizaria 8, 11, *12*, 20–2, *21*, 90
Rhizosolenia species 209–10
Rhodomonas salina, nucleomorph genome separation 194
rhodophytes (red algae) 8, 16–17, 203
    origin of apicoplasts 101, 125
    plastid genomes 95, **97**, 98
    secondary endosymbiosis *96*, 99, 193
Rho family proteins, *Dictyostelium discoideum* 220
rhoph1/clag gene family 39
ribonucleotide reductase, *Entamoeba* species 177–8
ribosomal DNA, processing in ciliates 68
ribosomal protein genes, *Entamoeba histolytica* 173
rice, gene transfer into nucleus 103
rif/stevor gene family 38, **39**
"ring of life" 118–19
RNA capping, *Entamoeba histolytica* 173
RNA editing, dinoflagellates 56, 58–9
RNA genes, Tritryps **159**
RNA interference (RNAi) 72
RNA polymerase gene, Foraminifera 83
RNA polymerase complexes, Tritryps 162
RNA polymerases, green plants 98
RNA processing, Tritryps 162–3
Robertinida 78, 85
Rotaliella elatiana, introns 84
Rotaliellidae, differentiated nuclei 80
Rotaliida 78, 85, 87
    introns 84
RUBISCO, diatoms 204, 208, 209

Saccharomyces cerevisiae 195
  ammonium transporters 207
Sagenoscena 21
saposin-like proteins, *Entamoeba histolytica* 176
scaffolds, WGS sequencing *145*, 146
scan RNAs (scnRNAs), ciliates 72–3
Schizoclades 9, 26
schizonts, malaria parasites 34
screening, HAPPY mapping 147
SDF-2, *Dictyostelium discoideum* 219, 221
secondary endosymbiosis 17, *96*, 99, 117, 192, *193*, 203
  apicoplast origin *124*, 125
  chlorophyll a/b lines 99–100
  Chromalveolata 100
secondary plastids, sequenced genomes **97**
secretin-type receptors, *Dictyostelium discoideum* 220
sequencing 143
  see also genome sequencing; genome sequencing
    projects
sexual cycle, diatoms 210
sexual reproduction
  ciliates 64–5
  diatoms 204
shellfish poisoning, dinoflagellates as cause 48
shells, Foraminifera 22
short IESs 68
short interspersed repetitive elements (SIRE) genes 160
shotgun library 141, *142*, 143
shotgun sequencing 141
SICAvar gene family 38, **39**
signaling molecules
  Dictyostelium discoideum 220–2
  Tritryps 165
signal transduction-related genes, *Entamoeba histolytica*
    175
silacateins 206–7
silaffins 206, 207
silica metabolism, diatoms 205–7, 209
Simpson, A.G.B. and Roger, A.J. *12*
SINEs (short interspersed nuclear elements), *Entamoeba*
    histolytica 172–3
single nucleotide polymorphism (SNP), *Thalassiosira*
    pseudonona 203, 204
SIRE retroelements, Tritryps 160
SITs, diatoms 206
size polymorphisms, *Plasmodium* species
  chromosomes 37
SLACS 160
sleeping sickness 24
slime moulds *see* Mycetozoa
slopalines 8, 19
SLRNA genes, Tritryps 159
slug form, dictyostelids 215, *216*, 217
small nucleolar RNA (snoRNA) genes, Tritryps 159

small subunit ribosomal RNA (SSU rRNA) sequences 10
  core Cercozoa 21
  Foraminifera 81, *82*, 83
  Haplosporidia 22
  microsporidia 182
  phylogeny 13, *14*
SNP (single nucleotide polymorphism) maps, *Plasmodium*
    species 42
soft walled Foraminifera 78, *79*
solitary amoeba, ancestry of dictyostelids 218
somatic nucleus, Foraminifera 80
Spirillina 88
Spirillinida 78, 85, 87–8
Spironemidae (Hemimastigophora) 9, 26
Spironucleus barkhanus, lateral gene transfer 114, **115**, 128
Spirotrichea 67
  chromosome processing 68, 69, 71
  cis-acting sequences **69**
  telomere repeats **71**
splicesomal introns, nucleomorphs 195
Spongomonads 8
Spongomonas species 21
sporangium, *Phagodinium* 26
spore formation, dictyostelids 215, *216*, 217, 219
spore structure, microsporidia *181*
sporozoites, malaria parasites 34
Spraguea lophii gene sequencing 185
stachel, plasmodiophorids 22
starch metabolism, nucleomorph genome 19
Stentor 65, 67
  macronucleus regeneration 67–8
Stephanopogon species 9, 24, 26
Stephanopyxis species *206*, 209–10
sterol biosynthesis, *Entamoeba histolytica* 174
Sticholonche 9, 22
Stichotrichia, DNA unscrambling 69–70
stigma, *Glenodinium foliaceum* 102
stop codons, dinoflagellates 51
strain choice, genome sequencing 139–40
stramenochromes 8, 18–19
stramenopiles 8, 11, 17–19, *18*, 110
  vestigial plastids 102
streptophytes 17
Stylonychia species *65*, 67
  cis-acting sequences **69**
  DNA unscrambling 69
  S. lemnae, chromosome fragmentation 70
  telomere repeats **71**
substitution rates, Foraminifera 88
subtelomeric regions
  Plasmodium species 35
  Tritryps 159–60, 162
subtelomeric rRNA operons, nucleomorphs 194–5
Suctoria, budding 67

sugar metabolism, Tritryps 163–4
sulfide dehydrogenase evolution 111–13
sulfur metabolism 113
supergroups 7, 10–11
  relationships 26
superoxide dismutase, *Entamoeba histolytica* 175
surface proteins, *Entamoeba histolytica* 169
SURFIN gene family 38
Symbiodinium species 48–9
  haploid DNA content **51**
  introns 52
  minicircles 101
  peridinin chlorophyll *a*-binding gene 50
syndineans 19
Syndinium borgerti, chromosomes 52
Synechococcus sp., partial urea cycle 208
synteny, Tritryp genomes 157, *158*
synteny analyses, *Plasmodium* species 39–40, 43
synurophyceans 18

tabulation, dinoflagellates 48
"TA" IESs, ciliates 69
Takayama species, tertiary plastids 57
TATA box sequences, *Plasmodium* species 41
taxon names, definition 11
TBP-related factor 4 (TRF4) 162
Teleaulax amphioxeia, as source of *Dinophysis* species plastid 58
telomerase, absence from *Entamoeba histolytic* 174
telomere addition, ciliates 70–1
telomere dependence, ciliate macronuclear chromosomal
    amplification 71
telomere repeats, ciliates **71**
telomeres
  Dictyostelium discoideum 218
  Tritryps 159–60, 162
telomere structure, *Plasmodium* species 35, **37**
Telonema 9, 26
temporary plastids, dinoflagellates 57–8
tertiary endosymbiosis 17, 56–7, 102
tertiary plastids, dinoflagellates 57
testate amoebae 3
tests
  Foraminifera 90
    morphology 78
Tetrahymena species 20, 64, *65*, 67
  chromosomal processing 68
  cis-acting sequences **69**
  conserved chromosome breakage sequence 70
  Dicer-like protein 73
  genome sequencing projects 71
  genome size 139
  telomere repeats **71**
  T. pyriformis, amitosis 66
  T. thermophila, genome sequencing 139

Textulariida 78
  paraphyly 85, 87
Thalassiosira pseudonana
  C4 enzymes 209
  carbonic anhydrases 208
  genome analysis 95, 198, 201, *202*
  genome characteristics 203–4
  nitrogen metabolism 207, 208
  nuclear-encoded proteins, import
      into plastids 205
  sequenced isolates 210
  silica metabolism 206–7
Thalassiosira puntigera 206
Thalassiosira rotula 206
Thalassiosira weissflogii
  C4-like photosynthesis 209
  carbonic anhydrases 08
Thassicolla species 20
thaumatomonads 8, 21, 22
theca, dinoflagellates 48
Theileria species 125
  1,4-α-glucan branching enzymes, absence 130
  genome analysis 131
  genome sequencing 139
Thermotoga maritima 114
thraustrochytrids 18
thyalkoid membranes
  origin 94
  structure in *Cerratium horridum* 101
thymidine kinase gene 132
Thysanomorpha 65
tobacco, plastid gene transfer into nucleus 103
toxin formation, dinoflagellates 48
Toxisarcon species, evolution 85
Toxoplasma species
  1,4-α-glucan branching enzymes 130, *131*
  T. gondii 125
    apicoplast 100–1, 128
    genome size *184*
Trachipleistophora hominis, HSP70 location 183
transcription
  Entamoeba histolytica 173
  Tritryps 162–3
transcriptional collision, microsporidia 186
transcriptional gene regulation, dinoflagellates 52
transcription factors, *Dictyostelium discoideum* 221–2
transcriptome projects, malaria parasites **36**
transmembrane protein kinases,
      Entamoeba histolytica 175
transmission electron microscopy 10
transporter gene SNPs, association with *Plasmodium*
      antimalarial resistance 42
transporters
  Dictyostelium discoideum 219

transporters (*Cont.*)
   Entamoeba histolytica 175
transposable elements
   Dictyostelium discoideum 218
   Thalassiosira pseudonana 204
transposon IESs 68
transposons, similarity to "TA" IESs 69
trans-sialidase (TS) genes 160
Treponemas agilis 110
TRF4 (TBP-related factor 4) 162
TribeMCL method 38
Trichia 15
Trichomonas species 109
   intradomain gene transfer 117
   T. vaginalis 23
      genome sequencing 139
      genome size *184*
      leucine-rich repeats 113
      N-acetylneuraminate lyase acquisition 111
Trichonympha species 23
Trichozoa 9, 23
Trimastix species 9, 23
Tritryp protein sequences, conservation in other
   organisms *162*
Tritryps
   drug targets 165
   gene content 165
      core proteome 161
      metabolism 163–5
      species-specific differences 161–2
      transcription and RNA processing 162–3
   genome organization
      gene clusters 157–*8*
      retroelements 160
      RNA genes **159**
      telomeres 159–60
   genome structure 155–7
   genome synteny 157, *158*
   hybrid genome of *T. cruzi* 161
tRNA genes, Tritryps 159
tRNA introns, nucleomorphs 195
tRNA tandem arrays, *Entamoeba histolytica* 172
Trochamminida 78, 85, 87
trophozoites
   Entamoeba histolytica 169, *170*
   malaria parasites 34
true fungi 13
Trypanosoma species
   genome sequencing 155
   genome statistics **156**
   genome structure 156–7
   retroelements 160
   T. brucei
      genome sequencing 144, 146

      genome size *184*
      gnd gene 103
      telomeres 160
   T. cruzi
      genome sequencing project 149–50
      hybrid genome 161
      lateral gene transfer 116
trypanosomatids 24, 155
trypanothione, presence in *Entamoeba histolytica* 175
trypanothione metabolism, Tritryps 164
tryptophanase, *Entamoeba histolytica* 176
*tuf*A gene 126
TWCA1 208
twin introns 100
type II fatty acid biosynthesis, Apicomplexa 131
tyrosine kinase-like enzymes, *Dictyostelium
     discoideum* 221

ubiquitin genes, Foraminifera 83
ultrastructural identity (UI) lineages 8–9, 10, 11
unequal macronuclear division, ciliates 66–7, 73
unilocular Foraminifera, evolution 85
3'-untranslated regions (UTRs), *Alexandrium tamarense* 50
urea cycle
   diatoms 208
   Dictyostelium discoideum 219
U-rich tracts, mammals 52
uridine kinase-uracil phosphoribosyltransferase
   (UK-UPRT) gene 132
Uronychia 65
Urostyla grandis 64
Uvigerina species 87

Valkampfia 23
Vampyrellids 9, 26
var gene families 38, **39**
var gene regulatory element, *Plasmodium* 41
variant surface glycoprotein (VSG) genes, Tritryps 155,
   157, 161
   T. brucei 160
vegetative growth, ciliates 65–8
vesicular transport model, diatoms 205
vestigial plastids 102
vinculin, *Dictyostelium discoideum* 222
VIPER retrotransposons, Tritryps 160
Viridaeplantae 8, 16, 17
   origin of plastids *96*
Vittaforma corineae 188
   gene sequencing 185
Vorticella 65

water moulds *see* oomycetes
whole chromosome shotgun (WCS) sequencing 140, 144,
   150

Dictyostelium discoideum 217
whole genome alignment, *Plasmodium* species 38–40
whole genome shotgun (WGS) sequencing 140, 144, *145*, 146, 149–50
   Entamoeba histolytica 171
   Thalassiosira pseudonana 203
whole genome SNP maps, *Plasmodium* species 42
Woloszynskia bostoniensis, nuclear genome 50

xanthophyceans 18–19
Xenophyophores 8, 22
Xenopus laevis, chromosomal processing 68

ycf genes, dinoflagellates 56
Yeast Artificial Chromosome (YAC)
   maps 141, 150
yeast genes, prokaryotic origins 118, 119
yeast nuclear genome, mitochondrial
   origin 126
yeasts
   Saccharomyces cerevisiae 195
     ammonium transporters 207
yir/bir/cir gene family 38, **39**

zoonoses, *Plasmodium knowlesi* infections 34